The Evolution of Political Ideology

Steven Charles Hertler
Aurelio José Figueredo
Mateo Peñaherrera-Aguirre

The Evolution of Political Ideology

A Natural and Civil History

palgrave
macmillan

Steven Charles Hertler
Saint Elizabeth University
Morristown, NJ, USA

Aurelio José Figueredo
University of Arizona
Tucson, AZ, USA

Mateo Peñaherrera-Aguirre
School of Animal and Comparative-Sciences Research
University of Arizona
Tucson, AZ, USA

ISBN 978-3-031-83940-5 ISBN 978-3-031-83941-2 (eBook)
https://doi.org/10.1007/978-3-031-83941-2

Niday Picture Library / Alamy Stock Photo

This Palgrave Macmillan imprint is published by the registered company Springer Nature Switzerland AG
The registered company address is: Gewerbestrasse 11, 6330 Cham, Switzerland

Steven Charles Hertler,
To my wife Danielle and our two boys, Dominic and Vincent

Contents

Notes on Contributors

Aurelio José Figueredo is Professor Emeritus of Psychology, Family Studies, and Human Development at the University of Arizona, USA. Dr. Figueredo also serves as Director of the Ethology and Evolutionary Psychology (EEP) Laboratory.

Steven Charles Hertler is a licensed examining psychologist and Assistant Professor of Psychology at Saint Elizabeth University, USA.

Mateo Peñaherrera-Aguirre currently serves as a research associate in the School of Animal and Comparative-Sciences Research at the University of Arizona, USA. His lines of scientific research include the evolution of lethal coalitional aggression in human and nonhuman animals, socioecological correlates of sociopolitical complexity, and multilevel selection.

List of Figures

List of Tables

Introduction: An Apolitical Book on Political Ideology

> The same political parties which now agitate the US have existed through all time. And in fact the terms of whig and tory belong to *natural* as well as to *civil* history. They denote the temper and constitution and mind of different individuals.
>
> —Thomas Jefferson[1]

Political Ideology's Natural History Expressed in the Course of Civil History

Thomas Jefferson, the motive force driving America's first opposition party, made the above-quoted observation on political orientation. In the *Republicans* and *Federalists* of early America, Jefferson saw the *Whigs* and *Tories* of England, which, in turn, can be broadly mapped onto related political orientations: libertarian *v.* totalitarian, liberal *v.* conservative, globalist *v.* nationalist.[2] These political orientations assume different forms and features depending on context and region, but, to Jefferson's point, derive from innate temperamental dispositions. More than 200 years later, a broad and deep body of evidence validates Jefferson's intuition, showing political ideology to be as much a matter of nature as of nurture:

- Taste bud density predicts conservatism, linking sensory processing to political orientation[3]
- Political ideology is reliably identified by olfactory cues that subconsciously guide attraction and assortative mating decisions[4]
- Among liberals, there is increased grey matter within the *anterior cingulate cortex*, a brain region associated with reward anticipation and impulse control[5]
- Democratic orientation is associated with greater activity in the left *insula*, a portion of the cortex associated with sense and emotional processing[6]
- Amygdala activation predicts political ideology and more broadly influences perceptual differences concerning the legitimacy of social systems[7]
- Political ideology tracks with *chronotype*, the sleep patterns associated with early and late rising[8]
- Political ideology is highly heritable, as per controlled twin studies[9]
- Political ideology correlates reliably with openness to experience, conscientiousness and other partially heritable personality traits[10]

As we subsequently explain at length, to the extent that emerging titles feature these endocrine, anatomical, and genetic findings, they neglect historical context. Demonstrating the point, in October of 2013,[11] Avi Tuschman presented his then newly published book, *Our Political Nature*, at Georgetown University's *Berkley Center for Religion, Peace and World Affairs*. Tuschman effectively said ideology evolved, but did not convincingly show his small audience of anthropologists and political scientists how it evolved. The question-and-answer session indicated interest, but was dominated by matters of historicity, signifying that it was not enough to say political ideology evolved; rather it was necessary to narrate its evolution from nonhuman primates and small-scale societies, through to modern polities tolerating organized opposition. Notwithstanding the receptivity on the part of the audience and the thoroughness on the part of the author, skepticism preponderated. Thus, as it stands, books on political ideology ignore either one or the other aspects of Jefferson's dictum. In other words, available books document either the *civil* or *natural* history of ideology: history and political science titles ignore genetics, neurobiology, and related biopsychological phenomena; while neuroscience and social science titles ignore historicity, civics, and related political contingencies.

By contrast, this monograph, *The Biology of Political Ideology*, is divided into two parts, as indicated by its subtitle: *A Natural and Civil History*. Addressing natural history, Part I evolutionarily explains why political ideology is substantially genetic, correlates with personality traits, and is represented in brain regions associated with risk and emotional processing. Addressing civil history, Part II traces the evolution of political ideology across anthropological transitions from non-human primates to small-scale societies, and across historical transitions from autocratic states through to democratic societies with political parties peaceably transferring power and tolerating opposition. In this way, we can offer an unprecedented synthesis explaining *why ideology has a natural history, and how that natural history unfolded through the course of civil history*.

In the remainder of this introduction, we expand on this abstract, discussing the parallel lines of literature that never seem to converge, ending with a more detailed description as to how we hope to yield a biologically informed political psychology that is not artificially dissociated from political history.

History and Political Science Titles

The information bulleted above, and related findings, is not typically featured in works of history and political science. For example, in a *Cambridge Companion to Liberalism*, Steven Wall warns readers of the heterogeneity of liberalism, and then proceeds to consider what all its various manifestations might have in common.[12] After defining classical liberalism, Wall contrasts it with modern liberal parties, traces its Enlightenment origins, and considers its relationship to civic republicanism, democracy, and conservatism. Wall and coauthors provide perspective on liberalism's Enlightenment origins, which were generally opposed to absolutism, moral policing, and traditionalism. Like Wall's edited volume, J. P. Diggins (1986) provides valuable insights into the interactions between persons and politics in *The Lost Soul of American Politics: Virtue,*

Self-Interest, and the Foundations of Liberalism. The same can be said of D. Womersley's (2006) *Liberty and American Experience in the Eighteenth Century*, D. Boaz's (2015) *The Libertarian Mind*, K. Minogue's (1963) *The Liberal Mind*, F. H. Knight's (1947) *Freedom and Reform*, R. C. Bannister's (1992) *On Liberty, Society, and Politics*, and Stears' (2006) *Progressives, Pluralists, and the Problems of the State.* Then there are the magisterial works of Quentin Skinner and J. G. A. Pocock. Many such sources contain marvels of acumen and synthesis; yet, even the most modern of these works does not reconcile explanations of political ideology with the above-listed biological correlates, or those many other points of fact populating the social science literature.

There are some few exceptions of which we are aware: Writer and journalist Colin Woodard (2016), in a history of American political party politics, writes instructively of *individual liberty* as vying with *freedom of community*. Moreover, Woodard (2016, p. 10) goes further, citing the works of famed entomologist and sociobiologist E. O. Wilson in support of the assertion that humans are "genetically hardwired" toward tribalism and group sacrifice, even while they "compete for survival and reproduction within groups via individual selection." Woodard (2016, p. 11) speaks of Wilson's *iron rule to genetic social evolution:* "Selfish individuals beat altruistic individuals, while groups of altruists beat groups of selfish individuals." Notwithstanding this insight, Woodard's *American Character* does not pursue this theme even within the introduction, and thereafter engages in a particularistic study of American politics without further reference to evolution. Another example is, *Political Realignment: Economics, Culture, and Electoral Change*, which begins by highlighting contemporary political volatility as expressed in the 2016 American elections, and analogous movements in Europe. It takes an appropriately dispassionate scholarly position, in that its author, Russell J. Dalton, explains broad patterns using historical analogies and statistical analyses. Dalton provides an insightful discussion of *cleavages*, fault-lines separating one political position from another. He distinguishes between social and economic cleavages, reviewing them as they are defined by social class, across historical eras, and within nation-states. In making this distinction, Dalton reviews Seymour Martin Lipset, Richard Hofstadter, and Ronald Inglehart alongside other political theorists, while plotting political parties graphically within a matrix formed by a social axis and an economic axis. Dalton does at length refer to the literature on personality, suggesting that openness to experience and other personality traits undergird cultural cleavages. In this respect, Dalton's book, like Woodard's, is uncommonly aware of relevant social science literature.

Social Scientific Approaches: Personality and Proximate Explanation

Despite these valuable examples of *consilient*[13] thought, political ideology's biological influence remains unacknowledged; or, if acknowledged, insufficiently warranted within works of history and traditional political science. The personality literature within the social sciences offers an alternative. This vast body of literature can be pursued in either of its two forms, which more or less follow the dichotomy between *lumpers* and *splitters*, or those that take a *personality style* approach v. those that take a *personality trait* approach. Representative of the personality style approach is *authoritarianism*, which captured the public imagination ever since Theodor Adorno published the *Authoritarian Personality* following the fall of Germany's Third Reich. As did Adorno, Altemeyer[14] cast authoritarians as illogical, shallow, and self-contradictory. Hetherington and Weiler used authoritarianism to explain correlations among policy positions on race, gay marriage, illegal immigration, and the use of force to resolve security problems. Stenner's, *The Authoritarian Dynamic*, consolidates literature on authoritarianism, describing environmental stressors as evoking latent dispositions. In consequence, the Authoritarian Dynamic is the consummate treatment of states interacting with traits to explain political behavior. With respect to the latent disposition itself, Stenner's book transcends historical particularism to seize upon the core elements of authoritarianism as they are manifest across time and culture: (1) moral absolutism, (2) intolerance, (3) punitiveness toward dissidents and deviants, and (4) animosity and aggression against racial and ethnic outgroups. Without using a biological framework to explain the ultimate origins of authoritarianism, Stenner nonetheless perfectly describes its relation to group-level existential threats psychologically stoked by immigration, military defeat, and declining group status. Putting these ideas together, we have a coherent explanation of an innate psychological disposition along with a description of how that psychological disposition is politically activated and expressed.

To illustrate a *splitter* approach that separately associates constituent traits to political ideology, we cite the work of Jeffery Mondak. Using a five-factor framework, Mondak (2010) reports that openness to experience and conscientiousness proved substantively associated with political ideology; and furthermore, these traits had opposite associations such that they served as anchors for each end of the political spectrum. Mondak reinvigorates the personality-based study of political ideology, which has a long research tradition extending to the early decades when the *five-factor model* of personality was newly articulated. Mondak is also specifically cited for his appreciation of personality's biological underpinnings.[15] Both political attitudes and actions, Mondak argues, are informed by personality traits, which are themselves, in large part, genetically informed. Mondak explicitly values insight into the origins of personality because such insight elucidates how and why personality informs ideology. Nevertheless, he confines himself to a three-part conjectural treatment of these origins, focusing on proximate genetic, rather than ultimate evolutionary determinants.

Evolution and Ultimate Explanation

Personality research, rich and fascinating though it is, lacks convincing causal explanations. The early authoritarian manifestation of personality research arose following World War II under the influence of psychoanalysis, and so broadly referenced psychoanalytic causality that is no longer convincing. The later trait-based manifestation of personality research arose in the 1980s under the influence of trait theory, and so remained causally agnostic. Apparently, there are surprisingly few modern publications grappling with issues of ultimate causation. Lakoff and Wehling's *Your Brain's Politics*[16] is one, wherein the authors use the cognitive science paradigm of *embodied cognition* in associating nurturant parenting with liberal attitudes, and strict parenting with conservative attitudes. Avi Tuschman, in his aforementioned book, brings together insights from primatology, genetics, neuroscience, and anthropology to explain the significance of personality traits in determining political ideology. In our estimation, the works of Faria (2019) and Haidt (2013) are among the best of these because both relate political ideology to the evolutionary push and pull between individual interests and national strength.

Filipe Nobre Faria's, *The Evolutionary Limits of Liberalism*,[17] distinguishes between group fitness and individual choice. The choices people freely make, Faria asserts, don't reliably redound to group fitness. When analyzed through a scientific, evolutionary lens, Faria (2019) claims, liberalism is strikingly unsustainable, notwithstanding its dominance over Western political thought for several hundred years past. Underlying this assertion is Faria's fear that Western liberalism will fall prey to competing national ideologies. What Faria (2019, p. 2) calls "the liberal satisfaction of preferences" is said to reveal "maladaptive tendencies which impact negatively on the fitness" of groups. "From an evolutionary point of view," Faria continues, "the liberal satisfaction of individual preference is not particularly desirable" because it operates "under a liberal meta-morality that potentially weakens groups in the evolutionary process of inter-group competition." By *liberal satisfaction of preferences*, Faria means to encompass both moral and economic decision making, and so is arguing that liberal freedoms in the capitalist or political sense can potentially damage group fitness, leading to "maladaptiveness and thus to institutional unsustainability." Faria asserts, "within political theory, no extensive and consistent research has utilized contemporary *multilevel selection theory* to analyze the theoretical underpinnings of liberalism." Faria is certainly justified in his claims to novelty and in his claims about the explanatory power of *multilevel selection theory*, an evolutionary meta-theory our book explains at length and relies on most fully to explain political ideology.[18]

Jonathan Haidt, in, *The Righteous Mind*, outlines a set of moral emotions, universally present in all, but proportionally different among liberals and conservatives. Haidt asserts that conservatives identify with morals outside the *care/harm* value set, to concern themselves, for instance, with whom their neighbors are sleeping. Haidt goes on to identify a *hivish* morality that is binding, and which instills trust, efficiency, and cooperation that can simultaneously benefit the ingroup while harming the outgroup. Haidt (2013) jointly discusses religion, morality, and political affiliation alongside an explanation of evolution and *group selection*, a component of multilevel selection, which again, is the primary biological theory we apply in this monograph. He outlines a newly expanded set of moral emotions, showing that conservatives identify with morals outside the *care/harm* value set, to concern themselves with expressions of *loyalty, purity*, and *sanctity*. Group selection causes others to rally around the flag, the leader, the nation, the cause. There is, too, a proper understanding of group selection taking place during a time of rapid, Neolithic evolution where the main selective pressures were other humans who punished non-reciprocators and non-cooperators in a process of self-domestication.[19]

Our Mandate and Its Fulfillment

We are neither political historians nor political scientists with a record of publications on political ideology. By contrast, we are psychologists, collectively publishing on war, primatology, history, violence, civilizational cycling, group dynamics, altruism, and personality. We think our approach, extending organically from our application of evolutionary meta-theories to these many topics, is well suited to study political ideology, which strangely has biological substrates antedating politics, while being shaped and expressed in the course of political history. Rather than being perceived as interlopers operating outside our area of expertise, we hope our synthetic approach brings novel insights. This book has the potential to frustrate the anthropologist or primatologist, as it does the political scientist, historian, biologist, geneticist, neuroscientist, or personality theorist, because it can never systematically or exhaustively explore information from these respective disciplines—that is not its purpose. Quite to the contrary, our paramount objective in this book is to narrate the origins and expression of political ideology, and is *not* either to catalogue biological findings or historical events. What we write about biopsychological phenomena and political history, we write because we believe it helps to advance this narrative; and what we forbear, we forbear because it is not necessary to advance this narrative.

Allow us to close with a description of this book's two parts, and ten aspirational goals which we outlined ahead of composition. To expand on Part *I*: In 1963, Niko Tinbergen published, *On the Aims and Methods of Ethology*,[20] wherein he asked four questions, which have since famously structured the study of behavior: Of any given behavior, Tinbergen asked, what is the (1) proximate causal mechanism, (2) development (ontogeny), (3) evolutionary history (phylogeny), and (4) ultimate adaptive function. While there are insufficient scientific data available at this point to answer each of these questions comprehensively, we approach the problem by working our way through several successive levels of biological organization of increasing complexity. Part I is divided into three sections: Section I, including this introduction and the first chapter, frames the book and articulates its organizing principles; Section II, comprised of the following four chapters, reviews proximate causes, encompassing causal mechanisms and behavioral development; and Section III, comprised of the next three chapters, reviews ultimate causes, encompassing adaptive functions and evolutionary history. Throughout this treatment, we attempt to tackle the problem of nested levels of biological complexity using the framework of *Multilevel Selection Theory*, which is introduced in the first chapter and revisited in the eighth. We propose that multilevel *selection* represents a theory uniquely capable of exposing the fault-line dividing *Right* from *Left*, and of making meaning from what might seem to be a hodgepodge of values, policies, and positions.

To expand on Part *II*: divided into four sections, this second part addresses questions of historicity, narrating the historical expressions of biological dispositions. Part *II* takes the abstract explanation of ideology explicated in Part *I*, and shows how it is manifest in different times and nations, and how it is channeled by various parties and persons. This second part is something of a historical narrative, with each chapter treating a stage on the way to the present—from leviathans that would brook no opposition, to proto parties, to mature stable states that channel ideological differences within constitutional frameworks, to the declining state that succumbs to rivals or rebels. In ranging widely across history, the chapters within this book's second part always keep to our main purpose, following evolved, temperamentally informed ideology as it is suppressed, overlaid, or fully active within various phases of group aggregation and state maturation. So when we discuss the life cycle of states, political parties, and realignments; when we pursue political parties and their emergence, demise, and corruption; or when we chart the rise, decline and fall of nations, we are always feeling for the pulse of political ideology as it is variously suppressed, polarized, or distorted.

Within the confines of a single volume, we seize on representative examples and core findings necessary to: (1) Explain political ideology in a manner consistent with its known heritability, and its anatomical and physiological differences observable via brain scan, EEG and autopsy; (2) account for the ultimate origins of ideological differences described in the psychological literature on personality and moral emotions; (3) supply an apolitical, scientific perspective inviting sympathy with competing ideologies, much needed in light of the hyper-partisanship now evident throughout the West; (4) interpret the underlying dynamics of group behavior depicted in the political science literature on populism and nationalism; (5) contextualize political histories of party formation, third party movements, rebellions, secession, and separation; (6) appreciate the shifting power dynamics between Left and Right as nations rise and decline, and as they enjoy peace and wage war; (7) understand ideology within cyclical history as it relates to the lifecycle of states; (8) relate demographic increase, stability, wealth, and other material conditions of life to the political expression of latent temperament; (9) mark the liberal and conservative temperaments in dialectic debate during the Enlightenment phase of constitution making wherein group selected values of ancient republics were reconciled with individually selected values of modern philosophy; (10) consider various forms of government by their ability to ritualize conflict, directing it into nonviolent channels and constraining it within productive political forms.

Notes

1. This quote was written in an 1813 letter to John Adams. The italicization of "civil" and "natural" was added.
2. All of these word pairs are overlapping instantiations of temperamentally informed political positions, though we provide detailed discussion as to their historical and philosophical differences as expressed within the political science literature.
3. Ruisch, B. C., Anderson, R. A., Inbar, Y., & Pizarro, D. A. (2021). A matter of taste: Gustatory sensitivity predicts political ideology. *Journal of Personality and Social Psychology, 121*(2), 394.
4. McDermott, R., Tingley, D., & Hatemi, P. K. (2014). Assortative mating on ideology could operate through olfactory cues. *American Journal of Political Science, 58*(4), 997–1005.
5. Schreiber, D., Fonzo, G., Simmons, A. N., Dawes, C. T., Flagan, T., Fowler, J. H., & Paulus, M. P. (2013). Red brain, blue brain: Evaluative processes differ in Democrats and Republicans. *PLoS one, 8*(2), e52970.
6. Haas, I. J., Baker, M. N., & Gonzalez, F. J. (2017). Who can deviate from the party line? Political ideology moderates evaluation of incongruent policy positions in insula and anterior cingulate cortex. *Social Justice Research, 30*(4), 355–380.
7. Nam, H. H., Jost, J. T., Kaggen, L., Campbell-Meiklejohn, D., & Van Bavel, J. J. (2018). Amygdala structure and the tendency to regard the social system as legitimate and desirable. *Nature Human Behaviour, 2*(2), 133–138.
8. Ksiazkiewicz, A. (2020). Conservative larks, liberal owls: The relationship between chronotype and political ideology. *The Journal of Politics, 82*(1), 367–371.
9. Alford, J. R., Funk, C. L., & Hibbing, J. R. (2005). Are political orientations genetically transmitted? *American Political Science Review, 99*(2), 153–167.
10. Fatke, M. (2017). Personality traits and political ideology: A first global assessment. *Political Psychology, 38*(5), 881–899.
11. https://www.youtube.com/watch?v=eo_wKbowynw
12. These additional liberal values, *community, democracy,* and *equality*, have variously been thought competitors to individual freedom. In answer to his question, Wall tentatively advances *a commitment to individual freedom,* though at the same time, he further cautions that definitions of

individual freedom can themselves vary. While individual freedom may well serve as a core value, community, democracy, equality, and a commitment to progress are recurrently associated with liberalism, either independently or through their relationship to individual freedom.

13. *Consilience* literally means the jumping together of knowledge and has been used to denote transdisciplinary efforts.
14. Altemeyer, B., & Hunsberger, B. (2005). Fundamentalism and authoritarianism. *Handbook of the psychology of religion and spirituality*, 378–393. Altemeyer, R. A., & Altemeyer, B. (1996). *The authoritarian spectre*. Harvard University Press. Altemeyer, B. (2004). Highly dominating, highly authoritarian personalities. *The Journal of Social Psychology, 144*(4), 421–448.
15. This book's author shares our ambition to move beyond purely environmental explanations of political ideology.
16. Lakoff, G., & Wehling, E. (2016). *Your brain's politics: How the science of mind explains the political divide* (Vol. 59). Andrews UK Limited.
17. This is the only book reviewed that explicitly takes a multilevel selectionist perspective, for, as explained by the author, multilevel selection theory offers an explanation of why "moral and behavioral pluralism is present in society." Faria specifically refers to multilevel selection theory on the second page of his introduction: "This book represents a unique research that uses the evolutionary model of multilevel selection to reassess the classical claims of public/rational choice theory regarding preference satisfaction in markets and liberal democracy."
18. Faria's book differs from our own in the following three ways: (1) it centers on liberalism rather than encompassing the broader political spectrum; (2) it focuses on the failings of liberalism whereas we focus on the dynamic tension created by opposing political ideologies; (3) it uses economic theory and philosophy whereas our book is more explicitly biohistorical.
19. Haidt's review of group selection, again a component of multilevel selection theory which we apply to political ideology within this monograph, is thorough and the concept is applied to political affiliation alongside moral reasoning and religious persuasion. That conservatives are group selected and liberals are individually selected is implied, seemingly suggested, easily discerned, but not, it seems, overtly stated. Still, this position has to be consolidated, more pointedly articulated, and fully explored, rather than being created *de novo*. As *The Righteous Mind* is of

such obvious relevance, Haidt's publications (2001, 2003, 2007) are of general utility. Haidt (2007) is a useful starting point in framing the present thesis. Therein, five innate intuitions are described as imparting culture-specific virtues: (1) harm/care, (2) fairness/reciprocity, (3) ingroup/loyalty, (4) authority/respect, and (5) purity/sanctity. As expressed in that paper, and during talks on its content, Haidt reviews ethnographic and individual differences across these five moral intuitions. Most importantly, the first two, dealing collectively with harm/care and fairness/reciprocity, appear to have a broad base within and between societies. Echoing Haidt, we see the first two moral intuitions as part and parcel of a general moral compass, but also that which is emphasized in liberal ideologies. On the other hand, the latter three forms of morality are more variably expressed, being consistently strong only among conservatives. Collectively, these have relevance, not to the individual and his proper treatment, but to the group and its cohesion and strength (Haidt, 2007). As one individual in communion with another, irreverence or irreligion has no direct or immediate consequences; yet, there may be deferred and indirect consequences to the conservative, which serves as the rationale behind its ostensibly irrational ire.

20. https://www.conted.ox.ac.uk/courses/samples/animal-behaviour-an-introduction-online/index.html

Part I

An Introduction & Organizing Theory

1

The Relation of Political Science to the Theory of Multilevel Selection

Steven Charles Hertler, Aurelio José Figueredo, and Mateo Peñaherrera-Aguirre

It would be no inapt definition of politics in our time to term it the search for social unity. Whatever political problems we may consider upon this fundamental question, we shall always ultimately be driven back. How far, and in what way, is our society one? How far is there an interest of the Whole, a monistic interest, which transcends the interests of the Many who compose that whole?

—Harold Laski, *Studies in the Problem of Sovereignty*

S. C. Hertler (✉)
Saint Elizabeth University, Morristown, NJ, USA

A. J. Figueredo • M. Peñaherrera-Aguirre
University of Arizona, Tucson, AZ, USA
e-mail: ajf@arizona.edu; mpeaher@email.arizona.edu

S. C. Hertler et al., *The Evolution of Political Ideology*,
https://doi.org/10.1007/978-3-031-83941-2_1

1 How Can Political Ideology Be Said to Evolve?

This volume traces the evolutionary roots of political ideology by examining empirical associations among genetics, endocrinology, neurophysiology, life history strategies, personality factors, and other individual difference variables. As will be established by the following four chapters, political ideology: (1) has genetic contributions; (2) correlates with known anatomical and physiological brain differences; (3) is affected by life histories; and (4) extends from personality traits and patterns. All of this is fascinating, but none of it is fully explanatory. Certainly, there is a measure of explanation at a proximal level, in, for example, reporting that conservatives have enlarged amygdalae. Yet, there is no contextualizing framework capable of ultimately explaining why these heritable and brain-based personality differences exist. Consequently, we begin this 20-chapter journey by introducing and explaining *Multilevel Selection Theory*, an indispensable theoretical framework capable of organizing all subsequent chapters as they variously review political ideology's evolution, both as an abstract feature of mind and as a contextual feature of society.

Before we begin, and to get off to a good start, let us first explain why we even think that it is reasonable to speak of political ideology *evolving*. As per Darwin's (1859) theory of evolution by natural selection, evolution occurred according to three related principles:

1. *Overproduction of offspring*, meaning that more offspring were produced in each generation than could be sustained by the local ecology's carrying capacity;
2. *Heritable variations among individuals*, meaning that the observed phenotypic (those characteristics of an organism arising from combined genetic and environmental influence) differences among individuals were at least partially heritable;
3. *Differential survival and reproduction among variants*, meaning that some of these individuals were better suited to the local ecological conditions, and thus possessed greater *fitness* in that environment. This process is called *natural selection*.

Subsequently, the *Neo-Darwinian Synthesis* of the early twentieth century combined Darwin's theory of natural selection with the findings of Mendelian Genetics. With the rise of population genetics, the evolutionary process was then operationalized as the *changes in gene frequencies from one generation to the next*. If this reasoning is valid, we must then ask how political ideologies could possibly be evolved in that manner. For example, many contemporary ideologies have scarcely even been in existence, at least in their present form, for more than one or two centuries. For *Homo sapiens*, a slow-breeding species, it is thus not highly probable for any genes coding for such specific ideologies to have evolved in that time frame. Chapter 2, *The Genetics of Political Ideology*, reviews political ideology's genetic loadings, while also describing their developmental influence on persons and their expressions within cultures. Suffice it to say at present, the plan of this volume is to, as evidence allows, walk the reader through various stages in this developmental and evolutionary sequence, and to tie these processes to the ultimate outcomes that influence political ideologies. As will be seen in Part I, genetically informed temperamental and cognitive variation evolved, which, as will be seen in Part II, interacts with culturally relevant content to inform political ideologies.

2 The Problem of Altruistic Cooperation and Social Order

Though initially envisioned as primer, this chapter became something of a narrative. It tells of how one question was approached by two disciplines. More specifically, it tells of the ways in which *political science* and *evolutionary biology* struggled to understand the ways in which individuals interfaced with groups. We first explain how evolutionary biology groped toward multilevel selection theory, thereafter describing political science's similar struggle to understand relations between subjects and states. Multilevel selection is a key theoretical model for ordering biological connections between genes, individuals, and groups, and one that can be extended to the ordering of those connections between *the one, the few, and the many* that political science has addressed in different ways across two centuries. We start with a brief bit of intellectual history that we think provides enough about multilevel selection's antecedents that readers will be better equipped to understand the abstract definition we thereafter provide.

A slate of theories, some of which we review, were in dialogue with one another, and sometimes vehement debate, about the nature of social cooperation. While consensus has not been obtained and debate remains,[1] we narrate how these theories eventually came to be clarified, reconciled, synthesized, and integrated into multilevel selection theory.[2] Altruism appeared antithetical to Darwinian evolution. Cooperation seemingly contradicted competition; and competition was a more obvious outgrowth of evolution. In the 1960s and 1970s several biological theories arose, all attempting to explain altruism and cooperation among individuals. It wasn't a Kuhnian process of anomalies accruing, causing a crisis, and ultimately ushering in a new paradigm. Anomalies were evident from the outset; they were recognized by Darwin, as when he struggled to explain cooperation among humans and animals, and especially when considering sterile worker castes within *eusocial* (many species of wasps and bees, and all species of ants and termites) insect colonies that forgo individual reproduction in favor of that of their queens (Shanahan, 2004). As passages quoted in the first chapter of our 2020 monograph, *Multilevel Selection Theoretical Foundations, Historical Examples, and Empirical Evidence*, show, Darwin effectively worked out these problems of cooperation on a conceptual level in a manner consistent with modern multilevel selection theory. Yet, his, and early biology's, limited understanding of genetic transmission made it impossible to fully reconcile these anomalies to the end of understanding problems of altruistic cooperation more fully. This only happened after the aforementioned *Neo-Darwinian Synthesis,* wherein evolutionary theory was reconceptualized alongside a modern understanding of genetics and genetic transmission to the end of creating a range of particularistic solutions, as reviewed immediately below.

Richard Dawkins' (1976) *selfish gene theory* reoriented evolutionary biology to the paramount importance of genetics within organisms.

Organisms were relegated to the role of *vehicles*; they were temporary vessels within which genes resided, with genes being that which endured and recombined within interbreeding populations. Yet, even before this clarification, genetic understanding since the Neo-Darwinian Synthesis allowed William Hamilton's (1964) theory of *kin selection* to explain altruism among relatives on the basis of shared genetics. A mother risks life and limb for her daughter because the mother's daughter shares half her genetic material. On this underlying genetic calculus, there is little distinction between these family members, and so every reason for self-sacrificial altruism to evolve. *Inclusive fitness* also explains altruism on genetic grounds,[3] but without the requirement that shared genetics come from recent common descent. In other words, one individual helps another in recognition of a shared phenotypic trait which signals a shared genetic basis, only that shared genetic basis does not necessarily come from a direct line of ancestry. Despite these distinctions, in both cases, shared genes facilitate altruistic cooperation. Robert Trivers' (1971) *reciprocal altruism* attempted to solve the problem of altruism where no direct or indirect genetic relationship was shared. Systems of cooperation can be established based on mutual gain, even when altruistic acts are not immediately or symmetrically recompensed. Gintis' *strong reciprocity* then stepped in with a complementary solution to this same problem. It explained how humans, even at great cost to themselves, were evolutionarily motivated to punish others into conforming and cooperating. Reciprocal altruism and strong reciprocity were respectively the metaphorical carrots and sticks explaining non-kin cooperation at larger levels of social aggregation among fully unrelated individuals.[4] Though by no means exhaustive,[5] these are major milestones within the field of social evolution. To summarize and simplify, these theories were attempting to explain social evolution along the following different *levels* of aggregation:

Kin selection	→ within families
Inclusive fitness/genetic similarity	→ across larger interbreeding populations
Reciprocal altruism/indirect reciprocity	→ among even less related individuals

3 Group Selection: The Midwife of Modern Multilevel Selection Theory

Emerging after the Neo-Darwinian Synthesis alongside kin selection, inclusive fitness, and reciprocal altruism, was *group selection,* the idea that natural selection can select for or against, not only individuals, but also groups. Just as strength evolves as a strong organism vanquishes a weak organism, cooperation evolves as a more cohesive group vanquishes a less cohesive group. And of course, no literal vanquishing is necessary; one group can simply have a higher birth rate and increase their demographic proportion relative to another group. Likewise, it is not only cooperation that can evolve, but other group-level traits capable of fostering martial, demographic, or territorial advantage. In this way, and as group selection claimed, evolution can operate at the level of groups, increasing or decreasing group numbers by selecting for or against traits.

As reviewed by Gilpin (1975), Okasha (2006) Haidt (2013), and especially Borrello (2010), there was controversy surrounding group selection. The controversy is traced to Wynne-Edwards (1962) who effectively framed group selection as a species-level evolutionary process tending toward the *good of the group*, protecting against, for example, demographic collapse[6] (Wilson, 2002). This is now described as *naïve group selection* in contradistinction to group selection as it is currently conceptualized. Through a tortuous process of criticism and resultant refinement that is beyond the scope of this chapter, group selection eventually became established on a firmer foundation.

With hindsight, the aforementioned theories of kin selection, inclusive fitness, and reciprocal altruism all explain the existence of hierarchically nested groups. Yes, they differ mechanistically in the weight assigned to genetic relatedness, and yes, they tend to explain differently sized groups, but all three are explaining the evolution of groups. In other words, kin selection strongly emphasizes genetic relatedness to explain cooperation among families, inclusive fitness less strongly emphasizes genetic relatedness in explaining cooperation among interbreeding populations, and reciprocal altruism dispenses with genetic relatedness altogether in explaining cooperation among large masses of unrelated individuals, but all three theories are instances of group selection—that is, evolution selecting for or against individuals, not solely based on their individual behavior or characteristics, but also on the behavior or characteristics of the group of which they are a part. This realization that such theories were not alternative competitors to group selection, but rather particular instantiations of group selection, only became known as group selection itself was assimilated as the main component to multilevel selection theory. Along with the individual, the levels within multilevel selection theory are the groups of various sizes discussed by group selectionists.[7]

Again, our interpretation, and we think one shared by David Wilson, Edward Wilson, Samir Okasha, and other multilevel selection theorists, runs thus: kin selection established the evolutionary logic of kin-based altruism. Nevertheless, it was recognized at length that a family is merely a lower-level, genetically related group, making it difficult to accept inclusive fitness and reject group selection. Reciprocal altruism was then another separate theory used to explain cooperation among non-kin, and also seen as a rival to group selection. Yet again, however, these reciprocal relationships, and the friendships and alliances they foster, merely represent non-kin groups. So, in effect, there were separate theories of cohesion and altruism for kin and kith, which, in hindsight, for many, appear only as components of group selection. Interested readers can trace group selection's labyrinthine intellectual history by reading the introductory chapters of Hertler, Figueredo, and Peñaherrera-Aguirre (2020) or Woodley of Menie et al. (2017). Beyond these two sources, readers can reference any number of books,[8] peer-reviewed articles,[9] and other sources.[10] Whether one finds this intellectual history of group selection intensely interesting, or oppressively dull, its further scrutiny is unnecessary to the goal of supplying a prerequisite intellectual armament,[11] enabling the comprehension of multilevel selection as it is explicitly or implicitly referenced in all forthcoming chapters.

4 Delineating Groups

The historical polemics regarding individual versus group selection, which led to the development of multilevel selection theory, unfortunately perpetuated an oversimplified dichotomy between individuals and groups as the two primary levels of selection. As explained in the preface to Hertler et al.'s (2020) monograph on multilevel selection, selective processes can occur and even co-occur at various levels of biological and social organization besides individuals and "groups." One important reason for this fact is that the word "group" is ambiguous, especially in complex fission-fusion societies (Hertler et al., 2020), such as those of humans. A "group" can mean any social aggregation of humans of varying levels of complexity. In complex fission-fusion societies, these levels are hierarchically nested, one within the other, like Russian Dolls. We have made two prior attempts to rectify these problems, one which illustrates fission-fusion, and one which delineates a more realistically complex range of groups. In discussing the transience of groups, and their fission and fusion up and down a ladder of aggregation, we featured the following graphic in our 2020 monograph on multilevel selection:

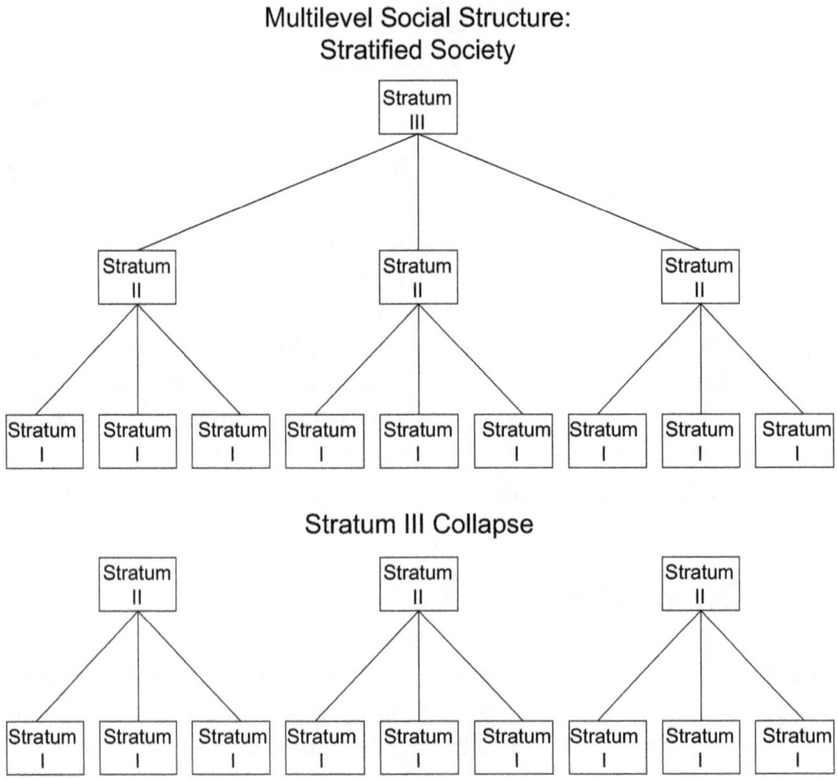

This schematic will be useful to create a delineated hierarchy within which individuals are nested inside multiple layers of aggregation, which are then embedded in biological and physical ecologies. We have also provided a taxonomy of groups. When updating the work of Urie Bronfenbrenner, Figueredo et al. (2017) suggested an *Evolutionary Ecological Systems Theory* that proposed a more biologically and anthropologically informed hierarchy of levels. Given its evolutionary inspiration, however, many of these levels no longer apply to the modern societies that are the main subject matter of the present volume. We have therefore updated this system to reflect the levels found within most modern societies as follows:

I. Individual
II. Immediate Kin (Consanguineal and Conjugal)
III. Extended Kin (Consanguineal and Conjugal)
IV. Friendship/Alliance Kith Networks
V. Local/Community-Level Organizations
VI. Province/Federated State
VII. National State-Level Organizations (Ethno-State)
VIII. Supranational State-Level Organizations

Thus, an adaptively relevant "group" may represent any level from the Levels II–VIII listed above. This alternative schema not only specifies what "group" individual interests might be conflicting with, but also implies that interests across these different levels might also be conflicting with each other. In the first case, we predict that individual interests will likely come into greater conflict with higher levels than lower levels. In the second case, there are many historical examples of either provinces or federated states coming into conflict with either national or supranational state-level organizations. Such conflicts often lead to either attempted or successful secession from the higher-level organization; and any violent confrontations that ensue are often referred to as "civil strife" or even full-scale "civil wars."

5 Walking the Line

At the risk of repetition, and before continuing on to consider implications, we provide the following consolidation: As per the dictates of multilevel selection, evolution proceeds by way of changes in gene frequencies. An additional implication of multilevel selection theory is that genes reside in other vehicles—not just individual organisms, but in groups of organisms. Just as organisms are bundles of evolving genes, so are families, tribes, and nations. Multilevel selection can be said to explain the multiple levels of selection upon which evolution acts. Traditionally, the organism was understood as the sole level of selection, but multilevel selection identifies genes as that which evolve, while relegating organisms to the role of a vessel or vehicle for propagating genes. Along with individual organisms, there are various groupings of organisms, such as families, packs, colonies, and troops; and among humans there are also sects, guilds, states, and empires. Along with the individual, these hierarchically nested levels of social organization are the various *levels* within multilevel selection theory; these are the vehicles through which gene frequencies increase or decrease.

Many have followed this line of logic to a blind end. They see the genetics residing in individuals, those individuals in groups, and then conceive of these groups in competition and at war with one another. So far, so good. Here is where the misstep takes place: *The groups themselves are then discussed as surviving and reproducing relative to the survival and reproduction of other groups*. The problem with this last leap of logic is that, when taken too far, it treats individuals as cells within an organism. However, individuals cannot be treated as such because this violates two realities. First, individuals can leave a group; they can emigrate and convert or defect and desert. Second, individuals are members of many groups simultaneously; they are nested within multiple groups at different layers of aggregation. They are simultaneously members of immediate and extended families, neighborhoods, states, and nations. Were these different layers of aggregation always in alignment, it might make a certain amount of sense to compare them to tissues and organs of an organism, but different layers of aggregation are not always so aligned, as in civil wars and intranational religious conflicts. This logical error of relegating individuals to the roles of cells within an organism is a holdover from *naïve group selection*, and is now described as *MLS2*; that is, multilevel selection two. Unlike *MLS1* which is the contemporary articulation of group selection discussed and applied in this book, MLS2 is theoretically possible, but not empirically supported. Furthermore, MLS2 thinking undermines the central tenet of multilevel selection theory, which, recall, is a gene-centric theory that relegates individual organisms to the role of vehicles; it follows that groups are also only afforded the status of vehicles. Just as genes are recombining as individual organisms breed and die, genes are recombining as groups form and dissolve.

While it is a categorical error to think of groups as tangible objects in the world, dismissing the importance of groups is an equal and opposite error. To assert that an individual's fate or fitness can be affected by the group in which he resides, is not tantamount to anthropomorphizing the group into a physical being.[12] Neither intangibility nor impermanence divests a group of its power to organize individuals and affect their fates, politically, economically, and with respect to survival and reproduction. Just as an individual affects whether or not his genes persist, so a group affects whether or not an individual (and ultimately the genetics of that individual) persists. A broad range of group behaviors and characteristics can alter the fate of groups and that of their individual members. Had the pass not been blocked at Thermopylae or the walls not defended during the Siege of Vienna, individuals within those groups would have perished and failed to propagate their genetics. The same could be said of the Battles of Marathon, Syracuse, Arbela, Metaurus, Arminus, Chalons, Tours, Hastings, and other *Decisive Battles of the World* collected by Sir Edward Creasy (2013). Most dramatically, think of the European discovery of the Americas, after which, with little reference to individuals involved, English demography seeded a continent, Amerindian demographics declined dramatically, Mestizo populations were created de novo, and African populations were separated, suppressed during transatlantic passages, bred, and miscegenated.[13]

6 Tension at Opposing Levels of Selection

With multilevel selection's history reviewed and its principles explained, let us more pointedly consider its implications for competition and cooperation. If genes are drivers and all else are vehicles, it follows that conflict and cooperation can happen even at the level of organisms. Indeed, such is the case. An entire literature exists on *selfish genetic elements*, which propagate themselves to the detriment of the organisms in which they reside (Burt & Trivers, 2006; Fishman & Jainike, 2014; Ågren, 2016; Lindholm & Price, 2016). By 2006 when Burt and Trivers (2006) published *Genes in Conflict: The Biology of Selfish Genetic Elements,* they made mention of thousands of articles with thousands of species implicated (Burt & Trivers, 2006). *Segregation distorters* (Robinson et al., 2008), *maternal effect killers, b-chromosomes* (Burt & Trivers, 2006), *virophages, polintons,* and *transpovirons* (Yutin et al., 2013), and other selfish genetic elements cause genomic imprinting (Werren, 2011), distorted sex ratios (Hurst, 1995), sterility (Ingvarsson & Taylor, 2002), cancer (Featherston & Durand, 2012), as well as alteration to genome size and architecture, and speciation events (Ågren & Clark, 2018). We need not delve further into the fascinating particularities of this literature, but instead focus on its significance. To that end, we provide the following excerpt from our 2020 monograph:

> Burt and Trivers, after characterizing the operation of selfish genetic elements across hundreds of pages, provide the following reflection: "The disunity of the organism is manifest once more" (Burt & Trivers, 2006, pp. 420–421). The weight of evidence and example leads to the insight that, what appear as unitary wholes are no more than tightly associated federations of cells, cohering as a complex whole with the aid of occasional coercion. In this way, the organism may just be the most tightly associated of all such groupings.

There is some "betrayal" so to speak at the organism level, but if tribes, nations, and other groups were as tightly bound as organisms, we might be able to think and speak of them as organism-like; as having an existence as a tangible and enduring entity. In other words, we could plausibly treat groups as those espousing naïve group selection and MLS2 have treated groups. Yet, there is still room for cooperation where cooperation can advance the survival and reproduction of organisms and their genes.

As we have done, many who write on multilevel selection cite the following Arab proverb: "I against my brother. I and my brother against my cousin. I, my brother, and my cousin against the world." It illustrates the nested levels of selection undergirding a hierarchy of allegiance. This proverb might be expanded, before coming to an end, by interjecting several more levels, such as that of the nation or cultural-religious population. Along these lines consider the following passage from Gowdy et al. (2016, p. 336):

> The same potential conflict between prosocial and disruptive self-serving behaviors occurs at all rungs of a multitier hierarchy. What's good for me can be bad for my family. What's good for my family can be bad for my clan. All the way up to what's good for my nation can be bad for the global village. The general rule is: Adaptation at any level of a multitier hierarchy requires a process of selection at that level and tends to be undermined by selection at lower levels. Another point is that the balance between levels of selection is not static but can itself evolve. When mechanisms evolve that suppress the potential for disruptive self-serving behaviors within groups, the group evolves to be a highly cooperative unit: a veritable superorganism.

This quote begins by describing lower-level aggregations, knit together on a genetic basis, but continues to describe higher levels of aggregation having no genetic basis. Genetic relatedness is integral to lower levels of selection; those relating to individuals and families. Largely on a genetic basis, we see multilevel selection play out in primate troops and small-scale societies. There is an innate tendency to look out for oneself and one's close kin, leaving less and less altruistic concern for those fully unrelated to oneself. Time, money, food, and other resources tend to be distributed on this basis. Often threat from a higher level quells squabbling at a lower level. For instance, families might reconcile when threatened by a rival family, while regional contests might be suspended by foreign invasion.

Several mechanisms stimulated larger aggregations of distantly related individuals. They were sometimes stitched together by ecological abundance which supported concentrated populations of persons; or by *ecological caging* wherein groups were penned into livable territory by harsh surrounding conditions. The punishment meted out by *strong reciprocity* coerces non-cooperators, while the promise held up by *comparative advantage* entices non-cooperators. Religion, perhaps especially universalizing and proselytizing monotheistic religions, created ties among large swathes of unrelated individuals. Then there are nations, formed around very broad-based common ancestry, that imperfectly align with states, which are sometimes, but not always, nation-states. These are also large conglomerations of unrelated or very distantly related persons. These higher levels of organization relate to the literature on *ultrasocial* societies. Ultrasocial societies are rare among species, being clearly evident only in eusocial insects and only emerged lately on the human evolutionary scene. There are precedents for human ultrasociality extending back to the upper Paleolithic with an acceleration of ultrasociality coming in the wake of the *Neolithic Revolution* once agriculture further fostered demographic expansion, division of labor, and sedentism (Gowdy & Krall, 2013; Gowdy, 2021). From thence came ancient empires and modern states, which are the highest levels within the multilevel selection framework (Turchin, 2013, 2016).

Naturally self-interested motives are countered by selfless motives, which are equally innate. These motives are both socialized and selected evolutionarily by other societal members who police and punish selfishness, and alternatively reward and revere selflessness. Now one can better see how multilevel selection theory is intimately related to altruism, and was first invoked to explain eusocial insects, and is routinely cited in the literature on altruistic punishment, and collective action problems. A group can be more successful than another group without coming into contact with that group, much less eradicating it. A highly group selected population can augment its numbers via self-sacrifice, solving collective action problems, differentiating labor, maintaining tight legal controls, and the like; all of which will tend to increase its numbers vis-à-vis other groups lacking these characteristics. This group will then expand and become more populous, resulting in directional selection within the human species, without a spear thrown or a shot fired (Gintis, 2017). At least two studies have begun to set the empirical footings upon which group selection theory can more stably rest, one of which shows group selection operating in Japan during the Warring States Period (1467 to 1567), and the other which shows group selection operating among Britannics from 1600 to the present (Woodley of Menie et al., 2017); and these are paired with evidence from animals, such as studies of hen breeding[14] and parasite propagation.[15]

7 Political Science and the Problem of Social Order

Hobbes' *Leviathan,* Mandeville's *Fable of the Bees,* Rousseau's *General Will,*[16] Tocqueville's *democratic associations,* and Montesquieu's *Spirit of the Laws* all attempted to answer, "the question of how men form themselves into association..." This vexed, perennial, and seemingly insoluble question "lies at the heart of Western political thought" (Runciman, 1997, p. 3).[17] Political scientists inherited the burden of this question from their philosophical progenitors, and are often seen writing in reaction to eighteenth- and nineteenth-century models emphasizing the state as an organically formed entity with a moral and legal existence in the process of becoming; and one separate, in some sense, from its subjects or citizens (Bevir & Rhodes, 2010; Adcock et al., 2007). Contra Bluntschli, Bosanquet, and Hegel, all of whom envisioned the state as a living, being, and growing entity existing above and separate from its inhabitants (Bevir & Rhodes, 2010; Farr & Williams, 2015), Harold Laski, F. W. Maitland, J. N. Figgis, Earnest Barker, and G. D. H. Cole founded *pluralism,* a major school of thought within political science (Runciman, 1997) emphasizing lower-order groups over and above sovereign states (Enroth, 2022). These are the individual citizens, ethnicities, elites, congregations, voting blocks, unions, parties, factions, clubs, and related interest groups underpinning the social whole. As seen in the example of Maitland, pluralists at most allowed that the state was a "highly peculiar group-unit" but still merely one species of association among many (Runciman, 1997, p. 95). Laski denies to the state monolithic allegiance, averring that monistic loyalties are fictions. Instead, "allegiance is divided between the different groups" to which one belongs (Laski, 1917/2014, p. 15). The state might not even command fealty over and above other levels. Religion and family are stiff competitors, respectively from above and below.[18] At the hands of pluralists, the state was shorn of its status as a preeminent whole. Neither the individual person nor the small association was in any way reduced to a cell or organ of a reified state.

However, Laski and fellow pluralists never really resolved this fundamental question. Without precisely coming full circle, their thinking progressed in reconsidering the seminal work of German legal scholar Otto von Gierke,[19] whose notion of *Genossenschaft* found unity in a form of bottom-up plurality (Runciman, 1997). Pluralists initially developed Gierke's ideas about interrelations among small groups, while deemphasizing Gierke's sympathies with German mystic nationalism in the cast of Hegel, who had seen man achieving "…existence, truth and morality only in his capacity as a member of the State." Most especially Figgis and Laski increasingly looked up the scale of aggregation. Figgis, in the later evolution of his thought, followed Gierke back to sympathies with something approaching a reified state (Runciman, 1997; Newman, 1993; Kavanagh, 1994; Lamb, 2004). Laski's reckoning with Gierke came after WWI, which was inaugurated by lower-level ethnic conflicts to the end of destroying three "multi-national dynastic empires": Habsburgs, Romanovs, and Ottomans. Contemporary political scientists have continued trying to resolve what the pluralists could not resolve. Bevir and Rhodes (2010) are prime examples of this continuing conundrum. Like the Pluralists, Bevir and Rhodes deny *autonomy, efficacy,* and even *preexistence* to the state that is in any way separate from its inhabitants and their behavior and thought. As with the pluralists, Bevir and Rhodes are writing in reaction to Victorian Era proto-political theorists, and their propensity to see the mother country or fatherland as having an independent existence.[20]

Reading pluralist literature or their philosophical progenitors, feels uncannily similar to our experience reading the biological literature. It is as if one is walking through the same town from a different direction, where the streets and stores have been renamed. Validating our experience is American sociologist Dennis Wrong, who, in *The Problem of Social Order* (1994), draws parallels between political science and biological evolution. Wrong (1994, pp. 5–6) first asked, *what are the relevant units?* "...individuals, dyads, families, small groups, local communities, politically organized associations, or 'imagined communities'" are among the "different levels" of social organization that "need at some point to be recognized and considered separately in their own right." "If the problem of order is seen as embracing units ranging from single individuals to organized political communities including millions of people," Wrong (1994, p. 6) continued, "it spans the entire continuum from face-to-face encounters between individuals to the relations between great powers in world politics." Further on in his book, when reviewing the intersection of political science with social Darwinism, and after commenting on Spencer and Sumner, Wrong (1994, p. 30) states:

> Yet the Darwinian focus on the individual as the unit of the species had also brought renewed attention to the problem of how these units succeed in subordinating their own strivings to a shared set of rules, the question first raised by Hobbes with such rigor and clarity in the seventeenth century.

8 Two Disciplines, One Problem

We have seen that evolutionary biology traversed the same ground in reverse. In contrast to political science, evolutionary biology was a latecomer to the problem of ordering social groups due to its focus on individual organisms since the 1960s. There were precedents[21] in Darwinian writings (1859, 1871), though only in the last quarter of the twentieth century did theoretical developments begin to explain the highest levels of aggregation.

With the guidance of Adcock et al. (2007), Hechter and Horne (2009),[22] Bevir and Rhodes (2010),[23] Enroth (2022) and related books on political science historiography, we came to see the dialectic between citizens and states as political science's *raison d'être*. We later indulge in a broader exploration of relations between citizens and states in Chaps. 12, 13, 14, but here focused on the pluralists as, in their example, we see a newly gelling discipline contrasting itself with prior philosophical theories tending toward the reification of the state. In reaction, pluralists went too far in the other direction, denying special characteristics to the state, and more generally, to higher-order groups. As said, where political scientists began with reified states and looked down the scale of aggregation to lower-level groups and individuals, evolutionary biologists began with individual organisms and looked up the scale of aggregation to higher-level groups and peoples. Just as pluralists went too far in reacting to reification, some evolutionary biologists went too far in reacting against traditional biological accounts focusing on the evolution of individual organisms. In doing so, naïve group selectionists and proponents of MLS2 committed an error akin to reification—they envisioned groups as organisms. Once again, we see two disciplines chasing down an elusive answer to the same question, beginning at different ends of the field and running in opposite directions.

As we have seen, evolutionary biologists were initially confused by the relation between organisms and their groups, but, at length, came to understand both as vehicles—temporary assemblages housing and perpetuating genetic information. Genes are transmitted across generations based on the efficiency of vehicles. They are better transmitted by neurophysically intact individuals and cohesive and altruistic groups. With this insight gained, evolutionary biologists could finally understand that, like individuals, groups are vehicles for genetic transmission, but unlike individuals, groups are intangible and less cohesive. In consequence groups should be acknowledged for their emergent properties without being afforded the status of superorganisms that themselves reproduced.

Evolutionary biology is no longer running up and down the scale of aggregation looking for answers. Multilevel selection cuts the Gordian Knot in recognizing both individuals and states, and all levels of aggregation in between, as vehicles for genetic information. With this in place, selection can be understood to operate at any and all levels of aggregation. In applying this insight, we can see that groups are neither agentic wholes nor reducible to the sum of their parts. Once again for the sake of clarity, groups, not excepting states, have emergent properties important to the propagation of genetics through the survival and reproduction of individuals within those groups. Thus, this book, in deploying multilevel selection theory, contributes a mechanistic framework structuring relations between genetics, individuals, and groups, thereby considering group attributes not reducible to group members.

Like post-pluralism political science, multilevel selection theory has a longstanding literature explaining how nested levels of identity and loyalty cause the fission and fusion of groups. Fission and fusion into larger or smaller groups can hang on questions of religion, ethnicity, and territoriality. This is why multilevel selection so cogently explains why France fought with England and Christians fought with Muslims. These national and religious ingroups fought their rival outgroups. Multilevel selection equally explains why, especially in the absence of external threat, political and religious ingroups fall out among themselves, as when France's Third Estate revolted against its First and Second, or why Christian powers fought among themselves after wresting the Holy Land from the Muslims. Outgroup threat fostered ingroup cohesion. When outgroup threat wanes, ingroup divisions emerge. This is because any large groupings, such as Christians or Frenchmen, are themselves composed of lower-level groups; and in turn, those lower-level groups are composed of even smaller groups, which are ultimately composed of individuals.

9 Conclusions

Multilevel selection, introduced in this chapter, is a framework for understanding how selection operates across at least the following four different levels of biological aggregation: (1) genes; (2) neurophysiology; (3) individual phenotypes; (4) groups. Genetic differences are the substrate of neurobiological differences that collectively inform those behavioral differences described as moral emotions, temperament, or personality, all of which comprise part of an organism's behavioral phenotype. These behavioral differences affect survival and reproduction. But this does not happen in a vacuum. As genes act within the confines of organisms, organisms act within the context of groups. While groups are not tangible, as are organisms, they are nonetheless aggregations with emergent properties fully capable of affecting the survival and reproduction of individuals, and their underlying genetics.

A large share of the power and novelty of this book is afforded by multilevel selection theory's application to ideology.[24] As readers will only see by the end of Chap. 8, multilevel selection explains the political along two lines. The first has been discussed implicitly: individuals form into groups for defense and resource acquisition or are parts of groups for reasons of relatedness, propinquity, or historical inertia. They then vie with other groups on that basis. Multilevel selection in consequence provides a framework for understanding interest group politics, such as when political cleavages align across classes, races, religions, sections, and clans. This perspective reverses Clausewitz's quip to read, *politics is a continuation of war by other means.* Only hereafter do we see how multilevel selection also explains political ideology as a feature of mind as it varies across individuals within any of these groups.

Notes

1. In applying Multilevel Selection Theory (MLS) without descending into an extended debate about its merits, we are following the precedent of evolutionary biologist David Sloan Wilson who provides a post-resolution account in 'Does Altruism Exist? Culture, Genes, and the Welfare of Others,' which was published by Yale University Press eight years ago. The entire intellectual history of MSL's theoretical progress is detailed in our 2020 monograph and is expertly narrated by Borrello (2010). MLS emerged alongside other theories attempting to explain social order, such as kin-selection theory. While some still see kin selection as separate and superior to MLS, many other theorists like ourselves see kin selection as one mechanistic account of how genetically related groups are formed and maintained. As D.S. Wilson and others see it, kin selection is not capable of explaining the behavior of large aggregates of distantly related individuals at higher levels of sociopolitical complexity. MLS theory is uniquely able to account for the behavior of the larger, state-level human societies that are the subject of the present work, which is about political ideology. We do not believe that such ideologies could have existed in the smaller scale pre-state societies such as hunter-gatherer bands. Again, this and other controversies about MLS defy condensation and require book length explanation. These books have already been written; we have written one. We want to signal to the reader that

differences of opinion exist which are discussed at length in our 2020 monograph and in Borello's history of group selection.

2. E. O. Wilson, Sober, and D. S. Wilson would readily agree to this statement, though others, like Stephen Pinker and Richard Dawkins would object strenuously; but this really only reflects the divide between those that are, and are not, adherents to multilevel selection theory.
3. See Park (2007) for an extensive exercise in differentiating kin selection from inclusive fitness.
4. Turchin's *ultrasociality* would then extend and further explain large-scale cooperation at the level of nations and empires.
5. Other noteworthy milestones in the field of social evolution include Price's (1970) *Price Equation*, West-Eberhard's (1967, 1970, 1975) elucidation of relevant ecological factors, E. O. Wilson's *Sociobiology* (1975), and Rushton's (1989, 1998) *Genetic Similarity Theory* (Hertler et al., 2020).
6. Wynne-Edward's theory of group selection is now often referred to as *naïve group selection,* with the adjective *naïve* denoting its "failure to incorporate plausible opposing forces to those of individual selection." This is quoted and reviewed in the second chapter of our 2020 monograph on multilevel selection.
7. We are eliding over decades of controversy while also offering an interpretive framework that is by no means universal, but one we think is empirically grounded and accurate.
8. Williams (1971/2008), Sober and Wilson (1998), Wilson (2002), Okasha (2006), Bahar (2017), Borrello (2010).
9. Smith (1964), Wade (1977, 1978), Wilson (1983), Grafen (1984), Boyd and Richerson (1990), Soltis et al. (1995), Henrich (2004), Richerson et al. (2016), Krasnow et al. (2015), Pruitt et al. (2017).
10. https://whyevolutionistrue.wordpress.com/2012/06/24/the-demise-of-group-selection/

    https://www.edge.org/conversation/the-false-allure-of-group-selection

    http://www.life.umd.edu/faculty/wilkinson/BIOL608W/WestGriffinGardner_08.pdf

    http://www.nytimes.com/2011/09/11/books/review/the-neighborhood-project-by-david-sloan-wilson-book-review.html?pagewanted=all

    https://www.prospectmagazine.co.uk/magazine/edward-wilson-social-conquest-earth-evolutionary-errors-origin-species

11. There were critics of group selection, and still are critics of multilevel selection. Criticism classically extended from unrealistic assumptions, but most often continues in some quarters because modern multilevel selection is not sufficiently differentiated from naïve group selection. Even as there remain prominent dissenting views (Pinker; Dawkins), consensus is being broached as to the validity of multilevel selection (Richerson et al., 2016; Wilson, 1975; Wilson, 2002). The conditions under which group selection operates are now more precisely defined and solidly established, and the concept itself is now differentiated from its earliest articulation and matured as one level within multilevel selection theory. Evidence is accruing and scientific opinion is following, even as it has not built to a consensus. Yet consensuses, like appeals to particular authorities, are ultimately poor ways to validate theories. Accordingly, this is why we included empirical evidence and historical examples in our 2020 monograph entitled, *Multilevel Selection: Theoretical Foundations, Historical Examples, and Empirical Evidence.* It would be folly to attempt any summary of two major parts of our prior monograph within one subsection of one chapter of this monograph.
12. The group is a hypothetical construct, but like other hypothetical constructs, is nonetheless able to impart fitness-relevant effects. The same is true for intelligence and personality; these are intangible constructs, but nonetheless are fitness relevant.
13. Higher levels of aggregation, such as modern states, were dependent on population growth and interconnectedness. Societies need to contact one another to define themselves in relation to one another. This is generally why prehistory contains anthropological accounts of lower-level groups because these were juxtaposed to one another with little opportunity for them to combine in conflict with larger groups.
14. Kerr (2009) describes breeding with chickens, noting that taking the most docile and fecund chickens from a coop and trying to breed them, though it fully corresponds with traditional evolutionary selection views, does not always work, as narrated by Gowdy et al. (2016, p. 337):

    > Imagine that you are a poultry breeder who wants to select for egg productivity. You monitor the productivity of all the hens and select the most productive hen from each cage to breed the next generation of hens. To your surprise, egg productivity goes down and hen mortality goes up. An ounce of multilevel selection thinking provides the answer: By selecting the most productive hen within each cage, you

> selected the most aggressive hens that achieved their status by suppressing the productivity of the other hens! The wiser strategy, and the one followed by the poultry industry, is to select the most productive cages rather than the most productive hen within each cage to breed the next generation.

Here individual and group selections are pitted against one another. A dominant, aggressive hen is most fecund relative to other hens within a given coop; but a hierarchically ordered coop is most fecund relative to other coops.

15. Using the example of a parasite trematode, the lanceolate fluke (*Dicrocoelium dendriticum),* Sober and Wilson show how self-sacrificial altruism evolves. This parasite creates a brain worm within ants which induces infected ants to rise to the top and lock its jaws around a blade of grass making it more likely that they will be eaten by a ruminant like a goat or sheep. The parasite then infects the ruminant liver. Thereafter, it is expelled in feces which are eaten by snails. Then the parasite comes out in a mucus envelope which is eaten by ants, and so the life of these parasites cycle between ants, ruminants, snails, and back again. Controlling ant behavior, making ants lock their jaws on tall blades of grass, is essential to completing the parasite's life cycle. But to make ants act in this uncharacteristic manner, requires one "volunteer" to become a "brain worm." That particular parasite that becomes a brain worm does not reproduce and so has no subsequent representation in the next generation. So, in light of this in isolation, it seems that brain worms should decrease and disappear. However, the key to understanding why this does not happen is to consider the individual ant with its population of 50-odd parasites inside, against other ants with their parasite burden. Within the ant, a brain worm has lower fitness when compared to its counterparts. However, the brain worm increases the chance of being eaten by a definitive host and so populations sequestered within ants with a brain worm have higher relative fitness than their counterparts without a brain worm. Consequently, self-sacrificing types that sometimes turn into brain worms will have lower than average fitness within their group, but higher than average fitness relative to groups with no brain worm types (because they are less likely to be eaten in the first place). And so, Sober and Wilson conclude, what looks to be something impossible to explain, becomes easy to explain.

16. Rousseau is of a different breed, but his concept of the general will is understood to dangerously concentrate passion at the level of the nation-state and amounts to a reified notion of will, as if the will of the state were akin to the will of an individual.
17. Runciman on the same page of the above-quoted book goes on to state: "An association is a group of human beings possessed of a distinct, formal identity based on the relation subsisting between its members, and it may fairly be said that it is around such groups that our political understanding has been constructed."
18. The following quote is further instructive of Laski's views, and is seen to correspond with those of Maitland, his fellow pluralist: "You must on this view admit that the State is only one of the associations to which he happens to belong, and give it exactly that preeminence-and no more-to which on the particular occasion of conflict, its possibly superior moral claim will entitle it" (Laski, 1917/2014, p. 19).
19. Born in 1841 and writing for several decades, Gierke was a deep thinker developing a philosophical position supportive of the legal question of whether an association or corporation could be held responsible as could an individual.
20. We see these changes in political science as probably reflective of democratization and also the earlier trend in the West wherein individuals became more overtly valued as individuals. These are the historical developments that might broadly inform these intellectual developments.
21. Darwin's most profound insights about groups, and what came to be known as group selection, was sparked by ratiocinations on eusocial insects and their altruistic ways. Selections of Darwin's thought on the matter can be seen in Shanahan (2004) and in the first part of our 2020 monograph on multilevel selection.
22. Hechter and Horne (2009) outline five theoretical approaches to social order: (1) *individual behavior* producing aggregate social order; (2) *hierarchies* imposing social order from above; (3) *markets* that produce interdependence from below; (4) *groups* that bind individuals and regulate behavior; (5) *networks* generating ties across groups. The fourth and fifth approaches are particularly relevant to our discussion of evolutionary biology. Durkheim, Tocqueville, and Goffman, solved the problem of social order by emphasizing *ties within groups,* while others solved this problem of social order, among them Gellner, Simmel, and Gluckman, by emphasizing *ties across groups* (Hechter & Horne, 2009)

23. These authors position themselves on the opposite end of the spectrum, as we see it, from where political science began, in that they promote a decentralized, bottom-up view of the state as can be seen in the following passage: "We argue that political scientists should think of the state as a series of contingent and unstable cultural practices, which in turn consist of the political activity of specific human agents" (Bevir & Rhodes, 2010, p. 1). Beyond their specific contribution however, Bevir and Rhodes provide a detailed history of political science's various stages of development and schools of thought.
24. Multilevel selection in no way replaces traditional explorations and explanations provided by political science and historical analysis. Indeed, this is why our book's first part is followed by its second part. Nevertheless, multilevel selection provides the biophilosophical glue that binds different levels of explanation together, and one that also allows connections to be made across species, levels of aggregation, and across large swathes of time.

Part II

Proximate Causation

2

The Genetics of Political Ideology

Aurelio José Figueredo, Steven Charles Hertler, and Mateo Peñaherrera-Aguirre

> The two parties which divide the state, the party of Conservatism and that of Innovation, are very old, and have disputed the possession of the world ever since it was made. This quarrel is the subject of civil history…Such an irreconcilable antagonism, of course, must have a correspondent depth of seat in the human constitution. It is the opposition of Past and Future, of Memory and Hope, of the Understanding and the Reason. It is the primal antagonism, the appearance in trifles of the two poles of nature.
>
> —Ralph Waldo Emerson[1]

A. J. Figueredo • M. Peñaherrera-Aguirre
University of Arizona, Tucson, AZ, USA
e-mail: ajf@arizona.edu; mpeaher@email.arizona.edu

S. C. Hertler (✉)
Saint Elizabeth University, Morristown, NJ, USA

S. C. Hertler et al., *The Evolution of Political Ideology*,
https://doi.org/10.1007/978-3-031-83941-2_2

1 Introduction

This chapter reviews literature connecting genetic differences to political dispositions. We begin with a primer on molecular genetics and functional genomics to contextualize politically relevant genetics within a broad biophilosophical framework. After this primer reviews indispensable complexities and important nuances, an overview of twin studies provides heritability estimates and evidence of additive genetic variance associated with political inclinations. We cite two reviews, and several specific studies, representative of a larger data set yielding moderate to high estimates, broadly indicating that, like personality and other individual differences, sociopolitical attitudes typically show an average heritability of 0.50. More briefly, we consider relevant single nucleotide polymorphisms identified by genome-wide association studies, an emerging approach with potential to corroborate behavioral genetic findings. After cataloguing a representative review of genetic evidence, we thereafter reflect on its significance by: (1) distinguishing between lower-order emotions and higher-order cognitions; (2) considering the lagged expression of genetic effects; (3) discussing models of political ideology to the end of more precisely specifying what is genetic; (4) recounting the way in which *states* can suppress the expression of *traits*; and (5) interpreting ideological differences in the context of the *Overton Window*. These five sections amount to a philosophical reflection on genetic evidence, contemplating what it means when political ideology is said to be heritable.

2 A Primer on Molecular Genetics and Evolutionary Functional Genomics

We start with an extremely basic account of how genes actually function. To do this, we must first understand how *polymers* work, without delving too deeply into the underlying biochemisty. The word polymer in Greek comes from roots meaning *many* and *parts*, respectively. Complex sugars and starches are polymers, being made of parts that are called simple sugars, or *monosaccharides*. *Disaccharides*, like common table sugar (sucrose), are composed of two such simple sugars, whereas starches (such as *amylose*) are composed of more of them arrayed along a chain. Similarly, proteins (*polypeptides*) are composed of multiple amino acids (*peptides*), also originally arrayed along a chain before folding themselves into a three-dimensional structure. Most importantly for present purposes, *polynucleotides* (such as RNA and DNA) are composed of individual parts called *nucleotides*. As with monosaccharides, there are various different types of single nucleotides. The order among these in a DNA molecule is called the *nucleotide sequence*. By involved mechanisms beyond the scope of the present primer, the nucleotide sequences in DNA molecules ultimately code for a specific sequence of peptides in any protein to be assembled. The general category of proteins includes that of *enzymes*, which are the biological catalysts that moderate all cellular and metabolic functions. It is thus that the macromolecular composition and the sometimes contingent *expression* of genes biochemically control all organismic functions. It follows that it is extremely imprecise to say that genes code directly for specific traits, as Gregor Mendel had assumed, and instead more correct to say that macroscopic traits emerge developmentally as a result of complex metabolic pathways in which hundreds and perhaps thousands of enzymes (and therefore genes) have been involved.

Nevertheless, the fundamental dynamics of these processes are quite commonly misunderstood. In what follows, we outline a genetically informed account of the differences between traits that are commonly described as being *genetic*, *environmental*, or *cultural*, as well as the relations between these overlapping ontological categories. Much of what follows is paraphrased from the seminal work of Mary Jane West-Eberhard (2003) in her monumental monograph, *Developmental Plasticity and Evolution*. To fully comprehend the nature of these interacting processes, however, it will first be necessary to define a few widely misunderstood technical terms.

Developmental is usually taken to denote changes that may occur within the lifespan of a single organism and this level is usually taken to be exclusive of genetic changes. Nevertheless, it has now been shown that many developmental changes involve *epigenetic* modifications that biochemically alter gene expression, so this blanket exclusion of genetic change must be amended to only exclude changes to the basic nucleotide sequence of the DNA, as opposed to other regions of the molecule or chromosome. Developmental changes may occur according to a genetically prepared template, but may also occur in reaction to environmental events. If such developmental changes are permanent, this process is usually called phenotypic *plasticity*, whereas if they are temporary and reversible with subsequent environmental change, the process is sometimes called *flexibility* in contradistinction. Phenotypic flexibility has also been referred to as *facultative* or *contingent* processes, given that it depends on immediate environmental circumstances. In addition, developmental changes might also fall into the subordinate categories of either *individual* learning or *social* learning. Individual learning occurs through direct interaction with the external environment, whereas social learning occurs through either interacting with or observing and imitating the behavior of other organisms, usually members of the same species. A special class of social learning phenomena has been called *cultural*, wherein behavioral or cognitive traditions may be passed on from individual to individual for multiple generations.

Genetic, on the other hand, is usually taken to denote permanent and stable traits that change either very little or not at all within the lifespan of a single organism, but may change over multiple generations via evolutionary processes. Traits typically called "genetic" are presumed to be *influenced*, although not completely *determined*, by the basic nucleotide sequence of each individual's DNA, although subject to the aforementioned epigenetic effects on gene expression, which might either up-regulate or down-regulate the output of gene products. As described above, such gene products are proteins, which include the enzymes that control all cellular and metabolic functions within the organism. In that sense, all biological processes are genetic, in that they are all partially dependent on genes and gene products to operate in context; but they are all also developmental, in that they are all partially dependent on environmental input and epigenetic regulation to work in context.

Even such qualified distinctions, however, are subject to further complications and interpenetrations. First, modern genetics does not actually predict that genes explicitly *code* for any specific phenotypic traits at the macroscopic level. That is a misunderstanding that traces all the way back to the work of Gregor Mendel (1866) himself, and has long since been corrected, even as the error persists within the popular culture. Since the discovery of the structure of DNA, genes have been redefined as segments of that polynucleotide, whose sequence of nucleotides codes for the amino acid sequence of a single polypeptide, or protein. Many of these proteins are in a special class known as enzymes, which catalyze and thus control all biochemical reactions within the metabolism of every living cell. Thus, genes are ultimately in control of all life processes, but only *indirectly*. There is quite a causal distance from these microscopic molecular processes and the macroscopic molar processes that we normally refer to as phenotypic traits. The mediating steps are covered by all kinds of intervening structures that are constructed sequentially during development from these first beginnings. These include some of the predictive factors mentioned above, such as characteristics of the organism's endocrinology, neurophysiology, life history strategy, personality, and other individual differences. Furthermore, genes act in concert with each other, through processes of mutual interaction and regulation, in the aggregate and as an integrated *system*. There are very few traits that are determined by single genes, and most of these fall within the province of lethal mutations that disrupt normal developmental processes. By analogy, replacing a single ingredient in a recipe, such as salt for sugar in a cake, can radically alter the outcome. Nevertheless, a cake cannot be reduced to the action of either salt or sugar alone.

Second, genes do not act in mutual isolation from their ambient environment, but are in a state of constant *transaction* with it (e.g., Johnson, 2007). Although some prefer to use the term *interactions*, these gene-environment transactions may or may not be modeled as statistical interactions. For example, gene-environment *correlations* are not interactions. These correlations appear when certain genes occur in greater frequencies in some environments than others. These correlations may be produced by at least three causal processes (Buss, 1987): (1) *selection*, in which organisms enact an active choice of specific environments; (2) *evocation*, in which organisms systematically elicit predictable reactions from certain environments, most notably social environments; and (3) *manipulation*, in which organisms modify their own environments, as well as those of others. On the other hand, gene-environment transactions may in fact be true interactions. When they are, they are often referred to as *epigenetic* processes. The existence of such gene-environment *interactions* implies that the same genes may function differently within different environmental contexts. Such interactions may include differential gene expression in the presence of other genes, as other genes constitute part of the environment of any gene, and this specific case is called a gene-gene interaction or *epistatic* effect. These interacting genes may be inside the same cell or inside the same organism, but may also reside within different organisms, especially in social species, and this latter process is called *social epistasis* (e.g., Woodley of Menie et al., 2022).

Third, as implicit in the immediately preceding point, a major portion of the evolutionarily relevant environment in certain species is the social and cultural environment (Avital & Jablonka, 2000; Hoppitt & Laland, 2013; Sumpter, 2010). Thus, any change in the social and cultural environment will necessarily produce corresponding changes in selective pressures acting on the genes. Symmetrically, some genes bias learning processes, and these may thus influence the behavioral development of any social and cultural traits to which organisms are exposed in their environment. Thus, although the "Cultural Anthropology" of the mid-twentieth century championed by such luminaries as Franz Boas, Margaret Mead, and Ruth Benedict, sought to draw a clear demarcation between genetics and culture, we now understand that genes and culture are hardly independent of each other (Lumsden & Wilson, 1981). This is because, within the common processes of natural selection, cultural traits self-evidently serve as part of the selecting environment for genetic traits, and genetic traits self-evidently serve as part of the selecting environment for cultural traits. It therefore follows that any changes in gene frequencies in a population will necessarily produce corresponding changes in selective pressure acting on the behavioral development of any social and cultural traits to which organisms are exposed within that same environment. An infamous quip encapsulating this principle was that "genes hold culture on a leash" (Lumsden & Wilson, 1981, p. 13). This reciprocal process has been called *gene-culture coevolution*, which describes how cultural evolution necessarily affects genetic evolution and genetic evolution necessarily affects cultural evolution (Boyd & Richerson, 2005; Lumsden & Wilson, 1981).

Fourth, West-Eberhard (2003) has theorized that the first step in the evolutionary process is often *phenotypic accommodation* to novel environmental challenges, and that this is only afterwards followed by *genetic accommodation* to fix the optimal phenotypic response to the new circumstances by means of *epigenetic selection* (see Figueredo et al., 2020). This latter process is the selection of genes which react in given and predictable ways to environmental triggers, rather than by causing the development of a fixed permanent and stable trait. Therefore, gene-culture coevolution may be considered a special case of genetic accommodation, in which changes deriving specifically from the *social* environment may produce selective pressures for certain gene-environment interactions (meaning systematic patterns in gene expression) and consequently for gene variants that are biased to respond in this adaptive manner. Thus, although such epigenetic processes are *contingent*, they are by no means arbitrary or chaotic and are therefore subject to selection.

In summary, there is a common overall pattern to the dynamics of any process of "selection by consequences" (see Figueredo et al., 2020). To understand this concept correctly, it is important to distinguish between the original selective pressures and subsequent evolutionary process that produced a certain response disposition in the *past* and the immediate reaction that the organism emits in the *present*, although that may be partially based on its evolved response dispositions. The first process is genetic evolution and the second is merely behavioral reaction. Similarly, it is important to distinguish between the immediate reaction that the organism emits in the *present*, and the effects of the consequences of those actions upon the survival and reproduction of that organism in the *future*. The first process is behavioral reaction (not yet become selective pressure, *per se*) and the second is, once again, genetic evolution. Thus, none of the hypothesized processes are teleological in nature,, and the temporal direction of causation always remains from the past to the present and from the present to the future. The full process is nonetheless a cyclical one between evolution and behavior. This is essentially the same argument for non-teleology put forward by Burrhus Frederic Skinner (1953) for the dynamics of operant conditioning.[2]

3 Genetic Influences on Political Ideology

Lewis (2016) outlines literature connecting genetics to political ideology, beginning with a seminal 1974 study by Eaves and Eysenck, and extending beyond Alford et al.'s (2005) landmark article in the *American Political Science Review* entitled, *Are Political Orientations Genetically Transmitted*? Eaves and Eysenck et al.'s (1989) were among the first to identify that, relative to dizygotic twins, monozygotic twins correlated more strongly on political positions regarding unemployment, abortion, ethnocentrism, morality, marital unions, and death penalty. Inspired by Eaves and Eysenck's work, subsequent studies gathered information on political attitudes across a large sample of twins. As Lewis' review cites, across many samples within many countries, "heritable effects on social and political attitudes has been reported virtually in every study." Závecz and Littvay (2022) provide a similarly serviceable review, also highlighting the importance of Alford, Funk and Hibbing, while identifying Martin et al. (1986) and Eaves et al. (1997) as pioneering precursors that were largely ignored.

Beyond such reviews, one can find evidence of genetic influence in large-scale studies. For example, a team of international collaborators, gathering survey data from 1970 to 2010 across Denmark, Sweden, and the United States of America, found significant mean differences across several political attitudes for which heritabilities ranged from 0.30 to 0.60. Also, after studying 1992 related participants Kandler et al. (2012) conclude that, "political attitudes are genetically but not environmentally transmitted from parents to offspring and that a substantial proportion of this genetic variance can be accounted for by genetic variance in personality traits." Hatemi and McDermott (2012b) produced heritability estimates across 26 items ranging from authoritarianism and traditionalism, to party identification and punishment attitudes. As Hatemi and McDermott report, political knowledge is highly heritable while party identification is not, though most items fall in the middle of these extremes. As the authors conclude, these attitudinal and temperamental political variables tend to average heritability estimates of 0.5, placing them on par with the heritability estimates associated with personality traits (Hatemi & McDermott, 2012b; Verhulst et al., 2010).

Our personal reading of these estimates is that the heritability of items rises as a function of their closeness to global, temperamentally informed individual differences. To further explore genetic influences on specific sociopolitical opinions, and in anticipation of applying evolutionary theory to prevailing conservative tenets reviewed in Chap. 8, we list ten below, citing evidence of their heritability (Kandler et al., 2012):

1. *Traditional heterosexual marriage*: Disgust sensitivity, a substantially genetic trait (Sherlock et al., 2016; Tybur et al., 2018), predicts a surprisingly wide array of ideological positions (Fournier et al., 2021; Curtis, 2011), and priming disgust responses among conservatives heightens bias against homosexuals (Terrizzi et al., 2010). Additive genetic effects explain thirty-six percent of the variance in attitudes towards homosexuality (Verweij et al., 2008), with other studies reporting similarly important heritability estimates (Eaves & Hatemi, 2008). Relatedly, conservatives engage in more traditional sexual practices with fewer partners (Hatemi et al., 2017), with such sexual variation being reportedly genetic (Verweij et al., 2009; Zietsch, et al., 2010).
2. *Pro-life stance*: Eaves and Hatemi (2008) report that similarities within families on the abortion question have little to do with the shared environment of homelife and instead, "the apparent effects of genetic transmission of liability are substantial."
3. *Stratification and hierarchy*: Preference for, and opposition to, inequality and associated political attitudes that enhance or diminish hierarchical differences, are moderately heritable (Kleppestø et al., 2019).
4. *Authoritarian and executive dominated power arrangements*: We see the familiar variable of disgust sensitivity emerging as an important underlying correlate again, with disgust sensitivity predicting preferences for authoritarian leaders.
5. *Tradition, deference and custom*: There is a shared genetic contribution that actuates ingroup preference and favors the maintenance of traditional norms (Lewis & Bates, 2014). Furthermore, empirical support was forthcoming for a latent genetic tendency to *obey traditional authorities*, that is expressed religiously, socially, and politically (Ludeke et al., 2013).

6. *Religiosity*: Political ideology is a species of ideology, and it is useful to draw parallels to, for example, religious ideology. The utility of this connection is supported by the correlation and apparent genetic association between social conservatism and religious ideology (Ksiazkiewicz & Friesen, 2017), with genetic factors explaining more than seventy percent of the variance across relevant scales (Bell et al., 2009).
7. *Controlled immigration*: Ethnocentrism and outgroup fear, possibly related to disgust sensitivity and fear of communicable disease (Navarrete & Fessler, 2006; Fournier et al., 2021), have a modest degree of heritability as per Orey and Park (2012). Hatemi et al. (2013) "find that the greater part of the relationship between fear and political attitudes is accounted for the genetic covariance between the two constructs." Likewise, Hatemi and McDermott (2012b) report that "social fear and immigration attitudes was due to a common genetic factor."[3]
8. *Group homogeneity*: There are within family similarities (Degner & Dalege, 2013) and genetic dispositions differentiating persons on an *ingroup affiliation factor* (Lewis & Bates, 2010). Bell et al. (2009) report moderate heritability estimates for politically oriented attitudes towards ethnic (0.42) and racial minorities (0.52).
9. *Strong local law-enforcement & capital punishment*: Hatemi and colleagues found moderately heritable attitudes towards capital punishment, with male siblings, male dizygotic twins, and male monozygotic twins respectively generating heritability estimates of 0.28, 0.41, and 0.52. Female siblings, female dizygotic twins, and female monozygotic twins respectively generated heritability coefficients of 0.31, 0.37, and 0.54.
10. *Military spending and armament*: Stam et al. (2012) show fear orientation, inclination towards preemptive war, and torture all to have substantial genetic loadings. Relatedly, McDermott and Hatemi (2017) found that, in both sexes but more fully among men, genetic factors explained, "individual differences in aggression for attitudes toward foreign policy or context-laden moral choices, such as sacrificing the lives of some for the greater good of many."

It is no coincidence that disgust sensitivity, latent conformist tendencies, fear and anxiety, and other factors are cited in relation to more than one of the ten policy positions above, as these ten policy positions are actuated by a restricted number of lower-order traits, as addressed in the subsequent section.[4]

Genome wide association studies (GWAS) and *genome-wide complex trait analysis* (GCTA),[5] along with the neurobiological differences reviewed in the subsequent chapter, provide important corroborating evidence, answering those who doubt the assumptions built into behavioral genetics studies, and so doubt the heritability estimates they produce. Hatemi et al. (2011) were among the first to identify candidate genes predictive of residence along the political spectrum. After controlling for participant age and sex, *NMDA* and glutamate-related receptor genes were variously associated with conservatism or liberalism. Hatemi et al. (2011) suspected that since NMDA receptors are related to cognitive flexibility under alternating schedules of reinforcement, it was feasible that these genes had an essential role in an individual's flexibility of political opinion in a changing social environment. This is one of several genes identified by genome wide association studies, though the promise of this technique runs ahead of its achievement.

4 Distinguishing Between Lower-Order Emotions and Higher-Order Cognitions

Claims about heritability are often met with incredulity. Skepticism can follow from fears of biological determinism, confusion about the meaning of heritability estimates, overreliance on environmental explanations, and related objections apt to be leveled at any claim that any trait is heritable. However, one objection can be uniquely raised about the heritability of political orientations: *How can political ideological variation have evolved prior to politics itself?*[6] At the outset of Chap. 1, we anticipated and addressed this concern, and presently provide what we hope will be a more complete answer to this apparent conundrum.

It is true, politics in its proper form was, if not the product of modernity, at least did not predate civilization. The answer, however, is that these ideologies are heritable because they are based on lower-level temperamental dispositions activated by politics and parties, but which evolved prior to their existence. Political ideological dispositions are grounded in lower-level traits predictive of approach and avoidance, fear and anxiety, as well as openness to novelty and change. Politics and political debate are often populated by pre-state society-level concerns writ large. Certainly, debates about supply-side economics and the Laffer Curve are rarified policy positions, but positions on economic matters such as redistribution may relate to lower-level inclinations toward fairness, altruism, charity, and the like. Furthermore, beyond the economic sphere, pre-state societal issues effectively inform the politics of defense, moral policing, criminal justice, and the like. Thus, on both these grounds, politics proper need not to have existed for genetic variation in political ideology to have evolved.

This reality can be seen in that political ideology is heritable, Smith et al. (2011) explain, through the medium of attitudes, which influence "chronic dispositional preferences for mass-scale social rules, order, and conduct." Underlying attitudes, much more than political decision making, are genetic; attitudes correlate in families from genetic associations much more than socialization (Hatemi et al., 2011; Klemmensen et al., 2012). Many authors understand these broad attitudes themselves as outgrowths of even lower-level, dispositions. As per Kleppestø (2021), genetic variation in political ideology extends from, "general socio-relational traits," which predict variation on the dimensions of, "in-group cooperation and submission to norms." Genes, as described by Hatemi and McDermott (2012b), "do not directly affect specific attitudes, but rather genetic propensity influences the disposition and operation of an emotive condition, which then manifests toward many targets…". Smith and Hibbing (2011) use individual temperamental differences to explain connections between genetics and voting behavior. Instead of understanding genes to directly dictate political positions on issues of the day, they explain genes to dictate individual differences in nervous systems, which in turn relate to political decision making. As an example, they report that electrodermal responsiveness has a substantial genetic loading, and correlates with heritable variation in attitudes. In summarizing two decades of reviews (Van Hiel et al., 2010; Hibbing et al., 2014; Jost & Amodio, 2012; Jost et al., 2017), Lammers and Baldwin note that "liberals and conservatives differ not only in their ideological leanings, but also in their epistemic, existential, and ideological motives, desires, and related aspects of psychological functioning." Illustrating one of the points above, Lammers and Baldwin continue thus: "A large literature ostensibly shows that differences between liberals and conservatives are not limited to political psychological judgement, but extend to physiology, neurological patterns, and basic and cognitive-psychological functioning." Though they go on to discuss small effect sizes, insufficient samples, and related methodological problems, Lammers and Baldwin (2022, p. 675) cite studies in evidence that, "Liberals have been shown to differ in their self-regulation; in how they deal with threatening stimuli; in their physiological reactivity in response to threat; in their communication preferences; and even in basic functioning of the brain."

Harden, in her book *The Genetic Lottery,* reviews the different causal attributions made by liberal and conservative leaning persons. Liberals are more amenable to explanations about luck and so they can more readily engage with Rawl's veil of ignorance, while conservatives believe more readily in Horatio Alger stories, emphasizing individual efficacy that makes one's proverbial bootstraps important. They then of course come away with very different explanations on accountability and culpability, and therefore find themselves on different sides of policy debates; for example, policy debates about wealth redistribution. Cognitions (*e.g.,* worldview, optimism, and pessimism), emotions (*e.g.,* anxiety and fear), attitudes (*e.g.,* toward newness and new peoples), and personality traits (*e.g.,* openness and agreeableness) are some of the many features of mind that evolved prior to politics but inform all that is political. So, in these many ways, we can see that genes influence temperament and lower-order cognitions that evolved prior to politics, which then are expressed within political contexts.

5 Development and Genetics: Lagged Effects

With the aid of longitudinal studies, the unique contribution of genetics is better understood. The power of the shared environment has always been stressed as pivotal, and indeed there is an effect of the shared environment shaping political ideology in adolescence and into early adulthood, but thereafter genetic influences emerge, which can be seen when comparing dizygotic and monozygotic twins. Dizygotic twins, after the effects of the shared environment fade, become less and less similar as their different genetics become expressed; whereas monozygotic twins continue to closely correlate along ideological dimensions. Hatemi and McDermott (2012b), the authors reporting these findings, also provide a graphical display of different cohorts, and highlight an inflection point after 20 and continuing through the remainder of adulthood. Prior to that inflection point, one sees correlational coefficients steadily rising from 0.3, 0.4, 0.5, to 0.6 on a composite measure of political ideology for both monozygotic and dizygotic twins; after that inflection point, the correlations remain in the 0.5 to 0.7 range for monozygotic twins, but fall to the 0.3 to 0.4 range for dizygotic twins. Again, it is as if children are increasingly imbibing the ideology of their shared environment as they age, and so becoming increasingly similar so long as they remain in that shared environment; but the effects fall away when that environment is no longer shared, and so the correlations lessen except where there is fuller genetic relatedness (Hatemi & McDermott, 2012b; Hatemi et al., 2009).

Reacting to such findings, Duckitt (2022, p. 185) reports that, "…family and parental influences on ideological attitudes largely disappear when offspring leave the parental home." "This suggests," Duckitt asserts, "that the parent-offspring ideological attitude correlations in adulthood primarily reflect shared genetic influences…" These are reminiscent of the fadeout effects found in *HeadStart* and adoption studies investigating the nature and nurture of intelligence. The intelligence literature is riddled with studies documenting intelligence quotient gains following adoption or early interventions. As reliably, these gains subsequently fade over time, with genetic influence expressing itself more strongly in later life. In other words, there is a partial effect of the shared environment that lasts only as long as the environmental enrichment lasts, with genetic influence more fully asserting itself thereafter (Rindermann, 2018; Warne, 2020).

There is one point of further concern about the developmental expression of ideology captured in the following quote: "If a person is not a liberal when he is twenty, he has no heart; if he is not a conservative when he is forty, he has no head."[7] Some think this is a product of worldly wisdom. The young idealist sheds his utopian hopes with the gall of experience. However, there is reason to suspect nonlinear genetic expression wherein genetic effects emerge in later life. Male pattern baldness is an accessible example. A child with strong genetic dispositions to male pattern baldness will nevertheless have a full head of hair. Perhaps in his early twenties and accelerating into his thirties, hair will thin and fall out, leaving a naked pate. Ideologically relevant personality traits can also be subject to nonlinear genetic expression, sometimes referred to as *heterotypic* expression (Corr & Matthews, 2009). Take the case of conscientiousness, a heritable personality trait correlated with conservative political ideology, and which augments with maturation. Full expression of genetic dispositions to conscientiousness might not be expressed behaviorally in children, teens, or even young adults, but will be at length thereafter. Through the medium of conscientiousness or otherwise, we should then be alert to lagged expression of genetic ideological differences when considering behavioral genetic research.

6 What Precisely Is Genetic?

As explained in our primer above, genetic influences are transmitted through a sequential cascade of effects, from protein synthesis to neuroendocrine physiology to complex psychosocial phenomena, acting and interacting throughout development. Thus, most of the genetic estimates documented apply to ideological differences, not specific party affiliations. We can see this distinction implicitly in the genetic estimates themselves in that party affiliation has the lowest genetic loading of all 26 factors reviewed by Hatemi and McDermott (2012b). Briefly, "ideologies are vast and expansive mental frameworks and language structures that hold together many different ideas" (Lewis, 2019, p. xv); ideologies are abstract mindsets creating coherence around policy issues (Levendusky, 2009); ideology is a worldview, a preconceived way of viewing the world that biases what is perceived and remembered, and how information is interpreted (Brennan, 2017).[8] Political ideology is one mechanism with which to bind persons together in political parties (Zmigrod, 2022), but it is not the only mechanism; and there are periods in history where party and ideology drift far apart, and others where they come into closer alignment. It may be a mistake, therefore, to judge the utility of this genetic literature from the vantage point of political parties; to insist for instance, that Republicans and Democrats be reliably distinguished genetically across time and place. As we do, Noel takes issue with those many authors who assume parties and ideologies are one and the same. An early example is Benjamin Constant who defined a party as a set of members, "professing the same political doctrine." Noel (2013, p. 14) condenses the differences in this quip: "…parties are about who stands with whom; ideologies are about what some group stands for." It follows that people of the same ideological disposition must share common cognitive features, but the same is not true of party members, who might share membership within a coalition in pursuit of different goals or the same goal for different reasons. Chapters 15 and 16 contrast ideology with party; suffice it to say at present that these are different, and that this difference is important in understanding the genetic estimates reviewed above.

After distinguishing between party and ideology, confusion can stem more fundamentally from disagreement about the nature of ideology itself. Misapprehension of genetic estimates follows equally from overly simplistic unidimensional models of political ideology, and needlessly complex multidimensional models of political ideology. McCarty et al. (2013) convincingly demonstrate the left-right dichotomy's empirical utility. Nevertheless, the left-right dichotomy simplifies reality. Of course, any model of political reality does so, but the left-right model uniquely conflates distinct dimensions in distributing all variation along a single continuum. There are several alternative, multidimensional models exploring the failings of the traditional left/right continuum (Wallace,

1908; Hermann, 1986; Lau & Sears, 1986; Iyengar & McGuire, 1993; Jost & Sidanius, 2004; Lammers & Baldwin, 2022):

- Eysenck's model combines an *authoritarian/democratic* axis with a *radical/conservative* axis.
- Nolan's Chart (1971) features a *libertarian/authoritarian* axis crosscutting a *left/right* axis to create conservative, liberal, libertarian, and authoritarian types.
- Rokeach's Chart (1973) shows freedom interacting with equality to create four quadrants: Socialism, capitalism, communism, and fascism.
- Pournelle's Chart (1964) features two axes, one varying from *enthroning reason* to *irrationality,* and the other varying from *state as the ultimate evil* to *state worship.*
- Mitchell's framework (2007) also provides two intersecting dimensions creating four quadrants*: libertarian individualists, republican constitutionalists, democratic progressives,* and *plutocratic nationalists.*

In many ways, such models provide nuance necessary to understand political differences between peoples and parties. Still, some of these finer distinctions accommodate macroeconomic preferences, free trade, or preferences relating to the scale and scope of government, but may be less necessary to explain genetically informed ideological differences of mind. Again, this is not to devalue these variables and any domains dedicated to their measurement to explain politics, but only to say they may be less relevant to genetically informed ideological variation. More relevant to genetically informed ideological variation are multidimensional models parsing between *preferences* and *power*; that is, distinguishing between pure ideological preferences classically encompassed by the right/left continuum, and varying tendencies to forcibly impose these ideologies on society.[9] This distinction, captured concisely by Nolan, and further discussed in Chap. 4, allows us to understand the apparent paradox of authoritarian communists and libertarian conservatives. Rather than deliberate further on the merits of various models, we merely acknowledge these distinctions here in this chapter, encouraging readers to think about their significance as we continue to develop our explanation of the evolution of a genetically informed ideology of mind.[10]

7 On States and Politically Relevant Traits

In the flush of success, having documented the heritability of political ideology, new terms, such as *political genotypes* and *phenotypes,* have been coined, followed by calls for the creation of a new subdiscipline of *genopolitics* (Wajzer, 2020; Alford et al., 2008; Littvay, 2012). However, critics seized on the inconsistent relationship between ideological traits and political behavior as evidence against claims made by behavioral geneticists (Charney, 2008; Wajzer, 2023). We can find an example of this inconsistent relationship between purportedly heritable ideological traits and political behavior in the research of Hatemi and McDermott (2012b): "…among populations where individuals have lost their job, suffered financial loss, or divorced, broad-sense heritability of economic policy attitudes, such as support for unions, immigration, capitalism, socialism, and federal housing, decreased to almost zero." Similarly Kaplan (2023) warns of waning relevance of genetic influences outside specified sociological parameters.

Debate between creators and critics of behavioral genetics data are playing out an old debate within psychology known as the *state-trait debate*. The state-trait debate arrayed social psychologists that emphasize the power of environmental determinants against personality psychologists that emphasize the power of temperamental determinants. Social psychologists would conduct naturalistic experiments showing subjects conforming to environmental pressures, while personality psychologists would counter with evidence of behavior following reliably from individual differences. Of course, both have a point because both environments and traits combine to determine behavior. Sometimes the power of environmental states will sway almost all to act in a certain way, while at other times, environmental determinants will be weak enough to allow individual differences to emerge.[11] The biological concept of a *reaction norm* mediates between these positions, integrating the sociological with the psychological. The reaction norm plots a trait on one axis and a state on another axis. Behavior is elicited at any given intersection between these axes. To illustrate the point, violence will follow when slightly provoking a highly aggressive individual, or when strongly provoking a slightly aggressive individual. However, the highly aggressive and slightly aggressive individuals will behave the same in some environments; both will behave peaceably when not provoked, and both will react violently when their lives are threatened. The reaction norm then helps us see that heritable traits as well as environmental influences are both important; without the former, one cannot understand why people act differently in the same situation, and without the latter, one cannot understand why different people act the same in certain situations. In a way, this is why this book does not merely end at Chap. 7, with the conclusion of its first part; because, as this section illustrates in microcosm, we really can't understand the significance of heritable ideological variation in a vacuum.

8 Debate Around a Shifting Center: The Overton Window

Beyond general cautions about complexity and context, consider the *Overton Window*,[12] a sort of strike zone, or prescribed area within which political debate takes place. In seventeenth-century England, Robert Filmer's version of absolute, divine-right monarchy (Ward, 2004) remained within the Overton Window, as did segregation during the American Civil Rights Era. In those times past, conservatives and liberals were arrayed across a divide centered on these issues. Of course, this is no longer the case. There are no serious conservative advocates or political platforms founded on divine right monarchy or segregation. In consequence, all would be skewed to one side of what was formerly a debate, irrespective of their genetically undergirded temperamental and attitudinal proclivities. Similarly, a generation ago a conservative might have argued over the fence with his liberal neighbor about gays in the military; a generation hence, the sons of these neighbors might both endorse the right of gays to serve in the military, but argue over the issue of gay marriage or adoption. The liberal son and his father, like the conservative son and his father, are arguing from their respective ideological orientations, but about different issues as the Overton Window moves across time and between generations.

Heritability is a much misunderstood metric. Heritability estimates can only be high when there is variation on a trait. A trait, such as having five digits on each hand, would not be reported as highly heritable, even though it is fully genetic, simply because there is no variation on that trait. Extrapolating from this general lesson, we will only witness high heritabilities on those traits for which there is variation, and, concomitantly, a genetic basis for that variation. In consequence, we have to acknowledge that there may be no timeless political issues reliably exposing heritable differences between liberals and conservatives. However, we should acknowledge this without then going so far in the other direction to conclude that these heritabilities are without meaning. In other words, this context dependent Overton Window issue should not be erroneously referenced to negate the meaning of heritable differences. Additionally, Overton Window issues should not be used to strip political ideology of any relation to content, by, for example, asserting that liberals stand merely for progress and change, and conservatives stand merely for reactionary resistance to such progress and change. Were this the case, one could never understand revolutionary conservatives like George Washington. Conservatives do broadly seek to conserve, and liberals do broadly seek to liberalize, but they also have content, morals, and values toward which they are tending; conservatives will not seek to conserve that which they deem dissolute or disordered, just as liberals will not seek to change that which they deem fair and just. In sum, the Overton Window frames cultural content to some degree, forming a range of debatable positions, and then heritable ideological differences predict which position along that range people occupy.

9 Conclusions

The facts and figures recording substantial heritability of political ideology presented within this first chapter are in some sense a prerequisite for all that follows. Were political ideology not substantially genetic, it could not be meaningfully explained by evolutionary theory. Our goal was to establish this prerequisite, and so we have neither reported novel research as have Hatemi and McDermott (2012b) or a comprehensive review such as that provided by Závecz and Littvay (2022). Consistent with this chapter's purpose, we first and most basically reviewed evidence of political ideology's heritability, along with the general genetic knowledge necessary to its assimilation. We have seen substantial genetic loadings for ideology.[13] This should be no surprise in light of Turkheimer's three laws of behavioral genetics (2000). We have taken pains to show these heritable political affiliations are higher-level manifestations of lower-level traits. Heritability rises the closer it stays to those cognitive expressions of ideology actuated by global temperamental features; heritability declines when measuring either individual issues or external affiliations. We did not evolve heritable variation in reactions to monetary policy, immigration, or incarceration, but on approach-avoidance, openness-conscientiousness, flexible and rigid thinking.

These first five chapters collectively construct a biology of political ideology, after which readers may come to see that political ideologies are not just absorbed environmentally through social transmission of political values; rather, political ideologies extend from genetic differences which give rise to neuroanatomical and temperamental differences which then form the foundations of political ideology. In turn, collectively, the biology of political ideology justifies its evolutionary explanation. These proximate realities about the genetics, neuroscience, and temperamental features of political ideology warrant ultimate evolutionary explanations of political ideology's existence. Keep in mind when reading on, especially when transitioning from Part I to Part II, that these biologically based dispositions are then variously manifest in different cultural contexts. Their significance and signal will be different in tribal v. feudal societies or between totalitarian regimes and representative democracies.

Notes

1. A Lecture delivered at the Masonic Temple, Boston, December 9, 1841. https://emersoncentral.com/texts/nature-addresses-lectures/lectures/the-conservative/

   The pith of this quote is contained in J. Purdy's, *A Tolerable Anarchy.*
2. "The manifestation of genetic influence on these preferences appears more complicated because of large-scale societies, institutions, and modern social structures, such as states and governments. The labels and meanings of issues, groups, and policies might change across time and

cultures, but the underlying connection between the core issues that are important to humans, including survival, reproduction, and defense, will remain. Indeed, genetic influences on attitude differences may be a remnant of ancient behavioral adaptation predating modern human society" (Hatemi & McDermott, 2012b).

3. The frequently cited relationship between conservatism and disgust sensitivity is described by Hibbing and coauthors to reflect a broader fear/anxiety orientation: "...physiological and psychological responses to features of the environment that are negative. Compared with liberals, conservatives tend to register greater physiological responses to such stimuli and also to devote more psychological resources to them" (Sherlock et al., 2016).
4. Furthermore, there are shared genetic influences on sociopolitical ideology and religious ideology (Friesen & Ksiazkiewicz, 2015).
5. For a review of the early promise and concerns about genome wide association studies, see Hatemi and McDermott (2012).
6. Especially for those that adhere to notions of the *Environment of Evolutionary Adaptedness*, these heritability estimates will be hard to assimilate. This term denotes an early belief among evolutionary psychologists that relegated all meaningful evolutionary change among humans to the Paleolithic Era, with any later changes during the Neolithic being thought cultural. This contrasts with later understanding that evolution accelerated during the Neolithic and continued to do so in its latter phases through to historical times and the present.
7. There are many variations on this quote and many uncertain attributions.
8. Brennan (2017, p. 37) further states, ideology undermines *epistemic behavior*; that is, they have compromised epistemological integrity and will allow bias to undermine rationalized pursuit of truths. This is a manifestation of *motivated reasoning*.
9. We provide our own multidimensional model in Chap. 3, one that goes beyond ideology, attempting to capture differences in life history and the broader propensity to control the external environment.
10. It might also help to distinguish between economic and social ideology, as they are conceptually, psychometrically, and possibly genetically distinct (Schwabe et al., 2016).
11. Relatedly, *within-group* ideological politics and *between-group* interest group politics are here relevant in that the former will accentuate and the latter will suppress genetically related attitudinal and temperamental differences.

12. See Karpenko (2019) for a discussion about how elite messaging can move the Overton Window via a process of propaganda, agenda setting, and messaging. This is important to note here because this sort of propaganda can temporarily or permanently but always by artificial means, push issues seemingly outside the Overton Window, which are actually inside it, were we to reference more directly the hearts and minds of persons within a particular sociopolitical culture.
13. We have also seen that political knowledge and participation are highly heritable; though this is a feature that will only be contextualized within Chap. 3.

3

The Neuroscience of Political Ideology

Steven Charles Hertler, Aurelio José Figueredo, and Mateo Peñaherrera-Aguirre

> Natural conservatism is a tendency of the human mind. It is a disposition averse from change; and it springs partly from a distrust of the unknown and a corresponding reliance on experience rather than on theoretic reasoning; partly from a faculty in men to adapt themselves to their surroundings so that what is familiar merely because of its familiarity becomes more acceptable or more tolerable than what is unfamiliar.
>
> —Lord Hugh Richard Heathcote Cecil[1]

1 Introduction

From *genopolitics* we now turn to *neuropolitics*, a term signaling subdiscipline status following the accrual of myriad neurobiological and physiological studies and findings (Connolly, 2002; Schreiber, 2017).

S. C. Hertler (✉)
Saint Elizabeth University, Morristown, NJ, USA

A. J. Figueredo • M. Peñaherrera-Aguirre
University of Arizona, Tucson, AZ, USA
e-mail: ajf@arizona.edu; mpeaher@email.arizona.edu

S. C. Hertler et al., *The Evolution of Political Ideology*,
https://doi.org/10.1007/978-3-031-83941-2_3

Neuropolitical literature documents self-regulation, conflict monitoring, emotional processing, and related physiological and anatomical individual differences predictive of political predilections (Kanai et al., 2011). For instance, augmented anterior cingulate cortical volume, along with amygdala, insula, and entorhinal cortical regions, are among the brain structures covarying with political ideology. Direct measures of brain differences, such as the increased insula size among Democrats, are better predictors of voting behavior than parental party affiliation and related sociodemographic data (Schreiber et al., 2013). As another example, take the right entorhinal cortex, a part of the hippocampus active in time-perception and memory that shows increased volume among conservatives (Haas, 2016).[2] This chapter also considers the smaller and newer body of neuroendocrine research as it relates to political affiliation. For example, McDermott et al. (2014) found that people could detect olfactory cues enabling them to assortatively mate based on political ideology. These anatomical and physiological differences predict perceptual and attitudinal differences that, in turn, predict political and ideological differences. As emphasized in this third chapter, neurobiological underpinnings of political differences expressed in modern nation-states extend from more fundamental and preexisting individual differences expressed throughout prehistory and ancient history. The neurobiological hardware we bring to the voting booth began evolving before humans were human, and continued evolving throughout prehistory into historical times—a point integral to producing a convincing evolutionary explanation.

2 Looking for Coherence Among Many Correlates

John Adams was among those obdurate realists that recognized rhetoric as rationalization.[3] The same is true of Franklin when he gave way to eating cod in violation of his vegetarianism: *If the larger fish ate the smaller, why may he* [Franklin] *not eat the larger*. Hume, preaching about reason being the slave of emotion, was of the same mind. These conservative Enlightenment thinkers respected reason, but understood its weakness in the face of passion. The validity of their position is seen in nearly every subsequent line of this chapter. As will be seen, there are very many politically relevant *limbic structures*, that is, subcortical brain regions associated with emotional processing, rather than rational thought. Lower limbic structures create emotional dispositions and perceptual biases, whether toward pessimistic fear or optimistic hope, to which rational argument is superadded. This is similar to the diagnostician that makes rapid instinctual judgments, using the remainder of the session to catalogue supporting symptoms.

In illustration of the above, consider that empathetic responsiveness within the *temporal–parietal junction* (Zebarjadi et al., 2023) and taste-bud density (Ruisch et al., 2021) are reported correlates of political ideology. Also, as mentioned in the introduction, political ideology tracks with *chronotype*, the sleep patterns associated with early and late rising (Ksiazkiewicz, 2020). Attention should also be given to hormones[4] such as *androstenone*, which predicts socioemotional responsiveness, reactions to facial images, but also political preferences (Friesen et al., 2020). Friesen and colleagues then extrapolate from their findings to explain how androstenone interacts with testosterone to create receptivity to tightly bound social orders:

> This relationship of androstenone to social order and hierarchy supports the hormone's connection to testosterone – that is, those sensitive to testosterone-related behaviors like aggression may desire strong leadership and tight ingroup coalitions to mitigate what they see as social disruption.

Still further, political ideology is reliably identified by olfactory cues that subconsciously guide attraction and assortative mating decisions (McDermott et al., 2014), with those on the opposite side of the ideological aisle being seen as less attractive as prospective partners (Easton & Holbein, 2021). Olfactory cues may be amplified by mobility, denser populations, social media, dating sites, as well as the concentration of people in jobs and colleges with similar values and ideologies (Klofstad et al., 2012, 2013). All of this creates more powerful assortative mating on the basis of political ideology. Then this could be compounded by internal migration, with the result being the amplification of partisanship.[5] Ideologically driven mate preferences signal, or at least are taken to signal, underlying moral emotions, as discussed by Bruchmann et al. (2018). These researchers had participants rate their own, and a prospective partner's, moral emotions. Participants rated prospective mates as having *individualizing* or *binding* moral values; thereafter, participants respectively and reliably identified prospective mates as supporting liberal or conservative candidates. So, participants were inferring ideological affiliations from moral emotions. It seems then that ideology serves as a cue for deeper moral emotional differences on which assortative mating takes place (Bruchmann et al., 2018). The variety of associations is expressed by Rabin[6] (2021, pp. 18–19):

> The connection between biopsychology and political orientations is becoming clearer with each new study. Liberals and conservatives think differently, feel differently, process emotion differently, experience sensation differently, see the world differently, process information differently, are motivated differently, experience humor differently, respond to new information differently, and have different moral foundations, and different conceptions of social justice, equality, empathy and fairness all based on differences in physiology, brain structure, and epigenetic outcomes.

There do seem to be, if not unity, some themes among these many correlates. To this point, and in contrast to most literature considering brain regions in isolation, Mendez (2017) articulates a "conservative complex" implicating the three aforementioned brain structures in interaction with one another. Reward anticipation and risk assumption, fear and anxiety, reactions to ingroup and outgroup members—these are the recurrent individual differences that reliably surface in this literature on the anterior cingulate, amygdala, and insula. Mendez envisions a "more nuanced lateralization of low intensity, rapid, automatic, and explicit recognition of fear and threat to the right amygdala," which is "paired with differences in the anterior insula causing exaggerated disgust responses, and the right ventrolateral PFC [prefrontal cortex] participates in overall self-restraint." Mendez suggests these structures appear to collectively "mediate conservative behavior, including political conservatism, and constitute a brake on change, maintaining stability, and protecting the status quo."

McRae (2022) reviews lesion and fMRI data alongside political messaging and political decision making as treated in 134 articles published between 2010 and 2021; likewise Schreiber (2017) provides a narrative summary of the development of neuropolitics which interested readers can reference for context. Yet, as before, our strategy is to provide a representative overview, giving preference to hard facts directly implicating brain structures relevant to political ideology and ideological differences.[7] Rather than randomly reviewing ever more politically significant brain structures and physiological findings, we organize findings into two additional sections, one focusing on generalized fearfulness, and the other on fearfulness expressed in the context of social groups.

3 Stress, Anxiety, and Fear: Amygdala Function and Physiological Differences

As per Rule et al. (2010), the amygdala, a limbic brain structure associated with fear and related emotional processing, is cross-culturally activated during voting. More than being activated by voting, increased amygdala volume predicts conservative political positions supportive of the status quo and those legitimating existing social orders (Nam et al., 2018). Conservatives show greater grey matter mass in the amygdala (Schreiber et al., 2013). "Conservatives," Rabin (2021) reports, "have a larger and more voluminous right amygdala and also give evidence of being less trusting than liberals."[8] Amygdala activation repeatedly predicts political ideology and more broadly influences perceptual differences concerning the legitimacy of social systems (Nam et al., 2018). Haas et al. (2020) cite a study by Pedersen et al. (2017) that potentially connects the amygdala to the hypothalamic-pituitary-adrenal axis through the bed nucleus of the stria terminalis (BNST). This axis is the tripartite regulatory system of adrenalin, which is released in response to stress and fear. In this way, increased amygdala volume stokes Hobbesian fears of disorder, thereby predicting conservative voting.

The amygdala and related neurophysiological differences predispose liberals and conservatives to differently process stimulation, anxiety, and fear.[9] As Claessens et al. (2022) state, "political ideology covaries with physiological differences, such as automatic eye-blink amplitude and electromyographic facial activity in response to aversive stimuli." As measured by eye blink amplitude, "defensive reaction was significantly stronger for participants who held right-wing views" (Jost & Amodio, 2012). Oxley et al. (2008) report high physiological responsiveness to randomized noises and threatening images to covary with support for "defense spending, capital punishment [and] patriotism," whereas low physiological responsiveness to these same stimuli covaried with support for, "foreign aid, liberal immigration policies, pacifism, and gun control." As reactivity-related individual differences are predictive of political preferences, so increasing reactivity by stress can shift participants factitively in the direction of authoritarian preferences, such as anti-immigration attitudes (Renshon et al., 2015; Mustafaj et al., 2022). Galvanic skin conductance, a physiological measure of anxiety, is said to be higher among conservatives, though the result is unreliable (Osmundsen et al., 2022). Salivary cortisol spikes more acutely among conservatives exposed to disappointing electoral news (Blanton et al., 2012). Preference for authoritarian leaders is associated with reduced stress recovery (Lepage et al., 2022). Also, pain sensitivity covaries with political preferences (Lee & Ma, 2023). As per Dodd et al. (2012) conservatives are simultaneously more alert and sensitive to threat and disorder. In other words, it is not simply a difference in reactivity, but there is this attentional element wherein conservatives are actively alert to the disorder and threat they fear.

4 Reactions to Novelty and Otherness

In the summer of 1954, social psychologist Muzafer Sherif collected a group of boys into a camp environment within an Oklahoma state park to study social interactions in a famed social psychology experiment known as *Robbers Cave*. Eleven-year-old males became *Rattlers* and *Eagles*, arbitrary designations paired with superficial badges that nonetheless mimicked race and religion in their ability to differentiate ingroup from outgroup (Sherif & Sherif, 1964; Perry, 2018). As Alford (2001) said, and as Sherif demonstrated, "group psychology is the state of nature." Haidt recognized the significance of this study for political ideology, as it is mentioned in *The Righteous Mind*. Parallels are neural as well as conceptual. Cikara and colleagues authored an article in the *Journal of*

Experimental Psychology subtitled, *Neural Representations of Generalized Group Concepts*, showing similar fMRI multivoxel pattern activation irrespective of whether participants thought about arbitrary group affiliations or those premised on political parties. Harris et al. (2022, p. 51) explain the significance of this study: "These findings suggest that arbitrary groups are represented in the brain similarly to partisan groups—providing the neural foundation of partisanship." Harris et al. (2022) add more about the group identity of political affiliation, citing studies that find people sacrifice financial incentives to harm political outgroups and will work with less competent partners rather than one from a rival political affiliation. Importantly, they draw parallels to religious group identities, suggesting that such findings are about group affiliation broadly, irrespective of the content around which those group identities are formed.[10]

This fundamental tribalism exemplified in the Sherif Study is an important baseline and backdrop, but there are also important individual differences in the features used, and in the extent to which, *us* and *them* are defined. We continue to consider ideological differences in response to novelty and otherness through the literature on the insula, executive functioning, and disgust sensitivity, all of which broadly suggest that conservatives tend to react more strongly to both otherness and novelty.

Democratic orientation is associated with greater activity in the left *insula*, a portion of the cortex associated with sense and emotional processing (Haas et al., 2017; Panish & Nam, 2024). The insula, or insula cortex, is a portion of the cortex located deep within the Sylvian Fissure. The insula is complex and can be parsed into regions or sections, some of which have been associated with emotions ranging from apathy and anxiety to disgust. Krosch et al. (2021) report insula volume and activity covaries with "political outgroup members, information about ingroup politicians, reactions to disgusting images and risky decisions." Graham et al. (2012) argue that emotional reactivity is more similar than is apparent at the implicit, visceral, and immediate level, with liberals cognitively moderating their emotional reactions. Ambiguous faces are more often perceived as dominant, threatening or angry by conservatives, and nondominant, nonthreatening, or joyful by liberals (Vigil, 2010; Giuseffi, 2012).

Additionally, there is literature relating political ideology to executive functioning. Executive functioning influences emotional expression and broadly connects to related literature on the anterior cingulate cortex.[11] Among liberals, there is increased grey matter within the *anterior cingulate cortex*, a brain region associated with reward anticipation and impulse control (Schreiber et al., 2013). Augmented anterior cingulate cortical volume predicts liberal political positions, as mediated through the more basic relationship of *conflict monitoring* (Kanai et al., 2011). Weissflog et al. (2013) administered a response inhibition test,[12] finding that liberals, and those with greater preferences for social equality and social change, have larger *event related potential* responsiveness within the anterior cingulate cortex. This study is predated by Amodio et al. (2007) who recorded event-related potentials from electroencephalograms during a *go/no-go* task. During this task, participants are habituated to a stimulus to which they respond. However, on select trials, an inhibition signal is presented, cueing examinees, contrary to the general expectation, to inhibit responses. The go/no-go task is similar to continuous performance tests used in clinical evaluations of attention. For example, the *Conners' Continuous Performance Test* presents randomized letters to examinees. Examinees click a button in response to all letters, except the letter *x*. So, in both cases such tests elicit a pattern of routinized behavior, and then randomly challenge respondents to flexibly deviate from that routinized behavior. Two findings were forthcoming. Amodio et al. (2007) found liberals more accurate in no-go trials, demonstrating more reliable response inhibition, meaning they were able to flexibly refrain from routinized responding when signaled by the no-go stimulus. Corresponding to better test performance, liberals also displayed "significantly greater conflict-related neural activity when response inhibition was required." In the paper, one can graphically see more peaked mean amplitude event-related potentials recorded within the anterior cingulate cortexes of liberals as they inhibited responding. Thus, accuracy coincided with neural discharge. This is taken as evidence of political ideology being partly mediated by *cognitive control* and *self-regulation* (Amodio et al., 2007; Jost & Amodio, 2012).[13] Arguably, these findings generalize to real-world conservative propensities to act traditionally and avoid change, and liberal propensities to deviate from traditional mores in response to environmental differences or changed situations. Again, as we would expect, lower-level relationships undergird these politically relevant associations. The anterior cingulate's political relevance is an expression of its more general relevance to outgroup reactions. For example, anterior cingulate activation is evoked in reaction to racial outgroups, and it is through this general reaction to outgroups that Kaplan et al. (2007) explain anterior cingulate cortical responsiveness to images of politicians among participants with rival party affiliations. In turn, these associations are related to ever more basic associations between anterior cingulate activation and fear responses (Frankland et al., 2004; Tang et al., 2005; Milad et al., 2007; Cullen et al., 2015; de Lima et al., 2022; Nagayoshi et al., 2022).

Lastly, there is literature relating political ideology to disgust sensitivity, with conservatives registering more active and strong disgust. Experimentally priming disgust augments the severity of moral judgment (Schnall et al., 2008). Disgust is also naturally elicited by pathogen prevalence. As seen in the title of Murray et al.'s (2013) article, *Pathogens and Politics: Further Evidence That Parasite Prevalence Predicts Authoritarianism*, there are associations between ecologies riven with parasites and minds susceptible to authoritarianism. This finding often goes unconnected with related political findings, though in the context of the behavioral immune system (Aarøe et al., 2017) the disgust literature can be, to some extent, connected to fear and anxiety as well as ethnocentrism. Disgust has recently been connected to the conservative dislike of abstract art. Long ago, Wilson et al. (1973) correlated conservatism with aversions to complex works of abstract art, a finding subsequently researched and supported (Feist & Brady, 2004; Wilson, 2013). Much more recently, Klebl et al. (2022), in their chapter entitled, *Disliking: From Adaptive Disgust to Ugliness*, state that "ugliness judgments may function to alert us to cues of pathogen presence and evoke disgust." Not only the anterior insular cortex, but the ventromedial prefrontal cortex and other parts of what is known as the *social decision-making network* are implicated in these aesthetic disgust reactions (Klebl et al., 2022). These conservative reactions to abstract art then instructively illustrate this section's emphasis on the dual threat of novelty and otherness.

5 Conclusions

Consistent with the genetic findings of Chap. 2, the neurobiological findings of this third chapter underscore fear, approach and avoidance, ingroup versus outgroup preferences, and related lower-order individual differences evolved in nonhuman primate and prestate societies.[14] There have been efforts to organize these findings into a coherent framework, as in the writings of Hibbing et al. (2014). These researchers, in trying to find some unifying order among these physiological differences separating liberals and conservatives, authored an article entitled, *Differences in Negativity Bias Underlie Variations in Political Ideology*. Negativity is put forth as an, "organizing element of the many differences between liberals and conservatives." Hibbing and coauthors state that "…compared with liberals, conservatives tend to register greater physiological responses to such stimuli and also to devote more psychological resources" to negativity.[15] Other researchers similarly report conservatives afford greater weight to negatively valenced information (Carraro et al., 2016). Conservative values of *moral traditionalism* and *ethnic separateness* were especially associated with biases toward processing negativity (Xu et al., 2022). Included in the issue of *Behavioral and Brain Sciences* in which the article was published, as is typical of that journal's format, were commentaries critical of this "negativity thesis." For example, Brandt et al. (2014) cautioned that the relationship's robustness is dependent on how negativity is defined, whereas Malka and Soto (2014) find the relationship relevant to social conservatism, but not economic conservatism. Lilienfeld and Latzman (2014) see conservatism as predictive of threat sensitivity, not generalized negativity. In turn, Bakker and colleagues cast doubt on the significance of threat sensitivity. Thereafter criticism centered once again on generalizability and also replication. Cross-cultural generalizability was called into question by Fournier et al. (2020), a research group failing to find strong correlations between political ideology and negativity (as measured via survey and skin conductance) across 17 countries spanning six continents. Furthermore, Johnston and Madson (2022) found no consistent relationship among a large US sample.

When defined broadly to encompass threat detection and when understood jointly as a cognitive and emotional bias, negativity is somewhat convincing as a unified explanation of political ideology. Even then, however, it does not clearly subsume related traits like fear and anxiety; at the same time, negativity does not clearly relate to disgust responses and reactions to novelty. Nevertheless, the search for an overarching trait that can subsume others is important; a trait that can parallel the scope of the right-left continuum itself. Attempts are made by personality psychologists. As will be seen in Chap. 5, attempts have been made to correlate ideology to single personality traits or global personality styles. Unfortunately, all these explanations, *genetic, neurobiological,* and *temperamental,* are proximate, and consequently limited in their causal explanatory power. As we have discussed previously, proximate explanations tell us important information about *how* something came about, but only ultimate explanations tell us *why* something exists. The important burden of ultimate explanation is assumed by Chap. 8. Our point at present is that it is hard to see unity, order, or purpose in genetic or neurobiological findings without specifying their adaptive significance within a compelling evolutionary context.

Notes

1. Lord Hugh Cecil, *Conservatism* (London: Williams & Norgate, 1912, p. 48).
2. As mentioned, the right entorhinal cortex, a part of the hippocampus active in time-perception and memory that shows increased volume among conservatives (Haas, 2016; Bergner, 2023). The entorhinal cortex, a portion of the brain's medial temporal lobe, is more specifically part of the cortical-hippocampus complex, which has been associated with memory and spatial mapping in animal models. However, the entorhinal cortex also seems active in time-perception and memory.
3. Diggins (1986, p. 109) cites Adams as recognizing that "...political language may be a poor guide to political reality..."
4. Hatemi and McDermott (2012) outline six substances (Serotonin, Dopamine, MAO, Oxytocin, Vasopressin, and Testosterone) whose known relations to lower-level behavior suggest their relevance to political ideology.
5. Such issues contributing to polarization are further discussed in Chap. 19.

6. Readers can see the original text wherein the paragraph contains, in its original form, all its many references. Each of these claims was associated with one or more citations.
7. Interested readers are also referred to Psychophysiology, Cognition, and Political Differences, a special issue of *Politics and the Life Sciences* (Mansell et al., 2021).
8. This finding might only relate to outgroup trust.
9. Pedersen et al. (2018, p. 43) also have an interesting perspective on the political significance of the amygdala.
10. One should recall this section hereafter, as all of these neuroscience findings are embedded within a multilevel selection framework founded on cooperation and competition between groups.
11. Executive functioning allows cognitive control of emotional impulse. It is interesting to consider how this interacts with culture. Cultures of honor value and reinforce raw emotional expression in certain contexts, which might need to be considered alongside one's raw ability to control emotions. Cultures of honor were common to the Democratic Party but have been more commonly associated with the Republican Party since Nixon.
12. This was a *go/no-go* task, as discussed through the work of the subsequent study.
13. Seemingly qualifying these findings, Buechner et al. (2021) report that liberals show greater cognitive flexibility and adaptation to changing circumstances, though conservatives are better at avoiding errors of impulse.
14. Inherent to social living and present to some degree in social species from insects to primates are issues of "leadership, political deception, law and order, military defense and immigration" (Claessens et al., 2022, p. 22). Similarly, Claessens and coauthors state, "…humans all over the world live in social groups, form coalitions, cooperate with one another, favor the in-group, adhere to social norms and laws…".
15. Similarly, Salter (2018) alludes to cognitive load and resource depletion as factors causing facultative shifts toward conservatism. We discuss facultative reactions to stress, pressure and risk, or what many refer to as existential threat. There are standing individual differences, but also responsiveness to environmental factors. Bachleda et al. (2020) similarly report the relevance of both states and traits in politically relevant negativity bias.

4

Childhood Adversity, Life History, and the Levels and Varieties of Political Engagement

Steven Charles Hertler, Mateo Peñaherrera-Aguirre, and Aurelio José Figueredo

> Only with difficulty does one draw a man out of himself to interest him in the destiny of the whole state, because he understands poorly the influence that the destiny of the state can exert on his lot.
>
> —Alexis de Tocqueville, 1948[1]

1 Introduction

As explained in Chap. 2, hereditary tendencies are only expressed through the processes of behavioral development, with genes being in a constant state of mutual causal transaction with the surrounding environment. In this chapter, we address the behavioral development of political engagement, irrespective of ideology. Whether right or left leaning, we ask, what

S. C. Hertler (✉)
Saint Elizabeth University, Morristown, NJ, USA

M. Peñaherrera-Aguirre • A. J. Figueredo
University of Arizona, Tucson, AZ, USA
e-mail: ajf@arizona.edu; mpeaher@email.arizona.edu

S. C. Hertler et al., *The Evolution of Political Ideology*,
https://doi.org/10.1007/978-3-031-83941-2_4

developmental and environmental factors cause some individuals, but not others, to invest time and energy in politics? Actively participating in politics, such as running for local school board elections, or supporting a candidate's campaign, are time-consuming activities, though so are the more prosaic tasks of reading newspapers, watching programs, registering to vote, and voting. Again, irrespective of party affiliation, why do some invest in these activities more than others? The social scientific literature documents an array of associations between economic distress and political participation; people become more involved in distressed macroeconomic circumstances. Then there is literature on *rational ignorance*, which explains declining engagement as a function of voter efficacy.[2] In other words, people vote when they think their vote counts, and they otherwise abstain. These are interesting phenomena, but they broadly affect all voters. This chapter consequently focuses on the individual differences that affect some voters over others. Deferring the discussion of details, we herein consider political engagement as an active attempt to shape the conditions to which one is subject.

2 Political Engagement: The Power of Environmental States

Explaining engagement, or lack thereof, is not for the faint of heart. Complexity abounds. But without making such distinctions, lack of engagement and forms of engagement can readily be conflated with basic ideological variability. On these grounds, political engagement warrants some attention.

Explanations of engagement vary in their locus, with some explanations emphasizing societies and social environments and other explanations emphasizing individuals and their character traits. Runciman provides two important societal determinants. Specifically, Runciman (2018) reports a *GDP* per capita threshold of $7000 for democracy to take root and remain stable. Mean age is the other important factor. Younger countries are less stable because less risk averse. Putting these together, we see older and richer countries as more stable, politically engaged, and politically enfranchised than younger and poorer countries.[3] Also consider the concept of *rational ignorance* wherein voters rationally remain ignorant because the costs of ignorance are small and diffuse, while, for lobbyists and special interest groups, benefits are large and concentrated (Faria, 2019). Rational ignorance then generates apathy among the electorate, concentrating power in the hands of elites. Then there is the *principal-agent problem.* In a representative democracy, citizens are *principals*, and representatives are *agents*. The principal-agent problem uncouples the interests of citizens from their representatives. Representatives become unaccountable, and as a result, citizens become uninterested (Faria, 2019). The principal-agent problem is one of many mechanisms undermining confidence in political systems and leaders, and which restrict political engagement. And, as Bevir and Rhodes (2010, p. 213) have observed, declining percentage of voters that vote, "signals discontent and distrust." Such explanations of engagement are more or less exclusively sociological, with all voters equally framed as flexibly and rationally responding to exogenous incentives. Endogenous traits, however, are also important, and emerge more frequently in contemporary social science literature.

3 Political Engagement: The Power of Endogenous Traits

As an example of endogenous determinants, take the issue of voter efficacy and participation discussed above. Even as belief in one's efficacy, the belief that one's vote and activity matters, is framed as an environmental predictor of voter turnout (Ksiazkiewicz & Friesen, 2017), both efficacy and turnout show independent genetic influence. Political efficacy is highly genetic and seems to closely covary with political participation (Klemmensen et al., 2012). An estimated 53% of variance in voter turnout is attributed to additive genetic effects (Fowler et al., 2008). These findings alternatively explain the positive correlation in voting between parents and their young adult children, which has heretofore been understood exclusively in the guise of social influence. Even the most sophisticated models and statistical analyses ignoring genetic differences explain, at most, a third of the variance in voter turnout. Of course, as discussed above, there are external conditions that stoke or suppress voting, but genetic differences should be taken seriously as an *internal constraint* (Fowler et al., 2011).

Bell et al. (2009) report, "a very high heritability estimate (62 percent) was found for general interest in politics." Genetic effects also explain why people generally vote or not, rather than some people voting sometimes and other people voting other times (Fowler et al., 2008).[4] Fowler explains that the *casual voter*, that turns out sporadically as political conditions change, is a comparatively rare type of voter. Reviewing data from the *American National Election Study* and the *South Bend Election Study*, Fowler concludes that most either vote or don't vote. "Turnout," Fowler (2011, p. 211) summarizes, "may be driven not by some thoughtful deliberative process, but from the expression of a trait that some people have and others do not." Fowler then interprets voting behavior as more broadly representative of cooperation: "…variation in behavioral types may be critical for cooperation more broadly conceived." Following Fowler's reasoning, Hatemi and McDermott (2012) understand the genetics of political engagement, sophistication, and political participation as politically relevant expressions of "altruism, cooperativeness, trust" and other manifestations of prosocial behavior. With "heritability ranging from 0.40 to 0.70," these prosocial behavioral differences are related to "polymorphisms in genes encoding the receptors of the neuropeptides oxytocin and vasopressin, as well as several genes in the dopaminergic system that interact with oxytocin and vasopressin…"[5]

Závecz and Littvay (2022) report that political knowledge, more than any political attitude, party affiliation, or underlying emotional stance, has the highest genetic contribution of all. Similar to the way Fowler tied voting behavior to stable traits, Brennan (2017) divides the electorate into those that do not care about politics, those that care about politics but are uniformed and reflexively partisan, and those that are rationally empiricist in their approach to politics.[6] Brennan finds reliably expressed voter ignorance across issues such as (1) budgeting for social security or foreign aid; (2) the basis for the Cold War; (3) congressional candidates in their districts; (4) who the United States of America fought in WWII; and (5) the meaning of *liberal.* Genetically derived variation in political knowledge may well relate to educational and intellectual constraints, with education and intellect predicting political engagement. In other words, education and intellect mediate the relationship between knowledge and engagement. Or one can say: *Understanding begets interest.* That is the lesson of correlations between intelligence and education on the one hand, and political knowledge and participation on the other (Aarøe et al., 2021). Higher levels of education make political participation more rewarding and practicable (Wolfinger & Rosenstone, 1980). Intelligence appears to have a unique effect on political participation. Specifically, controlling for personality, education, socioeconomic status, and related variables, intelligence separately predicts general interest in politics and voting behavior. Beyond participation and interest, intelligence predicts dispositions to free markets, equality and social liberalism; in turn, intelligence negatively predicts stereotypical thinking, dogmatism, and religiosity (Jones, 2006; Rindermann et al., 2012).[7]

In sum, it seems that there are likely portions of the population habitually disconnected from political life. They are chronically absorbed by subsistence needs, major medical issues, mental illness, or otherwise have developmental delays, educational gaps, or intellectual limitations precluding participation, inclination, or political thought. Beyond these, there are those that are capable but merely uninterested, or those that express interest when catalyzed by major events, such as war or populist movements. Leaving the fully unengaged aside, in the next section, we turn to levels and varieties of engagement, and consider what engagement is, from a biological perspective.

4 Political Engagement as Macrosystem Control and Extended Phenotypes

As has been seen, there are certainly some compelling explanations of political disengagement,[8] such as when the social scientific literature discusses low intelligence and restricted educational opportunity precluding political engagement, or when those with sufficient education and intellectual capacities choose not to engage as per the dictates of rational ignorance, are sapped of their motivation by the principal-agent problem, or are preoccupied by abject poverty. Hence, existing literature explains those that can't participate, and why some don't participate under certain external conditions; by contrast, it does not satisfactorily explain why those with equivalent ability and under equivalent external conditions, nevertheless show differential political engagement. The literature clearly states that different levels of political engagement under similar circumstances extend from genetic and neuroendocrinological differences, but are silent as to the origin and significance of those genetic and neuroendocrinological differences. We attempt to supply the deficit by contextualizing political engagement, framing it as a bid to control sociological systems that can affect individual and family fitness. In pursuing this goal across the remainder of this and the next section, we reference four theories: *Ecological Systems, Niche Construction, Extended Phenotype,* and *Life History*.

Thus, having explained what we mean by political engagement and disengagement, we turn back to the major theme of the present chapter, which was exploring the developmental and environmental factors that might influence some individuals, but not others, to invest significant amounts of time and energy in political activity. First, we start with the developmental question by expounding on the mutual and bidirectional interactions between the developing organism and its ambient ecology.

We understand political engagement to be a bid for control. In explaining our thoughts, we first refer to Uri Bronfenbrenner's *Ecological Systems Theory*. Ecological Systems Theory envisions individuals surrounded by concentrically nested environments, from the *microsystem,* to *mesosystem,* to *exosystem*, and *macrosystem*. Beginning with the family, parish and schools that surround individuals within their microsystems, there are successive levels of embeddedness, with culture, traditions and national attributes comprising the macrosystem. So when we see political awareness or political behavior, we understand this to reflect concern about, and attempts to control macrosystems. Macrosystems can affect microsystems and individuals. Concern and control at the macrosystem level has direct relevance to socioeconomic position, safety, health, and adaptive fitness. Some fear immigration will bring disease and depress wages, that war will carry off sons, that pro-life policies will interfere with family planning, that federal tax law will divest progeny of intergenerational inheritance of family farms and fortunes, that national gun laws will potentiate local school shooting, or that excessive redistribution will garner discretionary income. In these ways, people engage politically to control macrosystems that can affect microsystems.

Controlling systems, whether microsystems or macrosystems, is itself an *evolved adaptation;* it is an expression of *niche construction*. Niches are environments in which organisms live, and to which they must adapt. Niches can be passively experienced or actively constructed. In constructing a niche, an organism attempts to control the conditions it experiences and the selective pressures shaping its survival and reproduction. Animals construct niches when they burrow and build nests. The propensity to construct niches among animals can substitute for phenotypic adaptations: the squirrel's midden substitutes for the camel's hump, the decorated bower of the bower bird substitutes for the bird of paradise's colorful plumage, the beaver's dam substitutes for the turtle's shell. The squirrel, bower bird, and beaver are all, from an instinctively expressed *extended phenotype*, constructing niches, thereby changing the selective pressures to which they are exposed. Human niche construction is more elaborate and less fully instinctual. In virtually every way, we have constructed niches, discussed more generally as points of culture and components of civilization, that alter the evolutionary exigencies to which we were exposed in a state of nature. Houses, clothes, and climate control alter the thermoregulatory environment, farming and food storage alter nutritional exigencies, fire and weapons ensure we are not prey to predators etc.

We have attempted to chain political science's concept of political engagement to successively lower levels of behavior, first using psychological theory, and then using biological theory. In this way, we can contextualize political engagement, claiming that political engagement is an expression of a more fundamental propensity to control macrosystems and construct extended phenotypes, but we have not then explained why there is variation on these lower-level traits.

5 Life History Evolution

Thus far, political engagement has been reframed as one particular manifestation of a broader propensity to control macrosystems, construct niches, and express extended phenotypes. Again, this is context, telling us what type of behavior political engagement is, and why it is relevant to Darwinian fitness; but, again, we have not answered the question posed above, as to why there is variation within the population on these propensities. For the answer to this most fundamental question we turn to *life history theory*.

Life history theory represents a branch of evolutionary ecology that deals with the optimal allocation of bioenergetic and material resources among the different components of fitness. For example, fitness has been modeled as a multiplicative function of *longevity* and *fecundity*, in which many combinations of these two parameters can yield the same product (see Wolf & Figueredo, 2011; Figueredo & Wolf, 2009). The values that will be naturally selected are not arbitrary, however, but instead contingent upon the ambient ecology in which the organism is immediately situated. Thus, the theory implies that an optimal combination of life history parameters will be determined by selection.

Organisms make tradeoffs with respect to survival, evolving short or long lives, maturing rapidly or slowly, investing little or much bioenergetic resources in growth and maintenance. The same is true for reproduction: some organisms mate often, are not concerned about paternal certainty, and produce many offspring about which they care little and invest less. Alternately, other organisms mate exclusively and produce few offspring on which they dote much. Perhaps this analogy and brief explanation are enough to show that life history evolution examines biological tradeoffs among "essential patterns of organismal growth, maturation, reproduction, and survival" (Braendle et al., 2011). An organism's life history is then "…a composite of all the variables that contribute to the way in which an organism propagates itself" (Reznick, 2014, p. 268). "Time scales may be the single most valuable key to understanding" (Calder, 1984, p. 138) why organisms with *fast life histories* hasten through life, growing rapidly, mating early, and dying quickly, while organisms with *slow life histories* leisurely progress through life, growing slowly, mating later, and senescing gradually.

While there are species-typical means, there is some degree of life history speed variation around those means. This logic applies to humans. As a species, humans have slow life histories, but some have slower life histories than others. We see the consequences of life history speed variation all around us without appreciating either its unity or significance.

Early menarche, teen pregnancy, multiple partners, many births, restricted parental effort, rapid aging, and early death are basic biological signs of fast life histories. Fast life histories are also expressed behaviorally, with preferences for immediate consumption rather than delayed gratification, impulsivity, limited investment in training and education, less regard for rules and laws, and related behaviors and dispositions sometimes mistaken for social deviance or failings by social scientists, educators, and physicians.

Having provided this summary, we next want to acknowledge that there is actually some preexisting literature relating life histories to politically relevant variables. For example, life history evolution has been related to ideology through the medium of future orientation (Thornhill & Fincher, 2007) and system justification (Zhang & Zhong, 2019). Immigration and terrorism are existential threats that Blanchard et al. (2020) suggest elicited faster life history behavior, though this effect was found only among nonvoters. Gladden and Cleator (2018) relate life history directly to moral emotions and, secondarily, to conservatism. They report some association between slow life history and activation of both sides of the moral emotional dichotomy, meaning both *binding* and *individualizing* emotions. As Gladden and Cleator (2018) suggest, manifestation of the full range of moral emotions advances slow life history needs for affiliation and order. Though this is an interesting claim embedded in an interesting paper, we think life history is not a candidate explanation of ideology. There are slow life histories and fast life histories on both sides of the political divide. We think life histories are, by contrast, important to political engagement.

That life history speed maps onto political engagement is implicit in some extant political science literature. For example, Fowler and Kam (2006) report that future oriented planning and delay of gratification reliably differentiate voters and abstainers. Referring to a 2006 article written with a different coauthor entitled, *Patience as a Political Virtue: Delayed Gratification and Turnout,* these authors report:

> ... more-patient individuals were more likely to vote, confirming the proposition that more patient individuals are more likely to participate in elections. Since patience does not exist uniformly in a population, this connection between patience and turnout is supportive of our general

> claim that we can explain political behavior by identifying different types of individuals. Individuals vary in their patience, and, these differences in patience emerge at very young ages. As Fowler and Kam note, children as young as four years old have shown variance in willingness to delay immediate gratification, and these differences correlate with their dispositions as measured later in life. What is more, these differences explain a variety of other behaviors, including "drug addiction, educational attainment, savings and investment, and gambling" (2006, p. 16). By conceiving of a population of types in which some individuals are inclined to be patient and others are not, we can further understand a fundamental act of cooperation such as voting.

Fowler and Kam (2006) associate voting behavior and delay of gratification to a broad array of life history variables, even if they are not recognized as such. Delay of gratification itself is a classic time-relevant tradeoff discussed *ad nauseum* within the life history literature (Schechter & Francis, 2010; Kavanagh & Kahl, 2016; Stevens, 2017), as is savings (Brumbach et al., 2009), investment (Griskevicius et al., 2011), educational attainment (Chisholm, 1999), and cooperation (Figueredo et al., 2006). Life history variation is also implicit in descriptions of authoritarianism, as when Stenner (2005, p. 169) contrasts authoritarianism and conservatism:

> More generally, it is apparent that authoritarianism and political conservatism stand in very different relation to all of the variables in any way reflecting education or social class. Authoritarianism is reduced by cognitive capacity, years of education, attaining a college degree, family income, the household head's and the father's occupational prestige, and the mother's education. Political conservatism, by contrast, is increased by family income, the household head's occupational prestige, the mother's education (perhaps reflecting the social class of the family of origin), and attaining a college degree.

In distinguishing between conservatism and authoritarianism, Stenner, instead of referencing correlates of ideology, is referencing correlates of life history. As with Fowler and Kam, Stenner cites educational attainment, but also differentiates authoritarianism from conservatism via income, occupation, and prestige, all of which are downstream correlates of life history speed (Figueredo et al., 2004; Hertler, 2017).[9]

Connecting this section with the prior section, we suspect that slow life histories directly predict political engagement because slow life history strategists make long-term investments in education, offspring, and health that require control and predictability of environments. Political engagement is a bid for environmental control. Only with some degree of predictability and control are the long-term investments and delayed returns of slow life history strategists evolutionarily, socially, and psychologically compensated. As it stands, the relationship between life history and political engagement is only apparent because both life history and political engagement are related to time orientation and behavioral regulation. So while we are constrained to discuss life history's connection to political engagement through their joint relationships to time and behavior, again, we predict a direct connection awaiting to be established by future research.

On the other hand, we would do well to guard against the evident disdain of many social scientists for individuals that are politically disengaged. Viewing such persons with the contempt, expressed or implied, is hardly helpful to a scientific understanding. We must recall that faster life histories both evolve and develop within environments that are objectively difficult or impossible to actually control. Thus, if faster life history strategists decline to invest their limited resources in such pursuits, social scientists need to maintain some degree of scientific detachment regarding such adaptive decisions, however much they might conflict with pro-engagement values.

6 Conclusions

Having leveraged several theories to explain levels and varieties of political engagement, we provide some further discussion about how this all fits together. We furthermore think that this concluding section will show how engagement intersects with political ideology, in such a way that assuages doubts about the left-right ideological continuum, such as those discussed in the sixth section of the second chapter. Confusion ensues from conflating the left-right ideological continuum with two others. Consequently, we delineate three continua, *engagement, power,* and *ideology,* while relating them to the selective pressures that generate variability along them: (1) *phenotypic fitness continuum*, explaining political engagement; (2) *life history continuum*, explaining means of exercising political power; (3) *multilevel selection continuum,* explaining ideological position.

First, there is the *phenotypic fitness continuum*, explaining political engagement, with the emotionally unstable and intellectually underdeveloped on one side and the emotionally stable and intellectually developed on the other. The phenotypic fitness continuum explains why large swathes of people are apolitical. More broadly, it explains not only political engagement, but social engagement, and even financial success and achievement of other socially valued outcomes.[10] Those on the disengaged end of the phenotypic fitness continuum are disengaged through poverty that reorients focus to security and existential needs at the base of Maslow's Pyramid, through schizophrenia and other forms of major mental illness that distract and distort reality, through intellectual deficiency or educational impoverishment that simultaneously blunts interest in politics and understanding of political complexities. Complex trauma experienced in adulthood or adverse childhood experiences can restrict engagement. Those struggling with addiction or living in group homes might be among the politically unengaged. In short, political disengagement is one among other forms of general social disengagement common among those individuals featuring severe psychopathologies, cognitive impairment, financial impoverishment, and chemical dependence. As explained above, such individuals can hardly be expected to find themselves in a position to effect any social changes whatsoever, so it is hardly surprising that they are not motivated to try (although some of these conditions may partially overlap with fast life history speed, they are not by any means limited exclusively to, or highly correlated with, that evolutionary strategy).[11] Thus, political disengagement could very well be considered as a rational and adaptive allocation of limited time and energy resources among those who are objectively powerless for any reason.

If we establish that someone is, as are the vast majority of people, significantly far down the phenotypic fitness continuum to be politically engaged, then it will be important to determine where they fall on the remaining continua. There is the *life history continuum*, explaining preferred means of exercising political power, with both left- and right-wing authoritarianism on one side and left- and right-leaning libertarians on the other. If on the fast end of the spectrum, an individual will be present-oriented and be impatient of forms, precedents, and processes; they will likely show less consistent political engagement and express political opinions that are blunt and binary. By contrast, those on the slow end of the spectrum see the value in forms, precedents, and processes; they will likely show more consistent political engagement and express political opinions that are considered and informed. From this we can tell that we have a politically engaged person that is either authoritarian or not, but will still not know on which side of the ideological divide they reside. For that we turn to the *multilevel selection continuum*, and see group selected persons on the political right and individually selected persons on the political left. As detailed in subsequent chapters, the group selected are disposed toward the conservative end of the ideological continuum from their fears of societal decay and external conquest; they have stronger predilections toward order and stability which they prioritize above fairness and progress. By contrast, the individually selected are disposed toward the liberal end of the ideological continuum from their predilections for freedom and fairness, which they prioritize above order and stability.

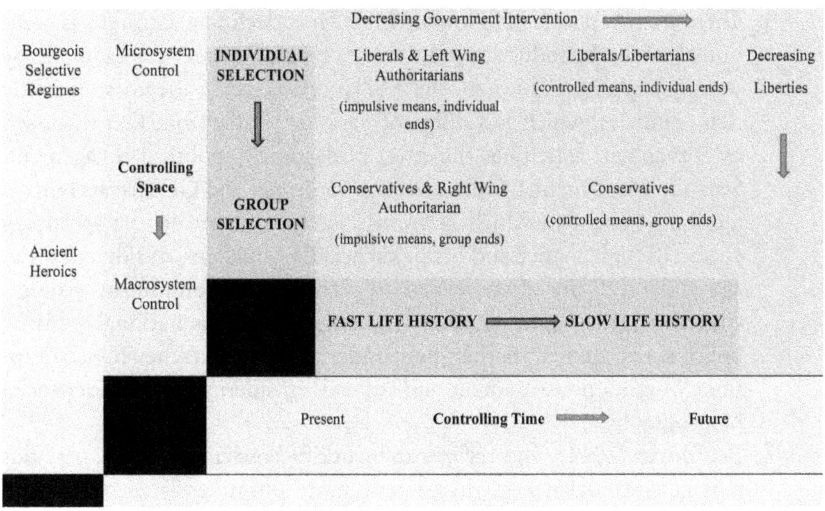

Graphically, here is how these three continua might be used to understand the full range of persons. Envision those far down the phenotypical fitness continuum, on the periphery of society, and so on the periphery of the graph above in all directions. All others can be placed in a Punnett square like the one above. Those sufficiently far down the phenotypic fitness continuum to express political engagement can be parsed into combinations of (1) group selected with fast life histories, being typical right-wing authoritarians; (2) group selected with slow life histories, being conservatives; (3) individually selected with fast life histories being left-wing authoritarians; and (4) individually selected with slow life histories being libertarians. In Chap. 5, we reorient toward the right-left ideological spectrum, and in Chap. 8 we explain how it maps on to the group-individually selected continuum, which might then be all the more clear for having distinguished among these three continua.

Notes

1. This is quoted from Chapter 24, page 247 of Hechter and Horne's (2009) *Theories of Social Order.*
2. Similarly, Faria (2019) refers to the *principal-agent problem* which uncouples the interests of the people from their representatives.
3. There are two points of follow-up here. First, declining earnings, as mentioned in the Introduction, can enhance participation; for example, by stoking populism. This is an effect of large declines in already sufficiently rich countries, which is a different national profile from that discussed by Runciman, which has the effect of depressing political engagement. Second, this Runciman's discussion of mean age and GDP treats both as randomly assigned, which is unfortunately common in the social sciences. Though chance and historical accident must be considered, mean age and GDP are often related to physical and community ecology, which in turn shapes population biology. Any cross national study of political engagement is then best undertaken when eschewing assumptions of random assignment and instead considering the broader social biogeography.
4. Boardman (2011) who reviews the work of Fowler and colleagues adds that genetic variance might become more prominently expressed and

important in differentiating people's political participation among the socially engaged (dubbed the *social trigger hypothesis*).

5. There is a related study that focuses on conservatism, but in reality captures an interaction with engagement: Kalmoe and Johnson (2022) accessed the Minnesota Twin Study to generate heritability estimates for conservatism. Rather than simply lumping all together, they parceled out participants based on political sophistication and found the following interaction: conservatism's genetic loadings increased as a functioning of political sophistication. For the most informed, conservatism's heritability was 74%, while for the least informed it was 29%. These authors consider this pivotal political sophistication through a traditional social scientific lens, as though it signaled a randomized lack of enrichment. We will develop a different explanation.
6. These Brennan respectively labels hobbits, hooligans, and Vulcans.
7. This latter literature set on intelligence and education relates to a more basic inability to engage, rather than the main body of engagement literature that relates to levels of participation among those with the ability to engage.
8. Social science courses and texts have a habit of discussing constructs without context. For example, personality courses are taught without ever explaining that personality traits are evolved, fitness relevant behavioral dispositions. Similarly, a text might discuss emotion without explaining that emotions exist and evolved because they evoke adaptive actions, as when disgust of pustules or putrefaction reflexively motivates avoidance so as to minimize risk of disease or poison. In some respects, we see the political engagement literature in the same light.
9. Income has the most tenuous and variable relationship with life history speed. Life history variation overlaps weakly with socioeconomic status. In turn, socioeconomic status correlates with voter turnout. The *discount rate,* defined as "the rate at which society as a whole is willing to trade off present for future benefits," increases as a function of socioeconomic variables. Nearly seven percentage points separate high school from college graduates, and the lowest from the highest earnings quartiles. Home ownership increases discount rates by nearly four percentage points. Marriage and age have weaker effects (Fowler & Smirnov, 2007; Fowler & Kam, 2006; Harrison et al., 2002). Fowler and Smirnov believe delay of gratification, a hallmark of slow life history strategists, is the causal variable connecting socioeconomic status and voter turnout.

10. Importantly, when we say phenotypic fitness, we are not using this in any eugenic sense, or even in any traditional Darwinian way referencing survival or reproductive success.
11. Several publications indicate that life history speed is weakly correlated with individuals' financial status and general cognitive abilities (Figueredo et al., 2004; Figueredo et al., 2014; Woodley, 2011).

5

Political Ideology as a Function of Personality and Individual Differences

Steven Charles Hertler, Aurelio José Figueredo, and Mateo Peñaherrera-Aguirre

> The reason people are not all on the same political page is not that some have seen the light and others have not, but instead because people exhibit fundamental and persistent psychological differences, differences that are largely rooted in biology. These differences are not about politics, but they matter for politics. By their nature, some people are psychologically predisposed to be liberals and others to be conservatives.
>
> —Mondak (2010, p. 131)

1 Introduction

We continue describing the theorized causal progression from molecular and behavioral genetics, to neuroanatomy and neurophysiology, through to behavioral development under varying degrees of environmental

S. C. Hertler (✉)
Saint Elizabeth University, Morristown, NJ, USA

A. J. Figueredo • M. Peñaherrera-Aguirre
University of Arizona, Tucson, AZ, USA
e-mail: ajf@arizona.edu; mpeaher@email.arizona.edu

S. C. Hertler et al., *The Evolution of Political Ideology*,
https://doi.org/10.1007/978-3-031-83941-2_5

adversity, coming now to this chapter, which wades into the depths of personality and related individual differences as they have been used to predict political ideology within the social sciences. This rich history begins with Adorno's *Authoritarian Personality* published in the wake of World War II, purportedly as a way to detect and explain proto-fascist tendencies. Altemeyer then revived social scientific interest in personality as a predictor of ideology using a refined measure of Adorno's excessively intricate scale.[1] Authoritarianism as a coherent collection of traits then was further improved by Stenner and related authors who addressed reliability issues, showing that inconsistent detection was a product of inconsistent *social activation*. In other words, it was not that authoritarianism itself was inconsistently present, but rather its detectability fluctuated with political climates that alternatively subdued and stoked authoritarian fears of disorder, dissolution, and weakness. Rather than complex trait conglomerates like authoritarianism, other personality psychologists looked to individual five factor personality traits. From these studies, as McCrae (1996) has noted, we get well-established correlations yoking political positions to underlying personality traits. For example, *openness to experience* statistically predicts less conservatism in that the highly open will invite experimentation and novelty, and, in turn, are not attached to custom and tradition, whether in politics or culture. High openness is inversely correlated with dogma, prejudice, and intolerance. By contrast, conscientiousness is often associated with basic conservatism, as well as ethics of personal responsibility, tradition, and virtue (Mondak, 2010). More recently, Haidt has looked to individual differences in *moral emotions* to explain the ideological divide. While Liberals are oriented toward *harm* and *fairness*, conservatives add to these, more strongly developed moral emotions of *authority, ingroup loyalty,* and *purity*—moral emotions which collectively elicit behavioral tendencies toward policing and judging the actions of others.[2]

Many of these theories are valid and compelling. They collectively establish political ideology as a biologically based, heritable worldview; an "applied political philosophy" (Heywood, 2021, p. 9)[3] that covaries with personality. Still, with the exception of Haidt,[4] none of these frameworks explain why personality and ideology covary, and why either are aspects of human psychology. Though this fifth chapter primarily reviews existing literature, it ends by discussing these individual differences as strategic variation evolved via a process of *negative frequency-dependent balancing selection*.[5] With this foundation, readers can then see, in Chap. 8, how this evolved variation is couched in the broader framework of multilevel selection theory.

2 Politically Significant Personality Styles: Authoritarianism

Theodor Adorno published the *Authoritarian Personality* (1950) after the fall of Germany's Third Reich, beginning the social scientific quest to associate ideology with personality.[6] Adorno cast authoritarianism as a conglomerate of nine traits, collectively denoting, "obedience to authority, moral absolutism and conformity, intolerance and punitiveness toward dissidents and deviants, animosity and aggression against racial and ethnic out-groups" (Stenner, 2005, p. 3). Yet, authoritarianism was hitched to the star of psychoanalysis, and waning interest in psychoanalysis stifled authoritarianism research. Moreover, there were significant empirical shortcomings plaguing Adorno's authoritarianism.

The so-called *Authoritarian Personality* is a *latent hypothetical construct,* and as such it requires empirical support by the various well-established means of *construct validation* (e.g., Campbell & Fiske, 1959; Cronbach & Meehl, 1955). Given that the *Authoritarian Personality* is a multivariate latent construct, measured by multiple, presumably convergent indicators, it must be shown that these indicators in fact correlate to the requisite degree. As with many hypothetical constructs in psychology, however, validation has progressed little beyond quasi-experimental studies supportive of predictive validity with respect to some expected behaviors, such as obedience to authority (e.g., Elms & Milgram, 1966). This nonetheless falls quite short of a full psychometric workup. For example, one of the central features of the generative theoretical perspective behind this construct is its purported relation to right-wing ideology. Problematically, measures most commonly used to assess authoritarianism, such as the *California F Scale*, do not appear to have much predictive validity when validated against measures of conservatism (e.g., Ray,

1979), which is odd given that the *F* is supposed to stand for Fascist. Possibly even more concerning, for it strikes at the very heart of the construct and not just at its relation to ideological orientation, is the F Scale's poor predictive validity, as seen in its failure to converge with empirically supported peer ratings and nominations (Ray, 1976; Titus, 1968).

Critics have also noted that there have been as many, if not more, left-wing, as right-wing, authoritarian governments. In response, newer research attempts to extract authoritarianism from the left-right continuum. There is the phrase *right-wing authoritarianism*; but the adjective *right-wing* was typically omitted because nearly everyone reflexively understood *authoritarianism* to be *right-wing* due to its historical association with fascism. As reviewed by Duckitt (2022), recent research on *left-wing authoritarianism* (Conway et al., 2018; Costello et al., 2022) comes down on Eysenck's (1981) side of the question, suggesting, contra Stone (1980), that left-wing authoritarianism is no myth. It may be that authoritarianism is an unvalenced tactic[7] without inherent ideological affiliation, supporting observations made in Chap. 4. This might have been obvious from the start by considering that the Bolshevik Revolution in Russia (1917) substantially predates the rise of either Fascism in Italy (1922) or Nazism in Germany (1933). As discussed in Chap. 4, authoritarianism might be an expression of fast life history tactics as authoritarianism relates to power grabs that favor immediate action to staid deliberation, and which disregards political forms, policies, precedents, and procedures. Thus, the authoritarian, as a fast life history strategist, is acting politically as they do personally; they are oriented toward the present, discounting the future, and displaying lower thresholds for impulsivity and aggression.[8] In consequence of empirical realities failing to buttress theoretical claims, subsequent research on politically significant personality styles has either attempted to *replace* or *rehabilitate* Adorno's work. An example of each attempt is provided below.

More psychometrically validated and less ideologically biased constructs have been proposed to replace authoritarianism (e.g., Bhushan, 1982; Heaven, 1985; Ray, 1984). Among these is the construct of *Social Dominance Orientation* (SDO). To provide validation of this construct, Fischer et al. (2012) conducted a cross-cultural meta-analysis based on SDO data collected from 27 societies. The study examined the association between SDO with participants' social values (aggregated at the national level) and various macroeconomic and sociodemographic indicators per country. Models revealed that SDO had negative and significant correlations with participant *egalitarianism* (negative views of social inequalities and endorsement of moral equality) and *intellectual autonomy* and *affective autonomy* (sense of meaning originating from their personal experiences, emotions, and attitudes) rather than their embeddedness within a collective. Analyses also indicated that SDO was positively and significantly correlated with participant *mastery* (desire to dominate their social environment), *hierarchy* (preference for social stratification), *embeddedness* (affiliation toward their ingroup generating a sense of belonging), and *discrimination* (against minorities and migrants). Fischer and colleagues also reported negative correlations between SDO and gross national income, democracy, and gender empowerment. These are all *convergent validity* indicators.[9]

Rather than replace it, Stenner's, *The Authoritarian Dynamic*,[10] tried to rehabilitate authoritarianism by defining it more precisely and describing how it interacts with eliciting environmental factors; though in doing so, to paraphrase Darwin, it, to some extent, still bears the indelible stamp of Adorno's lowly psychometric origins. Stenner describes environmental stressors that evoke latent authoritarian dispositions.[11] Effectually, the author is saying that authoritarian political behavior is rooted in a trait, activated by a state—trait authoritarianism, activated by social threat begets the *authoritarian dynamic*. *Social activation* seems to have been the missing element causing inconsistent detection of authoritarianism.[12] Inglehart's, *Cultural Evolution,* used the term *authoritarian reflex* in a similar manner. This authoritarian reflex is activated when territorial integrity is threatened by outgroups amid economic privation. Immigration and recession evoke authoritarian populism (Inglehart, 2018). Economic surplus seems to be a lubricant, decreasing tensions between immigrant and native populations, which, when not present, heightens existential threat, which, in turn, makes people hunker down and hold tighter to their smaller portion to the exclusion of those that might have otherwise shared in it. The *state-trait* debate is relevant here, where *states* represent the power of the environment stressed by social psychologists, *traits* represent the power of individual differences stressed by personality psychologists. Traits are like a spring that can be stretched or compressed by states. The spring, like the trait, was cast in a certain way, with certain metals, with material of a certain gauge, with a certain amount and spacing between coils, all of which affect how it responds to outside forces. The relationship between states and traits is reconciled by *norms of reaction* (Dingemanse et al., 2010; Morrissey & Liefting, 2016). As reaction norms have explained how traits interact with states, Stenner and Inglehart, applying this thinking, are attempting to explain how trait authoritarianism interacts with socially activating states.

3 Politically Significant Traits

From global personality styles we turn to personality traits. Books and chapters can stand in for this short section. One can speak of traits as they correlate with one or the other end of the political spectrum, political behavior at large, or issue positions. As traits can relate to multiple forms of political content, political content can relate to multiple forms of personality variation: Personality's political significance can be considered (1) individually or collectively, (2) at the factor or facet level, (3) in abstract or in context.

First, one can consider ideological correlations with individual traits, or with groups of traits. For example, Dark Triad traits, *psychopathy, narcissism,* and *Machiavellianism,* collectively correlate with the view that, "aggression and force are appropriate methods to achieve ideological goals" (Moss & O'Connor, 2020). Gøtzsche-Astrup (2021) found dark triad traits to interact with strong ideological commitments to predict political violence.[13] However, taken alone, trait narcissism is more or less equally distributed across the political spectrum (Hatemi & Fazekas, 2018), and only assumes relevance when considered as one of three dark triad traits (Arvan, 2013). Bell et al. (2021) found Machiavellianism and narcissism, two of the three dark triad traits, predictive of liberalism. Similarly, Krispenz and Bertrams (2023) found two of three dark triad traits, in this case narcissism and psychopathy, predictive of revolutionary tendencies designed to undermine existing orders. As can be seen, individual traits sometimes assume ideological significance only in the context of related traits.

Second, one can look to either the facets or factors within the Five Factor Model of personality,[14] which show negligible to compelling correlations to political ideology. Extraversion is an example of a politically insignificant trait. As Mondak (2010) says after conducting his own analyses, "the coefficients for extraversion provide no evidence whatsoever that this trait matters for ideology." Agreeableness is inversely correlated with support for capital punishment. It also seems that the highly agreeable favor international cooperative endeavors (Mondak, 2010). Similarly, neuroticism predicts support for capital punishment (Robbers, 2006; Kandola & Egan, 2014), though neuroticism's main contribution to political behavior may not be in its parsing between one and another political pole, but in determining one's level of participation in the political process more generally. McCann (2014) found positive correlations between neuroticism and liberalism only at the state level, as opposed to federal level elections. In these ways, three out of five traits either show negligible relationships with ideology, or strong associations with particular policy positions, but not to political ideology as a whole. By contrast, political ideology correlates reliably with openness to experience and conscientiousness (Mondak, 2010; Xu et al., 2013; Fatke, 2017). Conscientiousness is predictive of conservatism (Cotterill, 2023). In some studies it has been linked to isolationism, but also to support for using military force. Personal responsibility, tradition, and virtue are all valued by the highly conscientious (Mondak, 2010). High openness is inversely correlated with dogma, prejudice, and intolerance. Openness to experience, like all five factor traits, is comprised of lower-order *facets*, some of which, particularly *aesthetic sensitivity* and *intellectual curiosity*, show important independent relations to liberalism (Xu et al., 2021). Thus, conscientiousness and openness to experience have opposite effects on political ideology: Openness predicts left-leaning affiliations and attitudes, while conscientiousness predicts right-leaning affiliations and attitudes (Federico, 2022).

Third, one can focus on personality traits in abstract as was done above, or in the context of other politically relevant variables. When considering context, some previously dismissed traits take on new relevance. For example, agreeableness, in abstract, shows modest associations with ideology, but more consistent and clear associations with ideology when understood in the context of social class. Specifically, agreeableness produces stronger associations with liberalism within higher socioeconomic strata (Furnham & Fenton-O'Creevy, 2018). Similarly, interaction effects are reported for openness to experience, which is more strongly predictive of liberalism among the educated (Osborne & Sibley, 2015). Beyond single mediators or moderators, personality trait influence can be embedded in intricate webs of association, such as that provided by Schlenker et al. (2012). These authors relate lower-order traits like personality and moral emotions to higher-order beliefs like worldview and locus of control, which are jointly predictive of outcomes like happiness, adjustment, life satisfaction, and work ethic.

4 On *Moral Emotions* and *Visions*

Social scientists manufacture multiple terms for a single construct. To some extent this is true with respect to constructs explaining political ideology. Jonathan Haidt's *moral emotions* (Haidt, 2007; Darwin, 1871),[15] though they share insignificant conceptual and statistical overlap with personality traits (Kawamoto et al., 2017; Milojev et al., 2014), share significant conceptual and statistical overlap with personality styles (Sun & Smillie, 2023; McAdams & Mayukha, 2023), such as social dominance orientation, authoritarianism (Ballout et al., 2023), and psychopathy (Djeriouat & Trémolière, 2014).[16] The same is true of Thomas Sowell's *visions,* used to describe broad states of mind that are similar to the ideational differences arising from personality styles. These authors are focusing on different expressions of personality variation—Haidt on the emotional level and Sowell on the ideational level. It is important to understand personality broadly as genetically informed and brain-based differences, expressed *emotionally, behaviorally,* and *ideationally*. Even as we think Haidt and Sowell are respectively writing about emotional and ideational expressions of personality, we review these theories because of their important novel contributions, but treat them within this chapter because of their significant covariance with personality.[17]

In *The Righteous Mind* (2013), in earlier papers (2001, 2003, 2007), and during talks on the content[18] of these publications, Haidt reviews ethnographic and individual differences across five moral emotions: (1) harm/care, (2) fairness/reciprocity, (3) ingroup/loyalty, (4) authority/respect, and (5) purity/sanctity. The first two moral emotions have a broad base within and between societies, and jointly establish a general moral compass emphasized in liberal ideologies as *caring for others and for their fair treatment*. On the other hand, the latter three moral emotions are more variably expressed, being consistently strong only among conservatives. While liberals are oriented toward *harm* and *fairness*, conservatives add to these *authority*, *ingroup*, and *purity*—moral emotions which collectively elicit behavioral tendencies toward policing and judging the actions of others. These latter three moral emotions, "called *binding foundations* because they pull people together into cohesive groups and help to suppress individual selfishness," (Schlenker et al., 2012) are variously aroused by irreverence or irreligion, immigration and otherness, disorder and change. Collectively, these have relevance, not to the individual and his proper treatment, but to the group and its unity and strength (Haidt, 2007). The relatively restricted intensity of these latter three moral emotions allows liberals to agree with Jefferson when he said, "But it does me no injury for my neighbour to say there are twenty gods, or no god. It neither picks my pocket nor breaks my leg."[19] Haidt's framework improves upon predecessors by cutting nature at its joints, avoiding either the authoritarian research that confused power with ideological position, or the more atomized explanations of the personality trait literature.

Sowell uses the term *visions* or *social visions* to denote worldviews undergirded by fundamentally different views of human nature.[20] Pairings of philosophers illustrate the contrast: William Godwin v. the authors of the Federalist papers, Thomas Paine v. Edmund Burke, John Kenneth Galbraith v. Friedrich A. Hayek, Jean Jacques Rousseau v. Hobbes. Similarly, the American and French Revolutions respectively illustrate *constrained* and *unconstrained* visions. Sowell sees the American Constitution as essentially an outgrowth of a constrained vision, though he sees its erosion as coming from advocates of an unconstrained vision. Sowell then connects visions to political positions thus:

> One of the curious things about political opinions is how often the same people line up on opposite sides of different issues. The issues themselves may have no intrinsic connection with each other. They may range from military spending to drug laws to monetary policy to education. Yet the same familiar faces can be found glaring at each other from opposite sides of the political fence, again and again. It happens too often to be coincidence and it is too uncontrolled to be a plot. A closer look at the arguments on both sides often shows that they are reasoning from fundamentally different premises. These different premises - often implicit - are what provide the consistency behind the repeated opposition of individuals and groups on numerous, unrelated issues. They have different visions of how the world works.

Sowell sometimes uses other words such as man's *moral* and *mental natures* to stand in as synonyms for visions.[21] Sowell compares these worldviews to social maps. He notes that the world is too complex to know directly and fully so we have these visions functioning as simplifications. Visions are described as prefigured mental biases that color one's perception of reality, which is apt to be selectively ignored, distorted, or otherwise violated by the vision. Visions supplement knowledge just as *scripts* supplement memory. Sowell compares these worldviews to social maps, and seems to be invoking the concept of *social schemata*. Similar to visions, social schemata[22] are "mental templates" ordering new experiences through past expectations. Visions have a heuristic function that stands in for perfect knowledge; they simplify the world, as does ideology.[23] Especially in the latter phases of the above description, Sowell defines visions, not in a way that explains ideology, but in a way that is effectively the same as ideology. On the other hand, Sowell is correctly tracing opposing political ideologies to the more fundamentally opposite worldviews of which they are derivatives, and which we believe to be the ideational expression of personality. Personality orientations will lead people to emphasize different sides of the explanatory coin. Crime will be seen as a function of poverty by liberals and moral degeneracy by conservatives. Both are partially right as crime is a function of both nature and nurture, not one or the other exclusively. *Visions* or *worldviews* or *ideologies*, whatever we call them, imperfectly interpret reality, collapsing its complexity into comprehensibility. When personality is conceived of in this way, to actuate emotion and behavior and inform attitudes and cognitions, only then can personality be seen to structure one's entire worldview.[24]

5 Conclusions

As reviewed above, political ideology is partially reducible to *personality*, the name given to *behavioral, emotional,* and *ideational* individual differences. We have referenced politically significant personality styles, like authoritarianism, and politically significant personality traits, like openness to experience. As is political ideology itself, these underlying styles and traits are continuously distributed. For example, authoritarianism is the verbal anchor given to one pole of a continuum anchored on the opposite end by libertarianism; similarly, openness to experience is the verbal anchor given to one pole of a continuum anchored on the opposite end by closedness to experience.

We have yet to describe either the evolutionary causes of ideological or personality variation. This is the province of Chap. 8, which explains the evolutionary dynamics that create and maintain individual differences, distributing them across a range of values within populations. Suffice it to say at present that the personality continua reviewed in this chapter and the ideological continua considered throughout the book are more or less the same individual differences, simply seen from different disciplines and expressed in different contexts. Furthermore, we should state here that these individual differences evolved by the same mechanisms. Differences are distributed across a continuum, as opposed to centering on a single optimum. Meiotic and mitotic division, the number of muscles and bones, vision and hearing, digestive processes and blood circulation are among the innumerable traits for which an evolutionary optimum exists—the one best way used by all species members. Conversely, there is an increasingly vast literature on fitness *tradeoffs* which create phenotypic trait differences. Individual differences within species can be produced because selective pressures vary across space or time, but also because, within a single environment at a single time, selective pressures are pulling in opposite directions. Through this evolutionary logic, different personality types and traits are understood as different adaptive solutions catering to one or another selective push or pull. These individual differences are subject to *balancing selection*, sometimes more specifically to *negative frequency-dependent balancing selection,* a mode of selection wherein the fitness of a trait is the inverse of its frequency.[25] Again, these evolutionary dynamics, responsible for producing individual differences and consequent ideological variation, have a literature of their own, but are subsumed within multilevel selection theory as explained in the eighth chapter.

Notes

1. This is the test of *Right-Wing Authoritarianism*.
2. The reduced intensity of these latter three moral emotions allows liberals to agree with Jefferson when he said, "But it does me no injury for my neighbor to say there are twenty gods, or no god. It neither picks my pocket nor breaks my leg."
3. Heywood (2021) describes ideologies as frameworks from which people *critique existing orders, envision the future of society*, and derive *theories of social change*.
4. As discussed in the book itself, Haidt has actually mentioned group-selection (which is a central concept in multilevel selection theory) briefly on page 237 of *The Righteous Mind*.
5. Contrary to those many traits under directional selection that create a single, species-specific solution to an adaptive exigency, negative frequency-dependent balancing selection is a type of selective pressure that actively creates trait diversity. Some few traits, such as personality traits and their behavioral correlates, become less adaptive as they become more common. In consequence, as explained in the fourth chapter, negative frequency-dependent balancing selection selects for a range of traits that compete with one another.
6. The theory was paired with an inventory designed to detect and explain proto-fascist tendencies. Altemeyer then revived social scientific interest in personality as a predictor of ideology with a refined measure of Adorno's excessively intricate scale, which became known as the test of Right-Wing Authoritarianism.

7. Meaning separate from ideology, not independent of biological dispositions entirely.
8. With respect to cognition, authoritarian personalities conform to group norms and are intolerant of those that do not so conform; they are sensitive to threat, and seek *cognitive closure,* a sort of propensity to view things in clear and compartmentalized ways, eliding nuance and ambiguity (Federico, 2022; Kelemen et al., 2014). As per Altemeyer, cognitive failings and biases associated with authoritarianism include misattributions, tendencies to make conflicting statements, cognitive compartmentalizations that invite inconsistencies, and skewed interpretations of evidence. The literature is dotted with associations between reduced intelligence, dogmatism (Berggren et al., 2019), low educational attainment, and limited scholastic achievement, though it is a complex literature; one requiring caution about differentiating cognitive failings from cognitive differences (Sumner et al., 2023). Relative to discussions in Chap. 4, those with more significant cognitive failings would not be among the politically engaged. These less marked educational and cognitive differences associated with authoritarianism may nevertheless restrict engagement to more direct and less sophisticated means, while also limiting the apprehension of political complexities and gradations.
9. Contrary to the authors' predictions, SDO was not correlated with either unemployment or income inequality. A multilevel model (MLM) estimated the influence of democracy, gross national income, gender equality, and egalitarianism on SDO scores above and beyond the influence of the samples consisting of majority group individuals (compared to mixed groups or groups containing social minorities), and the scales' reliability estimates (Cronbach's α). The MLM revealed that higher reliability estimates produced lower SDO scores. In a similar vein, higher egalitarianism values negatively predicted SDO. In contrast, democracy, gross national income, and gender equality had no significant contribution to the model, supporting *discriminant validity.* Convergent and discriminant validation are the two major components of construct validation.
10. Stenner, further describes the nature of authoritarianism and its relations to group dynamics: Actuated by "persistent fears regarding societal chaos" (Stenner, 2005, p. 71), authoritarians implicitly prefer "autocratic social arrangements in which individual autonomy yields to group authority" (Stenner, 2005, p. 15). Rather than being idiosyncratic prod-

ucts of particular times or societies, authoritarians display an invariant "triad of racial, political, and moral intolerance." The locus of debate and groups implicated can change, but the personality dynamics remain the same. Stenner (2005, p. 18) refers to authoritarian's *groupiness,* meaning their dispositions to identify with large collectives, and to conform to overarching systems of order. Authoritarians *glorify* ingroups and *denigrate* outgroups; likewise, authoritarians dispense praise and punishment according to conformity to the dominant order and the common good. Authoritarianism imparts an "endogenous pre-disposition to be hyper-reactive to all that might threaten oneness and sameness" (Stenner, 2005, p. 71). Under *normative threat,* authoritarians express *defensive outputs* including "racial, political, and moral intolerance." By contrast, libertarians, located at the other extreme of the authoritarian continuum, and by nature disinclined to "groupiness," remain inattentive to the collective until it imperils the individual.

11. Stenner recognizes innate factors contributing to authoritarianism, though she continues to conceptualize authoritarianism as a failing, some sort of cognitive problem or negative outcome to poor child rearing.
12. This inconsistency in detection was responsible for waning interest in authoritarianism research.
13. In moving away from individual traits in preference to trait conglomerates, we might come full circle to a personality style approach such as that seen in the aforementioned work on authoritarianism. This is a valid criticism, though trait conglomerates like the dark triad were developed as apolitical descriptions of individual differences and so their ability to predict political ideology is, if not superior, at least less suspect.
14. Trait theorists analyzed adjectives descriptive of individual differences. Referred to as the *lexical model*, this approach assumed that personality variability must be encoded in language. Thousands of relevant adjectives were factor analyzed, and thereby condensed into a handful of traits, sometimes three, sometimes, five, sometimes sixteen. Five is the most commonly accepted number of factors: *Openness to Experience, Conscientiousness, Extraversion, Agreeableness,* and *Neuroticism.* Of these five, the last three do not show either meaningful or consistent relationships with political ideology, as understood as a cognitive complex extending across a spectrum.

15. We cite Darwin's, *Descent of Man,* because of frequent references to *moral sense* and *moral feelings* that have relevance to Haidt's use of moral emotions. Haidt clearly relates personality traits to moral emotions in his talks and writings, though we are not certain how he ultimately relates one to the other.
16. As McAdams and Mayukha (2023) state, "In a multitude of ways, human beings express and experience individual differences in their moral engagement of the world, all of which fall within the purview of personality psychology."
17. Duckitt (2022) attempts to bridge the literature on authoritarianism and social dominance, and then connect both to moral emotions. Duckitt asserts that authoritarianism positively correlates to loyalty, authority, and sanctity, whereas social dominance orientation is negatively correlated with care and fairness. First, Duckitt reviews the literature distinguishing authoritarianism from *Social Dominance Orientation,* which emerges as a distinct factor in most theories and studies. Where right-wing authoritarianism is predicted by low openness to experience and associated with outgroup threat; a social dominance orientation is predicted by low agreeableness and associated with ingroup competition. Authoritarianism, as in classic conceptualizations, extends from perceptions of a hazardous and erratic world. By contrast, a social dominance orientation relates to goal directed orientation to dominate within a competitive landscape or to express "superiority over others driven by a view of the social world as a ruthlessly competitive jungle..." (Duckitt, 2022, p. 183).
18. As *The Righteous Mind* is of such obvious relevance, Haidt's publications (2001, 2003, 2007) are of general utility. Once again, Haidt is perhaps one of the only researchers to connect political ideology to group selection, the theme we develop subsequently.
19. From Jefferson's "Notes on the State of Virginia," 1782 (monticello.org)
20. Sowell is very clear to state that he is simplifying the complex and dichotomizing a continuum.
21. When people clash over concrete policy there is usually not as vehement a misunderstanding; but when there is a conflict of vision this is where the most vehement clashes will occur and also when the role of vision will be most misunderstood.

22. https://www.oxfordreference.com/display/10.1093/oi/authority.20110803100515422
23. In an interview with Peter Robinson of the Hoover Institute, Sowell affirmatively replied when asked if his concept of conflicting visions extends farther back and can it be traced all the way through the Western tradition through to Greek philosophy. Peter Robinson also quotes Robert Kennedy as illustrating the unconstrained vision, and Sowell heartily agrees: "Some men see things as they are, and say why. I dream of things that never were, and say why not." https://www.youtube.com/watch?v=OGvYqaxSPp4
24. This is why political ideology predicts variables as diverse as "death anxiety; system instability; dogmatism; intolerance of ambiguity, low openness to experience, and uncertainty; need for order, closure, and negative integrative complexity; and fear of threat and loss of self-esteem" (Stankov, 2009; Jost 2013).
25. Like human personality researchers, animal personality researchers once thought that there was a golden mean or a most fit personality, deviation from which denoted decreased fitness (Dingemanse & Reale, 2013). At least by 1983 this view had been seriously considered long enough to earn a description; Lande and Arnold "The Phenotypic Selection Approach" (Dingemanse & Reale, 2013). This has been a sea change in thinking. The change has been profitable; it seems that this is an understanding of personality that truly is congruent with observed nature. Like human personality psychologists, animal researchers now recognize that the only optimum that is to be found is among the population at large. A population optimum suggests that there are multiple ways to survive and reproduce and that a range of behavioural strategies or personalities will evolve across the range.

Integrative Metacommentary: Part II. Proximate Causation

Aurelio José Figueredo, Steven Charles Hertler, and Mateo Peñaherrera-Aguirre

Proximate Causation

Part II's four chapters collectively describe processes of proximate causation, encompassing both mechanisms of causal influence and behavioral development. Chapter 2 covers the genetics of political ideology, starting with a primer on molecular genetics and functional genomics, including various gene-environment transactions and interactions, and afterwards reviews pertinent evolutionary genetic principles such as phenotypic and genetic accommodation as well as the principle of gene-culture coevolution. The chapter subsequently describes ten heritable features of political ideology that have been empirically established, thereafter drawing important theoretical distinctions between lower-order emotions and higher-order cognitions, innate and lagged genetic influences, states and traits, and finally defines the sociocultural Overton Window. Chapter 3

A. J. Figueredo • M. Peñaherrera-Aguirre
University of Arizona, Tucson, AZ, USA
e-mail: ajf@arizona.edu; mpeaher@email.arizona.edu

S. C. Hertler (✉)
Saint Elizabeth University, Morristown, NJ, USA

covers the neuroscience of political ideology, including elements of both neuroanatomy and neurochemistry that underlie the neurophysiologies of stress, anxiety, and fear, as well as reactions to novelty and otherness that have been hypothesized to influence political and ideological biases.

After reviewing these more fundamental biological bases of political ideology, Chapters 4 and 5 direct our attention to aspects of the internal psychological milieu and ambient social ecology in which behavioral development plays out. Chapter 4 starts by defining the construct of political engagement, then goes on to trace its roots to the various developmental and ecological influences that have been proposed by ecological systems theory, niche construction theory, extended phenotype theory, and life history theory. The phenotypic fitness continuum is then related to the life history speed continuum and to the multilevel selection continuum, ranging from individual to group selection. Finally, a two-by-two cross-tabulation is provided that presents a theoretically derived categorical breakdown of political ideologies according to these shaping selective pressures. Chapter 5 reviews personality and social psychological influences upon the development of political ideology. The construct of authoritarianism is reviewed and critiqued, followed by a brief survey of politically significant individual differences, such as openness to experience, conscientiousness, dark triad traits, moral emotions, and social visions.

The order of the chapters within Part II thus parallels the development of a living organism assembling and shaping itself over the lifespan. This progression starts at the smallest level, with the genes in a state of constant and continuing transaction with their environment, proceeds to the development of neural control systems that regulate emotional expression and evaluative functions, goes on to describe the evolutionary-developmental selective pressures that shape the individual organism's engagement with its surrounding social context, and then progresses to the traits of individuated personality that emerge from these biological substrates in mutual causal transaction with their social environment. Thus, Part II is organized according to increasing levels of biological complexity, but nonetheless begins to address two of Tinbergen's Four

Questions, namely those of mechanistic causation and individual ontogeny. These two questions have been subsumed within the more contemporary rubric of *proximate causation*, which is the theme of Part II. The remaining two of Tinbergen's questions under the contemporary rubric of *ultimate causation*, namely those of adaptive function and evolutionary phylogeny, were left to Part III.

Part III

Ultimate Causation

6

The Macroevolution of Sociopolitical Behavior and Within-Group Conflict in Nonhuman Primates

Mateo Peñaherrera-Aguirre, Aurelio José Figueredo, and Steven Charles Hertler

> In reality, there is nothing special about deadly human violence and war. Fundamentally the solution to the 'enigma of war' is that no such enigma exists. Violent competition, alias conflict—including intraspecific conflict—is the rule throughout nature, as organisms vie among themselves to survive and reproduce under ever-prevalent conditions of scarcity, which are accentuated by the organisms' own process of propagation.
>
> —Azar Gat[1]

M. Peñaherrera-Aguirre • A. J. Figueredo
University of Arizona, Tucson, AZ, USA
e-mail: ajf@arizona.edu; mpeaher@email.arizona.edu

S. C. Hertler (✉)
Saint Elizabeth University, Morristown, NJ, USA

S. C. Hertler et al., *The Evolution of Political Ideology*,
https://doi.org/10.1007/978-3-031-83941-2_6

1 Introduction

This chapter analyzes predictors of sociopolitical behavior in nonhuman primates, concentrating on within-group conflict frequencies among various taxa. Within the past 70 years, primatologists and evolutionary researchers have repeatedly identified associations between socioecological factors and the rise of complex cognition in nonhuman primates. For example, Jolly (1966) suggested that primate intelligence evolved in response to social pressures. Similarly, as per Humphrey (1976), complex cognitive abilities in nonhuman primates evolved in response to group living, with group living itself evolving in response to community ecological conditions. Subsequently, de Waal (2007), inspired by Niccolo Machiavelli's *The Prince* (1513), argued that chimpanzees' political behavior was not limited to dyadic agonistic interactions where individuals deceived and instigated others to obtain a particular goal. Instead, reciprocity, kin assistance, coalitions and alliances, and post-conflict resolutions (such as reconciliation) are social behaviors observed in chimpanzee communities (de Waal, 2000, 2007). The presence of sophisticated sociopolitical dynamics in nonhuman primates provided the empirical foundation for the *Machiavellian Intelligence Hypothesis*[2] (Byrne, 2018), wherein fitness outcomes were attributable to social abilities instead of merely to physical force during cooperative and agonistic interactions. Following this literature, several behavioral ecologists and primatologists have developed metrics destined to capture the degree of cooperation and competition within primate societies and their corresponding morphological and cognitive correlates.

The review of this literature, along with a series of novel analyses, provides comparative evidence of political behavior in nonhuman primates, laying the foundation for subsequent chapters on the evolution of human political nature. Even more than their ape and hominid ancestors, *Homo sapiens* have a particular evolutionary history, relevant to intelligence and social organization, that bears on the capacity to act politically. Most broadly, this chapter provides evidence supportive of the following conclusion: Humans are the most extreme case within a phylogenetic lineage already disposed toward Machiavellian intelligence and complex, violent social dynamics.

2 Neurocognitive Substrates of Sociopolitical Behavior in Nonhuman Primates

According to Burkhart and collaborators (2016), *general intelligence* in nonhuman animals refers to an individual's cognitive ability to obtain information based on its interaction with physical and social stimuli in the environment. Moreover, this knowledge organizes the individual's behavior in various settings (novel and familiar contexts) and facilitates solving physical and social challenges. The literature distinguishes between "*g*," referring to a domain-general cognitive ability estimated using individual-level data, and "*G*" describing a general cognitive dimension extracted from species-level behavioral indicators (Fernandes et al., 2020; Peñaherrera-Aguirre et al., 2023). For Burkhart and colleagues (2016), both *G* and *g* reflect behavioral flexibility and reasoning ability. Comparative psychologists and behavioral ecologists have identified *g* in numerous nonhuman species, including fish (Prentice et al., 2021; Woodley of Menie et al., 2023), avians (Burkart et al., 2017), rodents (Woodley of Menie et al., 2022), ungulates (González et al., 2019), carnivorans (Arden & Adams, 2016), and nonhuman primates (Fernandes et al., 2020, Woodley of Menie & Peñaherrera-Aguirre, 2022, 2023). Deaner and colleagues (2006) and Reader and collaborators (2011) reported evidence for *G* in nonhuman primates. Fernandes et al. (2014) expanded Reader and colleagues' database (2011) and conducted unit-weighted factor estimations and a Principal Axis Factor analysis. Their statistical modeling supported the presence of *G* across nonhuman primates, with this latent dimension loading positively and significantly onto tool use, extractive foraging, innovation, social learning, and tactical deception. This latent dimension explained 61.7% of the reliable variance, with factor loadings ranging from 0.688 (tactical deception) to 0.905 (tool use).

Shultz and Dunbar (2022) conducted a Principal Component analysis that revealed two components; the first dimension, *socioecological complexity,* explained 52% of the total variance and loaded onto tool use, extractive foraging, social learning, innovation, policing, deception, and coalitions. Component loadings ranged from 0.8 to 0.9. The second dimension, *reproductive cooperation,* explained 24% of the total variance and loaded onto cooperative breeding, alloparenting, collective action, and parental care. Component loadings ranged from 0.7 to 0.9.[3] Shultz and Dunbar also conducted multiple Phylogenetic Generalized Least Squares models (PGLS)[4] examining the association between the aforementioned dimensions with *G* scores derived from Deaner and colleagues (2006) and Reader and colleagues (2011). Analyses indicated that the socioecological complexity dimension correlated positively and significantly with the two *G* factor scores (Deaner and colleagues' $R^2_{adj} = 0.13$; Reader and colleagues' $R^2_{adj} = 0.14$). These findings provide additional macroevolutionary evidence associating *G* to behavioral indicators, such as policing, coalitions and alliances.[5]

3 Primate Societies and Levels of Competition

Plavcan and Plavcan and colleagues (1992, 1997a, 1997b, 2001, 2012) developed a competition index based on the intensity and frequency of within-group competition. Primate species featuring low-*intensity* conflict are characterized by the persistence of peaceful and mutually tolerant interactions among individuals, wherein a dominance hierarchy based on recurrent agonistic interactions is either absent or considerably tempered. In contrast, species featuring high-*intensity* conflict are characterized by despotic and intolerant interactions, wherein dominance hierarchies are based on persistent agonistic exchanges. Similarly, Plavcan (2001) classifies primate species into low-*frequency* and high-*frequency* conflict. Low-frequency taxa generally feature a single adult of one sex residing alongside multiple adults of the opposite sex, as in single male-multi female societies. High-frequency taxa generally feature multiple adults of the same sex residing alongside multiple adults of the opposite sex, as in multi-male-multi-female societies. To simplify and summarize, Plavcan (2001) classifies social conflict as (1) low frequency and low intensity; (2) high frequency and low intensity; (3) low frequency and high intensity; and

(4) high frequency and high intensity. Morphological indicators of intrasexual competition, such as residualized canine size dimorphism (adjusted for the allometric effects of body mass), positively correlate with the aforementioned levels of competition, providing additional statistical support via convergent validity (Plavcan, 2001).

4 Conservative, Revolutionary, and Bridging Coalitions in Nonhuman Primates

Coalitions and alliances are often classified into *conservative, revolutionary,* and *bridging* (Chapais, 1995; van Shaik et al., 2006). According to Chapais (1995), conservative coalitions and alliances are described as agonistic third-party interventions or coordinated aggression that consolidate individuals' rank order within the social structure. Consider three individuals: α (highest rank order), β (middling rank order), and γ (lowest rank order). A conservative coalition occurs when: (1) α initiates an agonistic third-party intervention towards γ, with β supporting α; (2) β initiates an agonistic third-party intervention towards γ, with α supporting β; (3) γ initiates an agonistic third-party intervention towards α, with β supporting α; and (4) γ initiates an agonistic third-party intervention towards β, with α supporting β. Conservative coalitions force lower-ranking individuals to yield to higher-ranking individuals (Chapais, 1995).

For Chapais (1995), bridging coalitions and alliances entail agonistic third-party intervention or coordinated aggression performed by a high-ranking and a low-ranking against a mid-level target. Bridging coalitions facilitate ascent in the dominance hierarchy. Bridging coalitions occur when: (1) γ initiates an agonistic interaction towards β, with α supporting γ; (2) α initiates an agonistic interaction towards β with γ supporting α;(3) β initiates an agonistic interaction towards γ, α with supporting γ; (4) β initiates an agonistic interaction towards α, with γ supporting α. According to Chapais (1995), bridging coalitions are often observed in primate societies featuring female philopatry, high levels of nepotism, and despotism. Within these social structures, high-ranking adult females born to dominant matrilines attack mid-ranking individuals who are often associated with more genetically distant matrilines. Recurrent bridging coalitions allow the offspring of dominant females to surpass the target's rank order in the dominance hierarchy.

In contrast, Chapais (1995) described revolutionary coalitions and alliances as agonistic third-party intervention or coordinated aggression of low-to-mid-ranked individuals against a high-ranking target. Revolutionary coalitions often lead to a reorganization of individuals' rank orders within the social structure. Revolutionary coalitions can occur when: (1) γ initiates an agonistic third-party intervention towards α, with β supporting γ; (2) β initiates an agonistic third-party intervention towards α with γ supporting β; (3) α initiates an agonistic third-party intervention towards γ, with β supporting γ; (4) α initiates an agonistic third-party intervention towards β, with γ supporting β. According to Chapais (1995), revolutionary coalitions are less stable than conservative coalitions, wherein it is common for a revolutionary coalition to transition into a bridging coordination.

5 Purpose of the Present Study: Methods

The current chapter examined the presence of a general within-group conflict factor based on the frequency of conflict, intensity of conflict, and coalitions and alliances across multiple primate species. The present study also predicted a positive association between endocranial volume and the aforementioned general within-group conflict factor.[6] The presence of a sizeable association between the latter variables justified the ancestral reconstruction of within-group conflict factor scores in 68 extinct and 1 extant (*Homo sapiens*) species of primates.[7] Moreover, we examine a predicted negative association between the reconstructed within-group conflict factor scores on the temporal persistence of extinct primate species, above and beyond the influence of temporal and phylogenetic autocorrelations.

6 Results: Within-Group Conflict Factors: Structural Validity

Horn's parallel analysis recommended the extraction of a single latent dimension. A Principal Axis Factor (PAF) analysis indicated this latent variable loaded positively onto the frequency of conflict (λ_F[8] = 0.995), coalitions and alliances ($\lambda_F = 0.638$), and intensity of conflict ($\lambda_F = 0.622$). This general factor explained 59.5% of the variance and exhibits high internal consistency (Cronbach's $\alpha = 0.79$). Part-whole correlations based on unit-weighted estimates supported the presence of a general within-group conflict factor that loaded positively and significantly onto frequency of conflict ($r = 0.898$, $p < 0.0001$), coalitions and alliances ($r = 0.810$, $p < 0.0001$), and intensity of conflict ($r = 0.803$, $p < 0.0001$). A similar pattern occurred with a restricted within-group conflict factor that loaded positively and significantly onto the frequency of conflict and coalitions and alliances ($r = 0.904$, $p < 0.0001$). An extension analysis revealed a positive and significant correlation between the aforementioned restricted within-group conflict factor and the intensity of the conflict indicator ($r = 0.562$, $p < 0.0001$). This result supports the initial inclusion of the intensity of conflict as part of the general within-group conflict factor.[9]

6.1 General Within-Group Conflict Factor: Predictive Validity and Criterion Validity

Several Phylogenetic Logistic Regression models[10] were conducted to examine the predictive validity of the general within-group conflict factor on several indicators of lethal aggression across primate species. Analyses indicated that the aforementioned latent dimension positively and significantly predicted male adulticide and infanticide. In contrast, there was no significant association between within-group conflict and female adulticide (for additional details, see Table 6.1). A lethal aggression index was generated by summing the dichotomous entries corresponding to female adulticide, male adulticide, and infanticide. A Phylogenetic Generalized Least Squares (PGLS) model estimated a positive and significant influ-

Table 6.1 Phylogenetic Logistic Regression models examining the influence of a general within-group conflict factor on female adulticide, male adulticide, and infanticide

Criteria	*b*-weight	Std. Error	*t*-value	*p*-value
Female adulticide	0.410	0.418	0.98	0.3265
Male adulticide	1.204	0.389	3.10	0.0020
Infanticide	1.401	0.432	3.24	0.0012

ence of within-group conflict on the latter lethal aggression index ($\beta = 0.487$, $p < 0.0001$), further supporting the latent factor's predictive validity. Lastly, a PGLS model revealed a positive and significant influence of polygyny on the general within-group factor ($b = 1.014$,[11] $R^2 = 0.384$, $p < 0.0001$), providing additional evidence concerning the latent dimension's criterion validity.

6.2 Indicators of Within-Group Conflict and Lethal Aggression

The Phylogenetic Logistic Regression analyses revealed a positive and significant influence of coalitions and alliances on male adulticide. Neither female adulticide nor infanticide reached statistical significance. Frequency of conflict positively and significantly predicted male adulticide and infanticide. A similar pattern emerged with intensity of conflict, wherein this predictor positively influenced male adulticide and infanticide. Neither frequency of conflict nor intensity of conflict significantly predicted female adulticide. For further details, see Table 6.2. Several PGLS models assessed the bivariate associations of coalitions and alliances, frequency of conflict, and intensity of conflict on the lethal aggression index. The models estimated positive and significant effects between the aforementioned variables and the lethal aggression index (coalitions and alliances: $\beta = 0.337$, $p = 0.0073$; frequency of conflict: $\beta = 0.459$, $p = 0.0002$; and intensity of conflict: $\beta = 0.419$, $p = 0.0007$).

Table 6.2 Phylogenetic Logistic Regression models examine the influence of coalitions and alliances, frequency of conflict, and intensity of conflict on female adulticide, male adulticide, and infanticide

Coalitions and alliances				
Criteria	*b-weight*	*Std. Error*	*t-value*	*p-value*
Female adulticide	0.511	0.633	0.81	0.4192
Male adulticide	1.549	0.591	2.62	0.0087
Infanticide	1.144	0.669	1.71	0.0872
Frequency of conflict				
Criteria	*b-weight*	*Std. Error*	*t-value*	*p-value*
Female adulticide	0.387	0.396	0.98	0.3285
Male adulticide	0.940	0.350	2.68	0.0073
Infanticide	1.298	0.366	3.55	0.0004
Intensity of conflict				
Criteria	*b-weight*	*Std. Error*	*t-value*	*p-value*
Female adulticide	0.252	0.430	0.59	0.5584
Male adulticide	0.870	0.385	2.26	0.0238
Infanticide	1.327	0.396	3.35	0.0008

6.3 Log Endocranial Volume and Within-Group Conflict

Several PGLS models evaluated the influence of z-log endocranial volume on various metrics of within-group conflict. Endocranial volume positively and significantly predicted primates' intensity of conflict, frequency of conflict, coalitions, and alliances, a restricted within-group conflict factor (comprised of coalitions and alliances and frequency of conflict), and a general within-group conflict factor. Of these, the intensity of conflict featured the smallest standardized estimate ($\beta = 0.362, p = 0.0375$), and the restricted within-group conflict factor had the largest ($\beta = 0.618, p = 0.0007$). Most models displayed sizeable joint phylogenetic signal values, further supporting using phylogenetic comparative methods.[12]

Several likelihood ratio tests examined whether the value of joint λ_B, computed for each model, was significantly different from 0 (indicating the absence of phylogenetic inertia) or 1 (the traits' residuals are phylogenetically conserved; hence, two or more species are expected to feature similar values due to shared ancestry). A value significantly similar to 1 indicates that the probability of traits' residuals change varies as a function of branch length, the longer a phylogenetic branch, the greater the likelihood that a specific values will vary. Most likelihood ratio tests were significantly different from 0 and 1. These results suggest that even though all models feature signs of phylogenetic inertia, the likelihood that the traits' residuals will change is not dependent on the length of the branches in the phylogenetic tree. These results are further described in Table 6.3.

6.4 Phylogenetic Path Analysis

Using various combinations of *z* log endocranial volume, the general *z* within-group conflict factor, and the *z* lethal aggression index, the following four models examined hypothesized direct and indirect effects: (1) *z* log endocranial volume and *z* within-group conflict predict *z* lethal aggression; (2) *z* log endocranial volume predicts *z* within-group conflict and *z* within-group conflict predicts *z* lethal aggression; and (3) *z* within-group conflict predicts *z* lethal aggression and *z* lethal aggression predicts *z* log endocranial volume; and (4) *z* within-group conflict and *z* lethal aggression predict *z* log endocranial volume. As indicated in Table 6.4, model 2 featured the strongest statistical support (*CICc weight* = 0.613), wherein *z*-log endocranial volume positively predicted the *z*-within-group conflict factor ($\beta = 0.614$, $p < 0.0001$[13]), and the *z*-within-group conflict factor positively predicted the *z*-lethal aggression index ($\beta = 0.324$, $p = 0.0131$).

6.5 Ancestral Character Reconstruction and Phylogenetic Bracketing

All phylogenetic bracketing models were computed between the unstandardized log endocranial volume and the unstandardized general within-group conflict factor scores, allowing the model to estimate the intercept instead of fixing it to 0 (as is required when computing statistical models with standardized variables). A Reduced Major Axis (RMA) regression

Table 6.3 Phylogenetic Generalized Least Squares models examining the influence of *z* log endocranial volume on *z* intensity of conflict, *z* frequency of conflict, *z* coalitions and alliances, *z* restricted within-group conflict factor, and *z* general within-group conflict factor

Criterion: *z Intensity of conflict*				
Phylogenetic signal	**κ[Fix]**	**λ_P[ML]**	**λ_P-*lb*(p-value)**	**λ_P-*ub*(*p*-value)**
	1.000	0.858	<0.0001	0.0001
Predictor	***β***	**Std. Error**	***t*-value**	***p*-value**
z Log endocranial volume	0.362	0.171	2.12	0.0375
Criterion: *z Frequency of conflict*				
Phylogenetic signal	**κ[Fix]**	**λ_P[ML]**	**λ_P-*lb*(*p*-value)**	**λ_P-*ub*(*p*-value)**
	1.000	0.894	<0.0001	0.0226
Predictor	***β***	**Std. Error**	***t*-value**	***p*-value**
z Log endocranial volume	0.598	0.197	3.04	0.0034
Criterion: *z Coalitions and alliances*				
Phylogenetic signal	**κ[Fix]**	**λ_P[ML]**	**λ_P-*lb*(p-value)**	**λ_P-*ub*(p-value)**
	1.000	0.000	1.000	<0.0001
Predictor	***β***	**Std. Error**	***t*-value**	***p*-value**
z Log endocranial volume	0.507	0.103	4.92	<0.0001
Criterion: *z Restricted within-group conflict factor*				
Phylogenetic signal	**κ[Fix]**	**λ[ML]**	**λ-*lb*(p-value)**	**λ-*ub*(p-value)**
	1.000	0.683	0.0021	<0.0001
Predictors	***β***	**Std. Error**	***t*-value**	***p*-value**
z Log endocranial volume	0.618	0.175	3.54	0.0007
Criterion: *z General within-group conflict factor*				
Phylogenetic signal	**κ[Fix]**	**λ[ML]**	**λ-*lb*(p-value)**	**λ-*ub*(p-value)**
	1.000	0.670	0.0003	<0.0001
Predictors	***β***	**Std. Error**	***t*-value**	***p*-value**
z Log endocranial volume	0.597	0.152	3.93	0.0002

Note. λ-*lb*: Lower bound relates to the value of 0; λ-*ub*: Upper bound relates to the value of 1

Table 6.4 Model comparison based on Phylogenetic Path models evaluating the contributions of *z*-log endocranial volume, the *z*-general within-group conflict factor, and the *z*-lethal aggression index[a]

Model	*k*	*q*	*C-statistic*	*p-value*	*CICc*	*ΔCICc*	*Likelihood*	*CICc weight*
2	1	5	2.08	0.3542	13.2	0.000	1.000	0.613
3	1	5	4.77	0.0922	15.9	2.69	0.260	0.159
1	1	5	4.97	0.0833	16.1	2.89	0.235	0.144
4	1	5	6.06	0.0484	17.1	3.98	0.137	0.084

[a]Phylogenetic Path models use a *d*-separation technique to estimate a *C*-statistic following the x^2 distribution, wherein $p < 0.05$ indicates a poor statistical fit (Gonzalez-Voyer & von Handerberg, 2014). In contrast to previous tables in this chapter, the term *k* represents the conditional independencies estimated in each model and is not an indicator of phylogenetic signal (κ). The term *q* indicates the number of parameters estimated in each model

model estimated a positive and significant association between log endocranial volume and the general within-group conflict factor ($p = 0.01$), as captured by the equation WGC = -2.410+(log ECV*1.565). The PGLS model also suggested that log endocranial volume positively predicted the general within-group conflict factor ($p = 0.0003$), generating the following equation: WGC = −1.389+(log ECV*0.801). A similar pattern emerged from a Phylogenetic RMA regression model ($p < 0.0001$), which generated the statistical equation WGC = 1.411+(log ECV*0.306).

Within-group conflict (WGC) scores with extant and extinct primate species revealed that the last common ancestor of all primates, considered in the reconstruction, featured low levels of within-group conflict. Within hominoids, the subtribe Hominina (*H. erectus*, *H. floresiensis*, *H. habilis*, *H. heidelbergensis*, *H. neanderthalensis*, and *H. sapiens*) exhibited greater WGC scores compared to the subtribe Australopithecina (*Ardipithecus ramidus*, *Australopithecus afarensis*, *A. africanus*, *A. garhi*, *Kenyanthropus platyops*, *Paranthropus aethiopicus*, *P. boisei*, *P. robustus*, *Sahelanthropus tchadensis*). The latter, in turn, displayed greater WGC values relative to the subtribe Panina (bonobos, *Pan paniscus*; chimpanzees, *Pan troglodytes*). The Ancestral character reconstruction also suggested that the last common ancestor of the tribe Hominini (including the subtribes Australopithecina, Hominina, and Panina) exhibited a WGC value that was slightly higher than the one for Panina and lower relative to Australopithecina and Hominina. Figure 6.1 describes this pattern in greater detail.

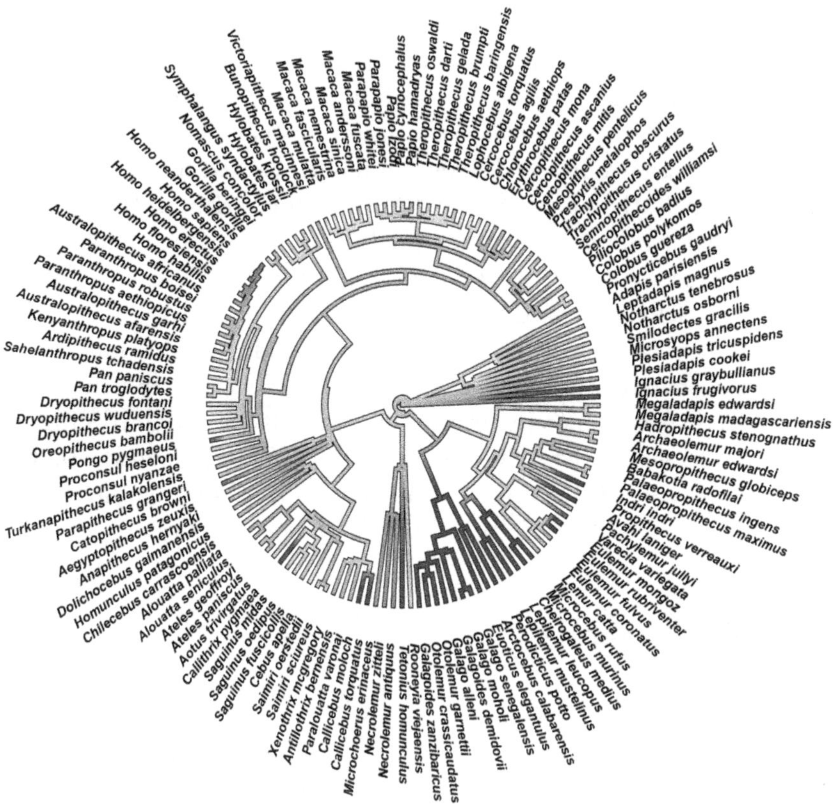

Fig. 6.1 Ancestral character reconstruction of a mean standardized within-group conflict factor (aggregating the values of three estimation procedures) across extant and extinct primate species

6.6 Within-Group Conflict, Temperature, and Species' Persistence

A bivariate examination identified a positive and significant association ($r = 0.663$, $p < 0.0001$) between log branch length and time range (calculated as the difference between the species' time since its first appearance and the species' time since its last appearance). Consequently, a unit-weighted estimation was conducted, revealing a species' temporal persistence factor that loaded positively and significantly onto log branch length ($r = 0.920$, $p < 0.0001$) and time range ($r = 0.904$, $p < 0.0001$). A PGLS model, limited to a subset of primate species that became extinct before the Holocene, examined the influence of average temperature values and the average general WGC factor on the standardized species' temporal persistence factor. The overall model reached statistical significance ($p < 0.0001$) and explained 19.2% of the variance. The analyses revealed that temperature did not predict the species' temporal persistence factor scores ($\beta = 0.069$, $p = 0.6985$). In contrast, the average WGC factor negatively influenced the latter criterion variable ($\beta = -0.669$, $p = 0.0055$). A rival model was computed, considering the interaction between average temperature and average WGC factor and the latter main effects. A model comparison indicated that the restricted primary effects model was statistically superior ($AICc = 69.641$, $\Delta AICc = 0.000$, $AICc\ weight = 0.770$) relative to a model also featuring the interaction term ($AICc = 72.061$, $\Delta AICc = 2.419$, $AICc\ weight = 0.230$).

7 Conclusions

Previous scientific publications have considered multiple ecological and behavioral correlates of coalitions and alliances in nonhuman primates (Chapais, 1995; de Waal, 2007; van Shaik et al., 2006). The present chapter supports a general within-group conflict factor (WGC) comprised of coalitions and alliances, frequency of conflict, and intensity of conflict, across extant primate species. Diagnostic statistics recommended the extraction of a general latent factor. WGC loaded positively and significantly onto coalitions and alliances and the frequency and intensity of conflict, accounting for a large proportion of variance. This latent dimension displayed predictive and criterion validity, as evidenced by the positive association between WGC values with polygyny and several indicators of lethal aggression. Several Phylogenetic Path models, and the corresponding model comparison, indicated that log endocranial volume positively and significantly predicted WGC, in turn this latent variable positively predicted a lethal aggression index. The positive and sizeable prediction of log endocranial volume on WGC, examined with three statistical procedures (RMA, PRMA, and PGLS), allowed for the ancestral reconstruction of this latent dimension across several extinct species of primates. This reconstruction revealed that the subtribe Panina (chimpanzees and bonobos) experienced a decline in WGC values relative to Australopithecina and Hominina. Even though Late Miocene and Early Pliocene species, such as *Sahelanthropus tchadensis* and *Ardipithecus ramidus*, evidenced a slight increase in WGC values, the speciation of *Australopithecus* sp., and *Paranthropus* sp., between the Early-to-Late Pliocene, is accompanied by a noticeable rise in WGC. *H. sapiens* and *H. neanderthalensis* featured the highest WGC values among extant and extinct primate taxa.[14]

Contra Rousseau, our state of nature was more often than not, *nasty, brutish, and short.* We are part of an ancestral lineage that exemplifies, not only social contention and complexity, but outsized cognition. We inherited a disposition part Hobbesian and part Machiavellian. While millennia would pass before these tendencies could be channeled into peaceful, political channels, this chapter's treatment of coalitions and alliances provides an evolutionary window into the rudiments of simple human political structures discussed in Chap. 7.

Notes

1. Gat, A. (2017, p. 246). *The causes of war and the spread of peace.* Oxford University Press.
2. Although several researchers have associated this hypothesis with the evolution and persistence of tactical deception in social species, initially, the MIH described the coevolution of social complexity and sophisticated cognitive abilities in nonhuman primates.
3. We say total variance as opposed to reliable variance because Principal Component (PC) analysis estimates the former, whereas Principal Axis (PA) analysis estimates the latter, having controlled for the error variances of each indicator. Thus, proportions of common factor variance reported by PA will generally be larger than those reported by PC analyses.
4. A phylogenetic comparative analysis that incorporates a trait's phylogenetic signal into the model when considering the influence of one or several predictors on a continuous criterion variable after accounting for

the underlying phylogenetic history (Peñaherrera-Aguirre & Fernandes, 2021).

5. Shultz and Dunbar (2022) acknowledged that, although the socioecological complexity dimension shared several indicators with Reader and colleagues' *G*, the similarity in the proportion of variance between their measures and the two G factors, indicates that the correlation between socioecological complexity and Reader and colleagues' *G* is a statistical artifact due to shared indicators among the two factors being correlated. Consistent with previous publications, the models revealed that logarithm of endocranial volume and that of neocortical volume positively and significantly predicted Deaner and colleagues' and Reader and colleagues' *G* scores. The PGLS models also indicated that logarithms of endocranial and neocortical volumes positively predicted the socioecological complexity dimensions It should also be noted that, in structural equation models, modeling variable *B* as an indicator of variable *A* is mathematically equivalent to modeling variable *A* as a cause of variable *B*, in that latent common factors are presumed to be the causes of manifest indicators. The distinction between these two scenarios is purely conceptual in that precisely the same empirical data would support both contentions.
6. Data on primates' coalitions and alliances, frequency of conflict, and intensity of conflict were obtained from Plavcan (2001), with data for *Gorilla beringei* extracted from Watts (1997) and Harcourt and Stewart (2007). Endocranial volumes were collected from Melchionna and collaborators' (2020) database. Similarly, information on polygyny and lethal aggression (female adulticide, male adulticide, and infanticide) was extracted from Gómez et al. (2021) online supplementary material. A phylogeny containing extinct and extant primate taxa was downloaded from Melchionna and collaborators' (2020) online data repository. Cenozoic temperature data (Celsius) were extracted from the paleobuddy R package (do Rosario Petrucci et al., 2022). Lastly, information on species' temporal persistence, calculated as the difference between time since its first appearance and time since its last appearance (in millions of years), was gathered from Herrera's online supplementary material (2017). Log10 transformations were implemented on variables featuring a considerable skew, as demonstrated by the presence of several outliers outside of a standardized range of –3.5 to + 3.5.
7. Principal Axis Factor modeling (*psych* package; Revelle, 2015) and Unit-weighted estimations were conducted to determine the presence of a

general within-group conflict factor comprised of coalitions and alliances, and frequency of conflict. Horn's parallel analysis (500 iterations based on a maximum likelihood estimation) was also performed to determine the recommended number of latent dimensions. The internal consistencies of the various common factors were examined with Cronbach's α. Phylogenetic independent contrasts (*ape* package; Paradis et al., 2019) were computed using the three conflict indicators. These vectors, controlled for phylogenetic effects, were later used in a Principal Axis Factor model and a Unit-weighted estimation. Tucker's Coefficient of Congruence was used to examine the structural similarity between the raw and the phylogenetically informed within-group conflict factors. Phylogenetic Logistic Regressions (*phylolm* package; Ho et al., 2016) were estimated between the general within-group conflict factor and various indicators of lethal aggression (female adulticide, male adulticide, and infanticide) to determine the predictive validity of this latent dimension. Additionally, a Phylogenetic Generalized Least Squares (PGLS; *caper* package; Orme et al., 2013) model explored the association between the within-group conflict factor scores and a lethal aggression index (computed as the sum of female adulticide, male adulticide, and infanticide). Several Phylogenetic Generalized Least Squares models were performed to examine the influence of log endocranial volume on (1) intensity of conflict; (2) frequency of conflict; (3) coalitions and alliances; (4) a restricted within-group conflict factor comprised of frequency of conflict and coalitions and alliances; and (5) a general within-group conflict factor encompassing all three indicators of social conflict. Four Phylogenetic Path analyses (*phylopath* package; van der Bijl, 2018) were estimated to explore the direct and indirect effects of log endocranial volume, the within-group conflict factor, and the lethal aggression index. As Phylogenetic Path models are based on the *d*-separation procedure, model comparisons are conducted by estimating each model's *C*-statistic (which follows a χ^2 distribution, wherein $p < 0.05$ indicates good statistical fit) and calculating a relative likelihood value (*CICc* weight). Lastly, three statistical procedures, Reduced Major Axis model (*lmodel2* package; Legendre & Oksanen, 2018), Phylogenetic Reduced Major Axis model (*phytools* package; Revell, 2012), and PGLS, were used to extract the various equations components needed to calculate the corresponding within-group conflict factor scores in a sample of 68 extinct and 1 extant (*Homo sapiens*) primate species.

8. The psychometric literature frequently uses the Greek term λ when referring to factor loadings, the present chapter will use the symbol λ_F when referring to these statistical parameters. In contrast the term λ_P is used when considering the phylogenetic and statistical parameter that indicates a trait's level of phylogenetic inertia.
9. Phylogenetic-independent contrasts (a procedure generating phylogenetic independent values that can later be analyzed using traditional statistical models) were computed on the latter conflict indicators. These variables were subsequently used to determine the structure of a within-group conflict factor net of phylogenetic autocorrelations. Horn's parallel analysis estimated a single latent factor. A PAF model revealed a general within-group conflict factor that loaded into the frequency of conflict ($\lambda_F = 0.864$), coalitions and alliances ($\lambda_F = 0.716$), and intensity of conflict ($\lambda_F = 0.459$). The model explained 49.0% of the variance, featuring an adequate degree of internal consistency (Cronbach's $\alpha = 0.71$). Tucker's Coefficient of Congruence index revealed that the raw and the phylogenetically informed within-group conflict factors were structurally similar ($CC = 0.989$, $p < 0.0001$). Part-whole correlations based on unit-weighted estimates supported the presence of a general within-group conflict factor, net of phylogenetic autocorrelations. This latent variable loaded positively and significantly into the frequency of conflict ($r = 0.845$, $p < 0.0001$), coalitions and alliances ($r = 0.816$, $p < 0.0001$), and intensity of conflict ($r = 0.723$, $p < 0.0001$). Tucker's Coefficient of Congruence index revealed that the raw and the phylogenetically informed within-group conflict unit-weighted factors were statistically identical ($CC = 0.999$, $p < 0.0001$).
10. A phylogenetic comparative analysis that considers the influence of one or several predictors on a dichotomous criterion variable after accounting for the underlying phylogenetic history (Peñaherrera-Aguirre & Fernandes, 2021).
11. A b-weight, instead of a β, is provided because polygyny is a dichotomous predictor.
12. This parameter is computed simultaneously based on both predictor and criterion variables. The models' joint Pagel's λ (λ_P) estimates ranged from 0.000 (coalitions and alliances) to 0.894 (frequency of conflict).
13. Significance tests estimated following the degrees of freedom formula n–k–1.

14. It is worth noting that although this chapter discussed a positive and sizeable association between log endocranial volume and within-group conflict in primates, additional studies are required to determine a predicted positive association between a General Cognitive Ability factor (*G*) and the aforementioned criterion variable. The limited number of primate species featuring both *G* and conflict data hindered the prospect of performing these analyses in the current study.

7

Evolutionary-Ecological Dynamics of Sociopolitical Complexity: Human-Animal Symbioses, Culture of Honor, and War

Mateo Peñaherrera-Aguirre, Aurelio José Figueredo, and Steven Charles Hertler

There is simply no proof that warfare in small-scale societies was a rarer or less serious undertaking than among civilized societies. In general, warfare in prestate societies was both frequent and important. If anything, peace was a scarcer commodity for members of bands, tribes, and chiefdoms than for the average citizen of a civilized state.

—Lawrence H. Keeley[1]

M. Peñaherrera-Aguirre • A. J. Figueredo
University of Arizona, Tucson, AZ, USA
e-mail: ajf@arizona.edu; mpeaher@email.arizona.edu

S. C. Hertler (✉)
Saint Elizabeth University, Morristown, NJ, USA

S. C. Hertler et al., *The Evolution of Political Ideology*,
https://doi.org/10.1007/978-3-031-83941-2_7

1 Introduction

Macro-coalitions within groups facilitated sociopolitical integration across groups. Thus, societies grew. Such an evolutionary transition was made possible through multilevel selective pressures promoting cooperation between these local political entities against supralocal political entities. In other words, operating under the assumption that *my enemy's enemy is my friend*, adjacent groups allied with one another against more distant outgroups, and, over time, merged into larger groups. It is important to understand the evolutionary transition to larger groups, because only in larger groups can sociopolitical complexity arise. Yet, as we also detail herein, with the growth of group size, came intensified violence and despotism. Several publications have reported such a curvilinear trend wherein warfare frequency and proportional killings increase from bands to tribes and eventually reach a zenith with simple and complex chiefdoms; thereafter, warfare frequency declines as complexity increases further among inchoate, ancient, modern, and contemporary states. In addition to reviewing those many publications studying interactions between warfare frequency, sociopolitical complexity, and lethal aggression, this chapter analyzes archaeological and ethnographic literature on lethal aggression in small-scale societies, providing empirical evidence concerning effects of direct and indirect influences of human-animal symbiotic associations, culture of honor, and external warfare between two distinct biocultural groups on sociopolitical complexity.[2] Precedents for such interactions are comparatively rare (Nawata, 2020; Peñaherrera-Aguirre et al., 2022; Steklis et al., 2023; Turchin, 2010; Turchin et al., 2021, 2022).

The rationale for including this seventh chapter is the same as the rationale for including the sixth: A more convincing ultimate explanation of modern political psychology comes from understanding the origins of brain and behavior as they are evolutionarily and culturally bequeathed from nonhuman primates and small-scale societies. It was through these transitions that selective pressures instilled genetic differences in the insula, amygdala, and other brain regions that give rise to those temperamental, perceptual, and emotional differences undergirding political orientation.

2 Lethal Coalitional Aggression: From Nonhuman Primates to Human Small-Scale Societies

As described in the previous chapter, coalitional cognition and behavior occur across various nonhuman animals, though the behavioral ecological literature indicates that *lethal* coalitional aggression is present in only a minority of nonhuman primate species (Gros-Luis et al., 2003; Southern et al., 2021; Talebi et al., 2009; Wilson et al., 2014; Wrangham, 1999). Lethal coalitional aggression is most clearly present among chimpanzees. Wilson and collaborators (2014) examined the influence of anthropogenic and socioecological factors on chimpanzee lethal aggression. Their database included information on: (1) intercommunity killings of weaned individuals; (2) intercommunity infanticides; (3) intracommunity killings of weaned individuals; and (4) intracommunity infanticides. Wilson and colleagues conducted a model comparison between alternative statistical analyses containing a combination of various predictors such as clade (western relative to eastern chimpanzees), mean number of adult males, mean population density, dimensions of the protected area, food provisioning, and ecological disturbance. The analyses indicated that the best model included the mean number of adult males and population density. Subsequent examinations determined that the number of adult males and population density positively predicted the frequency of killings per year for each chimpanzee community.[3]

Not only might drivers of lethal coalitional aggression among chimpanzees be all too familiar to historians, but cross-species comparisons find rates of lethal aggression in chimpanzees to parallel that of humans living in small-scale societies. Wrangham et al. (2006) reviewed ethnographic and primatological literature on lethal aggression in small-scale human subsistence societies (hunter-gatherers and farmers) and chimpanzee communities. According to the authors, the rate of chimpanzee lethal aggression (271 deaths per 100,000 individuals per year) was comparable in magnitude to human lethal aggression. The study also revealed that farmers (595 per 100,000 individuals per year) experienced greater killing rates compared to hunter-gatherers (164 per 100,000 individuals per year). Wrangham and colleagues also identified epidemiological differences. Whereas chimpanzees often targeted infants and adolescent-adult males, human warriors eliminated predominantly adult males (Wrangham et al., 2006).

Similarly, Walker and Bailey (2014) used ethnographic data on lethal aggression in 44 lowland South American societies to parse between the *Parochial Altruism Model* and the *Chimpanzee Model*, two alternative hypotheses developed to describe the evolution of lethal coalitional aggression in small-scale societies. Their Generalized Linear Mixed models revealed that external warfare featured a higher body count than internal warfare. In a similar vein, the number of victims was higher during external warfare relative to within-village conflicts. Furthermore, Walker and Bailey reported higher body counts where perpetrators relied on treachery to eliminate their targets. Revenge (as a response to previous instances of lethal aggression, cuckoldry, or sorcery accusations) was the primary catalyst behind lethal aggression. Although these results provided strong evidence in favor of the Parochial Altruism Model, Walker and Bailey's (2014) analyses also supported the Chimpanzee Model, wherein males established coalitions against other factions, the efficacy of which can be seen in their low cost (attackers died during only 2% of engagements) and high impact (lethal aggression led to the weakening or fragmentation of the rival faction).

3 Leveling Mechanisms and Anti-Hierarchies in Small-Scale Societies

Cross-species comparisons indicate that despotism is more pronounced in specific clades of nonhuman primates and human societies than in others. For example, elevated rates of despotism can be observed in several great apes and old-world monkey species (parvorder *Catarrhini*) as well as many human chiefdoms and inchoate states. Lesser apes, such as gibbons and siamangs, as well as most human forager bands and some tribes, feature relatively lower levels of despotism (Boehm, 1999; Flannery & Marcus, 2012; Turchin, 2015). Some evolutionary researchers hypothesize that the socioecological characteristic that distinguishes egalitarian human groups is the existence of leveling mechanisms, operating as social sanctions, destined to curb the rise and persistence of despots and upstarts. Such anti-hierarchical social reorganizations depend on low- and mid-ranking individuals cooperating against despotism by dominant actors. Boehm and collaborators (1993) described eight behavioral and social sanctions directed toward upstarts and defectors. These punishments included: (1) *Public Opinion*, manifested as informal disagreements or in a formal forum destined to discuss underlying grievances; (2) *Criticism*, aimed at a leader's actions or intentions; (3) *Ridicule*, mockery destined to belittle or question an individual's status; (4) *Disobedience*, refusal to follow an instruction or plan; (5) *Deposition*, replacing or removing an individual from a position of leadership; (6) *Desertion*, abandoning the upstart or despot; (7) *Exile*, expelling the problematic individual from the group; and (8) *Execution*, relying on capital punishment to eliminate the despot or upstart.

Boehm et al. (1993) examined the latter leveling mechanisms in a sample of 48 egalitarian societies. The data included North America, Mesoamerica, South America, Africa, the Mediterranean and Levant, Asia, and Oceania societies. Desertion occurred at the highest frequency (35%), followed by Public Opinion and Execution (21% each). Deposition, Disobedience, and Criticism featured slightly lower frequencies (19%, 15%, and 13% respectively). Lastly, Ridicule and Expulsion displayed the lowest frequencies (10% and 4% respectively). Leveling sanctions were often implemented in response to aggressive behavior and dominance, followed by partial or complete ineffectiveness of leaders. Other transgressions included selfishness, resource monopolization, and unkindness toward others. The predominance of Desertion, Public Opinion, and Execution as the primary forms of social sanctions across egalitarian groups suggests that affected parties prefer abandoning the group when dealing with dominant, aggressive, or incompetent leaders. If desertion is impossible, expressing discontent in a private or public setting is adopted as an alternative strategy. A statistically significant negative correlation between Desertion and Execution ($r = -0.38$, $p = 0.0077$) indicates that eliminating an upstart occurs less frequently if people can depart from the community. As the remedy of Execution did not correlate significantly (range $r = -0.039$ to -0.246) to any of the other social sanctions, this implies that leveling mechanisms do not constitute a system of proportional response, wherein individuals exhaust nonlethal alternatives before adopting more severe punishments. Moreover, the weak negative correlations among all forms of sanctions (mean $r = -0.048$, $p = 0.7460$) indicate that the preference of egalitarian societies for certain types of punishment is independent of the preexistence and use of other sanctions.

4 Culture of Honor Theory and Lethal Aggression

Culture of Honor Theory (Nisbett & Cohen, 1996) states that, without formal institutions guaranteeing personal safety and property rights, a culture of honor develops to counter individual or collective victimization (Nisbett & Cohen, 1996; Uskul & Cross, 2020). Cross-cultural variation in reputational concerns, revenge-seeking behaviors, and lethal aggression is, in part, attributable to socioecological factors. Specifically, according to Nisbett and Cohen (1996), honor and retribution are especially prevalent in herding societies, relative to farming societies, due to a higher risk of poaching domestic animals. Supportive of these predictions, Linquist (2015) analyzed the influence of socioecological correlates of culture of honor across 18 small-scale societies based on data curated as part of the *Human Relations Area Files*. Analyses suggested that within-group antagonism and the parental endorsement of aggressive behaviors were more prevalent in herding than in agricultural societies. Similarly, ethnographic examinations in four East African tribes find greater aggressiveness and self-reliance among herders compared to farmers (Edgerton, 1971; Goldschmidt, 1965).

Nisbett and Cohen (1996) also hypothesized that culture of honor would be absent or attenuated among hunter-gatherers whose foraging economies do not yield surpluses that could incentivize raiding. However, in some ways contrary to Nisbett and Cohen's hypothesis, Figueredo and colleagues (2004), after administering a culture of honor scale to inhabitants of Mexico and Costa Rica, found high-density foragers (fishers) had a culture of honor intermediate between herders and farmers; a finding consistent with related studies (Allen & Jones, 2016; Keeley, 1997; LeBlanc & Register, 2004; Otterbein, 2004) demonstrating relatively high frequencies of lethal competition among prehistoric and extant hunter-gatherers. Concerning the elevated rates of lethal aggression in tribal societies dependent on slash-burn horticulture rather than herding, Nisbett and Cohen (1996) argued that low soil productivity, environmental unpredictability, resource scarcity, intrasexual competition, and the lack of strong institutions regulating conflict contribute to the persistence of lethal aggression in tribal horticultural societies. In contrast, farmers who depended on stable agricultural systems, found in predictable ecologies, and capable of producing economic surpluses, experienced lower rates of lethal aggression (Nisbett & Cohen, 1996). Moreover, stable agriculture is generally found in sizeable sociopolitical organizations featuring institutions destined to regulate competition and protect farmers from raiding and theft (Nisbett & Cohen, 1996). Culture of Honor Theory, as can be seen from these several deviations from its predictions, did not consider the multilevel structure of human sociopolitical systems. To remedy this omission, the following section discusses the reported variation in lethal retribution emerging within and between human communities.

5 Controlling Lethal Retribution

According to the ethnographic literature, revenge killings occur in small-scale societies, either as blood feuds and capital punishment within a group or retaliatory raids and massacres between groups (Hertler et al., 2020; Kelly, 2005; Otterbein, 2004). In terms of within-group dynamics, after a homicide is perpetrated, grievers may solicit monetary compensation or demand that the killer be ejected from the community (Boehm, 1999; Dunbar et al., 1995). No further actions are implemented if the reparatory sanction is deemed adequate. Alternatively, if the affected party remains unsatisfied, the conflict may escalate. Grievers could target the killer without discussing the subject with other members of the group, an action that could trigger a blood feud within the community (Boehm, 1999; Peñaherrera-Aguirre et al., 2020).

To avoid escalation, it is not unusual for the affected party to meet with influential individuals from the same group, sometimes even including the murderer's kin and allies, to discuss the killer's execution (Boehm, 1999; Hertler et al., 2020). Whereas revenge killings within a group often restrict their actions toward a specific target, perpetrators of between-group raids and massacres target any individual in the rival community (Kelly, 2005). This process of *social substitution* in retributive aggression can perpetuate violent feuding, and is often further complicated by the problem of equivalence—the difficulty of the aggrieved group finding a revenge target of equal social value in the aggressor group. This problem of equivalence is especially pronounced in smaller-scale groups.[4] A well-functioning culture of honor might reduce aggression (Figueredo et al., 2001) by solving the problem of equivalence, thus thwarting ongoing cycles of retribution, while maintaining adequate deterrence against future intergroup aggression (Daly & Wilson, 1988). More often, rather than settling a dispute, intergroup retaliatory killings produce additional grievances, leading to future incursions (Gat, 2009; Hertler et al., 2020; Kelly, 2005).

Cross-cultural analysis provides unique insight into the coevolution of revenge killings and sociopolitical dynamics. After analyzing socioecological and feuding data from the *Human Relations Area Files,* Otterbein and Otterbein (1965) found feuding correlates with *patrilocality*, wherein coalitions of related males (fraternal interest groups) seek retribution when their interests are affected. Otterbein and Otterbein (1965) also examined the association between external warfare and feuding as a function of the level of political integration: feuding and external warfare increase as political complexity increases, and decrease as political complexity decreases (Otterbein & Otterbein, 1965). More recently, Nawata (2020) extracted multiple socioecological variables from the *Standard Cross-Cultural Sample* to determine the mediating influence of warriors' social rewards (non-material benefits such as prestige, status, and praise; Nawata, 2020) on the association between an honor culture and intergroup conflict. The author's analyses confirmed the existence of three latent dimensions: *honor culture, social rewards for warriors,* and *frequency of intergroup conflict.* According to Nawata, bivariate analyses indicated that honor culture was positively associated with social rewards for warriors, and intergroup conflict, but not with political integration, social stratification, and community size. In turn, social rewards were positively associated with intergroup conflict, and, to a lesser extent, with community size and political integration. Lastly, intergroup conflict was correlated with political integration, social stratification, and community size, albeit these effects ranged from small to medium in magnitude. A well-fitting structural equations model, wherein social rewards mediated the relationship between honor culture and intergroup conflict, confirmed Nawata's predictions. An alternative model that included ingroup loyalty as another covariate also fit the data. The analysis revealed sizeable and significant associations between honor culture, social rewards, and intergroup conflict. These results suggest the potential presence of a higher-order latent dimension. The model also indicated that ingroup loyalty was significantly correlated with intergroup conflict but not with honor culture or social rewards.

6 Lethal Aggression and Structural Inequality Across Sociopolitical Organizations

According to Turchin (2015), despotism and inequality have varied throughout human evolution, following a nonlinear pattern. Turchin's description of the initial stages of sociopolitical evolution is consistent with Boehm's theory (1999) of anti-hierarchical systems in egalitarian groups, wherein, relative to several *catarrhine* primate species, human bands of hunter-gatherers use social sanctions that operate as leveling mechanisms, curbing the rise and persistence of upstarts and despots. The sociocultural evolution of reward systems organized in terms of the acquisition and adscription of non-heritable prestige allows high-status individuals to attain an array of benefits, from obtaining mates and allies, to exerting non-coercive influence in social and political arenas (Fried, 1967; Macfarlan et al., 2014). According to Flannery and Marcus (2012), coercive gift-giving and indebted servitude raised inequality in small-scale societies. These socially and politically stratified communities established institutions legitimating high-status positions across generations, thereby facilitating the rise of simple and complex chiefdoms (Flannery & Marcus, 2012; Fried, 1967). Illustrating the curvilinear relationship discussed in this chapter's introduction, only as societies expanded much further, escaping the intermediate sizes that featured the highest rates of violence, would there be investment in several ultrasocial institutions promoting cooperation and proscribing social parasitism among distantly related individuals (Peñaherrera-Aguirre et al., 2020; Turchin, 2015). Turchin (2015), reflecting on this arc of human social evolution, noted that archaic states reached levels of structural inequality that exceeded those of several nonhuman primate species, a process that persisted until the emergence of modern and contemporary constitutional democracies.

We see this same curvilinear pattern through the work of Gómez and colleagues (2016) who conducted a phylogenetic and cross-cultural examination of lethal aggression in human and nonhuman mammalian clades. The authors' estimates indicated that the percentage of lethal aggression varied as a function of sociopolitical complexity. Prehistoric bands of hunter-gatherers featured a lower percentage of killings relative to prehistoric tribes, which in turn, had lower values than historic chiefdoms. Historic states displayed percentages slightly higher than prehistoric bands but considerably lower relative to prehistoric tribes and chiefdoms. Regarding contemporary ethnographic data, Gómez and collaborators reported that contemporary bands exhibit more lethal aggression than contemporary tribes. This result remains inconsistent with previous publications indicating that tribal societies experience higher killing percentages relative to bands of hunter-gatherers (Pinker, 2012). Despite this statistical anomaly, Gómez and colleagues reported that contemporary states featured lower killing percentages than extant small-scale societies. These results strongly suggest that the total portion of killed individuals declined in more complex sociopolitical systems.[5]

The association between conflict and the level of sociopolitical complexity differs depending on whether one considers the frequency of intergroup competition (Keeley, 1997; Otterbein, 1989) or the intensity of lethal aggression, operationalized as the percentage of individuals killed (this indicator does not distinguish between intragroup and intergroup competition; Gómez et al., 2016). Figure 7.1 summarizes this information, stating that *continuous warfare* follows a negatively

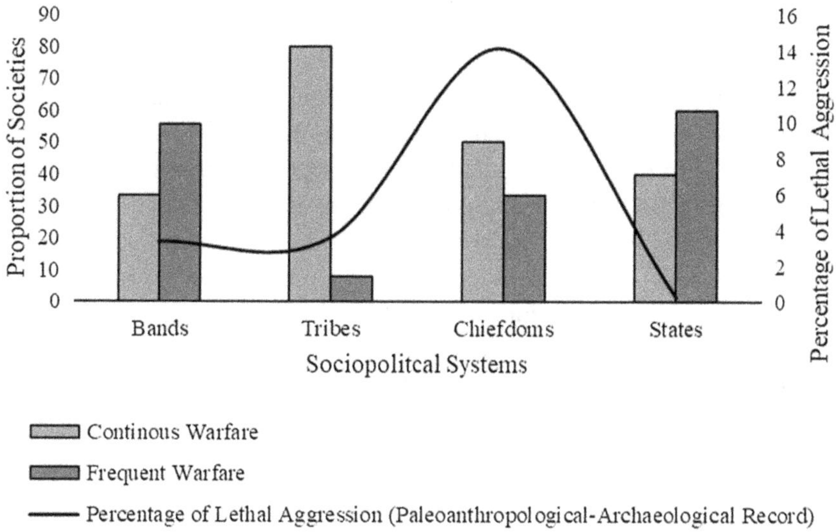

Fig. 7.1 Proportion of societies and percentage of lethal aggression in the paleoanthropological and archaeological record disaggregated by sociopolitical system, based on data collected from Keeley (1997) and Gómez and colleagues (2016)

accelerated curvilinear pattern (inverted u-shaped), with a larger proportion of tribes and chiefdoms displaying continuous warfare. *Frequent warfare* follows a positively accelerated curvilinear pattern (u-shaped) as a function of sociopolitical complexity, with a larger proportion of bands and states featuring frequent warfare. Also, tribes, and chiefdoms experienced a higher percentage of lethal aggression.

7 Sample and Statistical Analyses

Going beyond available information to provide further clarity on these issues, data from Murdock's *Standard Cross-Cultural Sample* (SCCS) were collected.[6] The SCCS contains 1781 indicators of ethnographic information on 186 societies. Following Nawata's cross-cultural study (2020), the present chapter explored the existence of a *Culture of Honor, Reward System*, and *External Warfare* latent dimensions. Measurement models also explored the presence of three general factors: *Parasite Stress,*

Homicide, and *Internal Conflict*.[7] Unit-weighted factor estimations were used to determine the corresponding factor structure of each latent dimension, with part-whole correlation being mathematically equivalent to Principal Axis Factor loadings (Figueredo et al., 2013). All External Warfare and Internal Conflict indicators are operationalized as frequency scales rather than percentages, prevalence, or incidence rates of killings. After determining the statistical adequacy of the latter common factors, a Sequential Canonical Cascade (SEQCA) model[8] was computed to explore the direct and indirect effects of *Biogeography*, *Agricultural Dependence*, *Animal Herding*, *Equestrian Culture*, *Culture of Honor*, *Reward System*, *Homicide*, *Internal Conflict*, and *External Warfare*, on *Sociopolitical Complexity*.

The statistical examination detected an *Animal Herding* factor that featured a high internal consistency and accounted for 68% of the variance, with factor loadings ranging from 0.800 to 0.840. A *Parasite Stress* factor also displayed an adequate level of internal consistency and explained 77% of the variance, with factor loadings ranging from 0.370 to 0.705. The *Culture of Honor* dimension featured an adequate level of internal consistency, with this latent factor explaining 96% of the variance and factor loadings ranging from 0.670 to 0.744. In a similar vein, the *Reward System* dimension displayed an adequate level of internal consistency. This latent factor explained 90% of the variance, with factor loadings ranging from 0.571 to 0.868. The analyses also detected a *Homicide* dimension that displayed high internal consistency and explained all of the observed variance with high factor loadings. An *Internal Conflict* dimension was detected with this latent factor, featuring high internal consistency and explaining 85% of the variance. Factor loadings ranged from 0.592 to 0.902. The present examination also supported the presence of an *External War* factor that exhibited a high internal consistency and explained 88% of the variance. The factor loadings ranged from 0.659 to 0.877. These results, alongside the measurement models for the *Culture of Honor* and *Reward* dimensions, are consistent with those reported by Nawata (2020). A *Sociopolitical Complexity* factor was extracted, this dimension exhibited adequate internal consistency and explained 96% of the variance. Factor loadings ranged from 0.689 to 0.739 (Tables 7.1 and 7.2).

Table 7.1 Unit-weighted factor loadings were estimated using several Principal Axis Factor analyses for the corresponding cross-cultural constructs

Dimension	SCCS ID	Descriptor	Loading
Animal Herding	SCCS1486	Goats Present	0.840
	SCCS1487	Sheep Present	0.800
	SCCS1514	Cattle Present	0.830
Parasite Stress	SCCS1253	Pathogen stress: Leishmanias	0.409
	SCCS1254	Pathogen stress: Trypanosomes	0.370
	SCCS1255	Pathogen stress: Malaria	0.705
	SCCS1256	Pathogen stress: Schistosomes	0.590
	SCCS1257	Pathogen stress: Filariae	0.654
	SCCS1258	Pathogen stress: Spirochetes	0.649
	SCCS1259	Pathogen stress: Leprosy	0.677
Culture of Honor (based on Nawata, 2020)	SCCS1774	Revenge Related Norms	0.670
	SCCS1765	Emphasis on Courage of Boys in Late Childhood	0.744
	RSCCS625	High Value Placed on Males Being Aggressive, Strong, and Sexually Potent	0.733
Reward System (based on Nawata, 2020*)*	SCCS1773	Prestige of Warriors	0.571
	RSCCS905	Rewards (Special Gifts, Praises, or Ceremonies, Not including Ritual Purification for a Man Who Has Killed an Enemy in Battle or Otherwise Shown Skill in War)	0.833
	RSCCS903	Prestige Associated with Being a Soldier or Warrior	0.868
Homicide	SCCS1665	Individual Aggression—Homicide	0.840
	SCCS1675	Socially Organized Homicide	0.840
Internal Conflict	RSCCS891	Frequency of Internal War	0.627
	RSCCS773	Internal Warfare (between Communities of Same Society)	0.902
	RSCCS767	Conflict (Social or Political) in the Local Community	0.592
	RSCCS768	Conflict between Communities of the Same Society	0.896

(*continued*)

Table 7.1 (continued)

Dimension	SCCS ID	Descriptor	Loading
External Warfare (based on Nawata, 2020*)*	SCCS1778	Frequency of Interethnic Violence/ Attacking	0.659
	RSCCS892	Frequency of External War— Attacking	0.877
	RSCCS893	Frequency of External War—Being Attacked	0.782
	RSCCS774	External Warfare (with Other Societies)	0.744
Sociopolitical Complexity	SCCS235	Mean Size of Local Communities	0.739
	SCCS1740	Levels of Political Hierarchy	0.702
	RSCCS765	Political Participation: Range of Community Decision Making	0.689

Table 7.2 Internal consistency values (Cronbach's α) and proportion of variance explained by each latent dimension examined in the present chapter

Dimension	Cronbach's α	Eigenvalue	R^2
Animal Herding	0.840	2.035	0.678
Parasite Stress	0.766	2.456	0.769
Culture of Honor	0.770	1.539	0.964
Reward System	0.741	1.772	0.901
Homicide	0.827	1.411	1.000
Internal Conflict	0.831	2.106	0.851
External Warfare	0.811	2.369	0.879
Sociopolitical Complexity	0.714	1.514	0.961

8 Results

The omnibus test indicated that the model was statistically significant and explained 46% of the variance. In the first step of the Sequential Canonical Cascade model, *Agricultural Dependence* was not significantly predicted by *Biogeography*. In turn, *Animal Herding* was significantly predicted by *Agricultural Dependence* and *Biogeography*. Having an *Equestrian* culture was positively and significantly predicted by *Animal Herding* and *Biogeography*. *Agricultural Dependence* did not have any influence on this step of the model.

Parasite Stress was positively predicted by *Animal Herding* and *Biogeography*. An *Equestrian Culture* and *Agricultural Dependence* did not contribute to this step of the model. A *Culture of Honor* was positively and significantly predicted by an *Equestrian Culture*, neither *Parasite Stress*, *Animal Herding*, *Agricultural Dependence*, nor *Biogeography* significantly predicted this criterion variable. A *Reward System* associated with warrior prestige and status was positively predicted by a *Culture of Honor* (a finding consistent with Nawata's 2020 model) and *Agricultural Dependence*. Neither *Parasite Stress*, an *Equestrian Culture*, *Animal Herding*, nor *Biogeography* significantly predicted this criterion.

Homicide was positively predicted by *Reward System*. Neither *Culture of Honor*, *Parasite Stress*, an *Equestrian Culture*, *Animal Herding*, *Agricultural Dependence*, nor *Biogeography* significantly predicted this criterion variable. *Internal Conflict* was positively predicted by *Parasite Stress* and *Biogeography*. *Homicide*, *Reward System*, *Culture of Honor*, an *Equestrian Culture*, *Animal Herding*, and *Agricultural Dependence* did not contribute to this step of the model.

External Warfare was positively and significantly predicted by *Homicide* and *Reward System* (this latter finding remains consistent with Nawata's 2020 model). *Internal Conflict*, *Culture of Honor*, *Parasite Stress*, an *Equestrian Culture*, *Animal Herding*, and *Agricultural Dependence* had no significant effect on the criterion variable. In the last step of the model, *Sociopolitical Complexity* was positively and significantly predicted by *External Warfare*, having an *Equestrian Culture*, and *Agricultural Dependence*. In contrast, *Internal Conflict*, *Homicide*, *Reward System*, *Culture of Honor*, *Parasite Stress*, *Animal Herding*, and *Biogeography* had no significant effect on this step of the model (Table 7.3).

A Hierarchical General Linear Model assessed the contribution of *Biogeography*, *Agricultural Dependence*, *Herding Animals*, *Parasite Stress*, *Culture of Honor*, *Reward System*, *Homicide*, *Internal Conflict*, and *External Warfare*, on *Sociopolitical Complexity* (see Table 7.4 for further details). The model reached statistical significance and explained 55% of the variance. *External Warfare*, *Parasite Stress*, *Domestic Animals*, and *Agricultural Dependence* had positive and sizeable effects on *Sociopolitical Complexity*. In contrast, Culture of Honor had a negative and significant influence in

Table 7.3 Sequential Canonical Cascade Model examining the influence of *Biogeography, Agricultural Dependence, Animal Herding, Equestrian Culture, Parasite Stress, Culture of Honor, Reward System, Homicide, Internal Conflict,* and *External Warfare*, on *Sociopolitical Complexity*

Model	Effect Size (E)	C. I.	F Ratio	df1/df2	*p*-value
Overall (V = 2.072)	0.46	0.00,1.00	1.51	70/252	0.01
Variables	*Effect Size (sR)*	*C. I.*	*F Ratio*	*df1`df2*	*p-value*
Y variable: Agricultural Dependence					
Biogeography	0.45	0.00,1.00	1.38	7/39	0.24
Residual Mean = -21.06; SD = 12.01; Skew/Kurtosis = -0.13/-0.87; Range = -47.22–3.64					
Y variable: Animal Herding					
Prior Y variable					
Agricultural Dependence	0.37	0.08, 0.60	1.72	1/38	0.002
X variable					
Biogeography	0.61	0.35,1.00	4.14	7/38	0.002
Residual Mean = -6.57; SD = 4.24; Skew/Kurtosis = 0.25/-0.79; Range = -15.62–0.17					
Y variable: Equestrian Culture					
Prior Y variables					
Animal Herding	0.36	0.07, 0.59	1.00	1/37	0.003
Agricultural Dependence	-0.21	-0.48, 0.09	3.52	1/37	0.07
X variable					
Biogeography	0.58	0.31,1.00	3.70	7/37	0.004
Residual Mean = -7.71; SD = 4.44; Skew/Kurtosis = -0.18/-0.83; Range = -17.39–-1.17					
Y variable: Parasite Stress					
Prior Y variables					
Equestrian Culture	-0.15	-0.43, 0.15	3.23	1/36	0.08
Animal Herding	0.59	0.35, 0.75	48.54	1/36	<0.0001
Agricultural Dependence	0.04	-0.26, 0.33	0.24	1/36	0.63
X variable					
Biogeography	0.61	0.47,1.00	7.57	7/36	<0.0001
Residual Mean = -3.71; SD = 2.32; Skew/Kurtosis = 0.29/-0.92; Range = -7.99–0.32					
Y variable: Culture of Honor					
Prior Y variables					
Parasite Stress	0.05	-0.25, 0.34	0.13	1/35	0.72

(*continued*)

Table 7.3 (continued)

Model	Effect Size (E)	C. I.	F Ratio	df1/df2	*p*-value
Equestrian Culture	0.32	0.03, 0.57	4.99	1/35	0.03
Animal Herding	-0.23	-0.50, 0.07	2.61	1/35	0.12
Agricultural Dependence	-0.12	-0.40, 0.19	0.64	1/35	0.43
X variable					
Biogeography	0.30	0.00,1.00	0.63	7/35	0.73
Residual Mean = -18.84; SD = 1.54; Skew/Kurtosis = -0.18/-0.77; Range = -42.94–3.38					
Y variable: Reward System					
Prior Y variables					
Culture of Honor	0.58	0.34, 0.75	21.42	1/34	<0.0001
Parasite Stress	-0.04	-0.33, 0.26	0.11	1/34	0.75
Equestrian Culture	-0.13	-0.41, 0.17	1.09	1/34	0.30
Animal Herding	-0.04	-0.33, 0.26	0.10	1/34	0.76
Agricultural Dependence	0.33	0.03, 0.57	6.88	1/34	0.01
X variable					
Biogeography	0.13	0.00,1.00	0.16	7/34	0.90
Residual Mean = -16.97; SD = 9.47; Skew/Kurtosis = -0.20/-0.74; Range = -39.03–-3.10					
Y variable: Homicide					
Prior Y variables					
Reward System	0.33	0.04, 0.57	4.93	1/33	0.03
Culture of Honor	0.28	-0.02, 0.53	3.46	1/33	0.07
Parasite Stress	-0.14	-0.42, 0.16	0.90	1/33	0.35
Equestrian Culture	0.05	-0.25, 0.35	0.13	1/33	0.72
Animal Herding	0.10	-0.21, 0.38	0.41	1/33	0.53
Agricultural Dependence	0.04	-0.26, 0.33	0.07	1/33	0.79
X variable					
Biogeography	0.19	0.00,1.00	0.22	7/33	0.90
Residual Mean = -8.00; SD = 4.71; Skew/Kurtosis = -0.14/-0.80; Range = -18.37–0.11					

(continued)

Table 7.3 (continued)

Model	Effect Size (E)	C. I.	F Ratio	df1/df2	*p*-value
Y variable: Internal Conflict					
Prior Y variables					
Homicide	0.07	−0.24, 0.36	0.42	1/32	0.52
Reward System	0.15	−0.16, 0.43	2.01	1/32	0.17
Culture of Honor	0.19	−0.11, 0.46	3.42	1/32	0.07
Parasite Stress	0.49	0.22, 0.69	22.18	1/32	<0.0001
Equestrian Culture	−0.11	−0.40, 0.19	1.16	1/32	0.29
Animal Herding	−0.16	−0.44, 0.15	2.32	1/32	0.14
Agricultural Dependence	−0.04	−0.26, 0.34	0.17	1/32	0.68
X variable					
Biogeography	0.56	0.33,1.00	4.18	7/32	0.002
Residual Mean = -23.17; SD = 12.88; Skew/Kurtosis = -0.17/–0.81; Range = -51.91–-4.92					
Y variable: External Warfare					
Prior Y variables					
Internal Conflict	0.12	−0.19, 0.40	1.05	1/31	0.31
Homicide	0.28	−0.02, 0.53	6.14	1/31	0.02
Reward System	0.33	0.04, 0.57	8.69	1/31	0.006
Culture of Honor	0.16	−0.15, 0.43	1.93	1/31	0.17
Parasite Stress	0.18	−0.12, 0.46	2.60	1/31	0.12
Equestrian Culture	0.20	−0.10, 0.47	3.22	1/31	0.08
Animal Herding	0.02	−0.28, 0.32	0.03	1/31	0.86
Agricultural Dependence	0.16	−0.15, 0.43	1.94	1/31	0.17
X variable					
Biogeography	0.53	0.23,1.00	3.18	7/31	0.01
Residual Mean = –37.42; SD = 21.11; Skew/Kurtosis = –0.15/–0.87; Range = –83.34–-8.23					
Y variable: Sociopolitical Complexity					

(continued)

Table 7.3 (continued)

Model	Effect Size (E)	C. I.	F Ratio	df1/df2	*p*-value
Prior Y variables					
External Warfare	0.42	0.14, 0.64	11.47	1/30	0.002
Internal Conflict	-0.02	-0.32, 0.28	0.03	1/30	0.87
Homicide	-0.20	-0.47, 0.11	2.52	1/30	0.12
Reward System	0.01	-0.29, 0.31	0.01	1/30	0.91
Culture of Honor	-0.24	-0.50, 0.07	3.67	1/30	0.06
Parasite Stress	0.10	-0.20, 0.39	0.68	1/30	0.42
Equestrian Culture	0.31	0.01, 0.56	6.35	1/30	0.02
Animal Herding	0.14	-0.16, 0.42	1.34	1/30	0.26
Agricultural Dependence	0.31	0.01, 0.56	6.30	1/30	0.02
X variable					
Biogeography	0.21	0.00,1.00	0.42	7/30	0.88
Residual Mean = -24.29; SD = 13.77; Skew/Kurtosis = -0.14/-0.88; Range = -54.51–-4.55					
Note. N = 47.					

Table 7.4 Hierarchical General Linear Model examining the influence of *Biogeography, Agricultural Dependence, Animal Herding, Parasite Stress, Culture of Honor, Reward System, Homicide, Internal Conflict*, and *External Warfare*, on *Sociopolitical Complexity*

Variables	Effect Size (sR)	C. I.	F Ratio	df1/df2	*p*-value
External Warfare	0.34	0.15, 0.51	19.85	1/78	<0.0001
Internal Conflict	-0.15	-0.34, 0.06	3.65	1/78	0.06
Homicide	0.07	-0.14, 0.26	0.74	1/78	0.39
Reward System	0.14	-0.06, 0.34	3.55	1/78	0.06
Culture of Honor	-0.24	-0.42, -0.04	9.92	1/78	0.002
Parasite Stress	0.29	0.10, 0.46	14.41	1/78	0.0003
Domestic Animals	0.42	0.30, 0.54	5.07	1/78	0.0002
Agricultural Dependence	0.20	0.00, 0.39	7.08	1/78	0.009
Biogeography	0.12	0.00, 0.34	0.38	1/78	0.91
Multiple R	0.74	0.50,1.00	4.61	20/78	<0.0001

Note. N = 99

the analysis. The model also revealed that neither *Internal Conflict*, *Homicide*, *Reward System*, nor *Biogeography* significantly affected *Sociopolitical Complexity*.

9 Discussion

The present chapter provides a cross-cultural examination of biogeographical, socioecological, and cultural factors favoring the evolution of human sociopolitical complexity. Measurement models supported the presence of several general latent factors, from *Parasite Stress* and *Internal Conflict* to *External Warfare* and *Sociopolitical Complexity*. A structural model examined via a Sequential Canonical Cascade Analysis (SEQCA) ordered the influence of these dimensions on *Sociopolitical Complexity*. The model revealed that societies featuring *Agricultural Dependence* also relied on *Animal Herding*. In turn, herding animals positively predicted having an *Equestrian Culture*. Consistent with previous epidemiological publications, *Parasite Stress* was attributable to *Animal Herding* and biogeographical effects, such as living in a tropical region. The current analyses also indicated that *Culture of Honor* was not significantly predicted by *Animal Herding* but by an *Equestrian Culture*. These results offer a more nuanced perspective concerning the evolution of *Culture of Honor*. The SEQCA replicated Nawata's (2020) results wherein a *Culture of Honor* positively predicted the presence of a *Reward System* based on the attribution and acquisition of prestige and status to warriors. These social and political incentives predicted societies' scores on the general *Homicide* factor. Although, at first glance, *Homicide* could be considered a manifestation of *Internal Conflict*, the lack of a significant association between these latent dimensions does not support the inclusion of homicide indicators into the *Internal Conflict* factor.

Moreover, *Parasite Stress*, rather than any other prediction in the analysis (including the significant albeit moderate effect of *Homicide* and *Culture of Honor*), had a positive and sizeable influence on *Internal Conflict* above and beyond the latter variables. The SEQCA also suggested that *Homicide*, in addition to Nawata's *Reward System*, positively predicted *External Warfare* in the Standard Cross-Cultural Sample. Lastly, the statistical model revealed that *External Warfare*, *Equestrian Culture*, and *Agricultural Dependence* positively predicted *Sociopolitical Complexity*. Moreover, the positive influence of *External Warfare* on *Sociopolitical Complexity* persisted when conducting a similar statistical examination with a larger dataset.

10 Conclusions

Thus far, we have seen that violence is part of our phylogenetic heritage. At the same time, we have seen that small hunter-gatherer bands, using leveling mechanisms, somewhat successfully controlled intragroup despotic violence, even as they were less successful in controlling intergroup violence. As group size increased to the middling levels of chiefdoms, despotism increased without any compensatory decrease in intergroup violence. Retributive justice within these cultures of honor often failed to stem cycles of violence, especially where horsemanship was prevalent. Levels of despotism and violence would only abate in the historical times, and then only appreciably with the social controls found in more thoroughly complex modern societies, which, unlike this and the prior chapters on evolutionary history, are only considered in the historical chapters of Part II.

Notes

1. War before Civilization.
2. It bears mention that we are not contending that this is an inevitability of societal evolution. There are many ecological and other contingencies acting as determinants or contributory factors. Despotism and democracy can be seen at variously sized social structures. Also of note, this progression is implicitly historically documented in chapters nine through nineteen.
3. Of note, these results indicated that non-anthropogenic factors were the main contributors to chimpanzee lethal aggression.

4. If a revenge target of lesser value than the original victim is attacked, then the retribution will be deemed (by one's group) insufficient; if a revenge target of greater value is taken, then the retribution will be deemed (by the other group) excessive and disproportionate.
5. However, due to the nature of the paleoanthropological and archaeological record, it is not feasible to classify each instance of lethal aggression accurately as an outcome of coalitional within-group or between-group competition. This methodological obstacle precludes more rigorous examinations of the latter data to determine whether warfare, operationalized as lethal competition between two or more distinct biocultural groups (i.e., external), among prehistoric small-scale societies eliminated a higher percentage of individuals relative to historical states.
6. These data are currently stored and curated as part of the DPLACE online data repository (Kirby et al., 2016).
7. Internal consistency values (Cronbach's α) were computed for each latent dimension, with estimates above a threshold of 0.70 considered to display adequate levels of internal consistency.
8. A SEQCA operates as a system of Hierarchical Linear models, established as a series of equations wherein an outcome variable, in a pre-specified stage of the analysis, functions as a predictor in the ensuing phase of the model (Peñaherrera-Aguirre et al., 2022). Consequently, the order in which the variables are included in the analysis must follow a pre-specified and theoretically based order (Figueredo et al., 2017).

8

Faction and Party in Modern Constitutional Republics: The Politically Relevant Dynamics of Multilevel Selection

Steven Charles Hertler, Aurelio José Figueredo, and Mateo Peñaherrera-Aguirre

> The present constitution will not ensure the greatness of France, the sole object of my preoccupations and supreme goal of my life. The regime of parties is harmful. It prevents the execution of a great external policy, just as it compromises internal peace and stability
> —Charles de Gaulle

S. C. Hertler (✉)
Saint Elizabeth University, Morristown, NJ, USA

A. J. Figueredo • M. Peñaherrera-Aguirre
University of Arizona, Tucson, AZ, USA
e-mail: ajf@arizona.edu; mpeaher@email.arizona.edu

S. C. Hertler et al., *The Evolution of Political Ideology*,
https://doi.org/10.1007/978-3-031-83941-2_8

1 Introduction

This chapter analyzes the dynamics of political factions and parties in the modern state, using the United States of America (USA) as a convenient model for a constitutional republic. Thus, it is unclear how much of what is said here about the differences between Conservative and Liberal, Right and Left, etc., might or might not be generalizable to other such polities. Nevertheless, one has to start somewhere.

The founding of the USA is something of a time-lapse. What took so long to grow up organically and unconsciously in England unfolded rapidly in the latter part of eighteenth century America. Founders were first architects of a constitutional cage capable of controlling political competition between interest groups and then became partisan actors within that constitutional cage engaging in political competition with ideological opponents. In their role as constitutional architects,[1] Washington, Adams, Jefferson, Madison, and other founders were erecting a framework, first to minimize between-group competition and then to control what remained by countering interest against interest across multiple levels: within and among branches of the federal government, between state and federal authority, across classes by marrying republican protections to democratic forms. The success of the founders in controlling competition between groups allowed latent within-group ideological differences to emerge, upon which ideologically divergent parties formed.[2] James Madison embodies this bifurcation. Madison was first working with Hamilton as coauthor of the Federalist Papers; and thereafter working against Hamilton as the action arm of the burgeoning Republican opposition party. After the war he was father of the constitution; after the constitution, he was penman of the Republicans.[3]

The rapidity of this turnabout illustrates the dual nature of the political. First, there is interest group politics. This is effectively between-group competition as it existed prior to state level societies. States grew up around groups, monopolized violence, thereafter rendering between-group competition nonviolent—in other words, political. Rich and poor wrangled about redistribution and taxation, state and federal governments disputed sovereignty, Protestants tried to limit Catholic rights in eighteenth century England, Blacks struggled for equality in nineteenth century society, guilds erected barriers of entry into their fields, unions fought employers over wages, all of these are rightly political. However, they are transmutations of violent and martial between-group competition. Once these interest groups are bound within, and constrained by, the state, they continue to compete, but do so nonviolently by means of political infighting. Rather than slitting throats and battling on the field as in pre-state societies, these interest groups try to grasp levers of power, gain influence, obtain resources, and otherwise use state power to promote group interests. Second, there are ideological affiliations cutting across any such groups. This is the ideology of mind that explains political debates taking place across kitchen tables and water coolers.

In sum, references to *political* or *politics* routinely blend these distinct manifestations, the first of which is between-group competition among interest groups, the second of which is within-group competition among ideological positions. Mapping onto these different manifestations of the political are two distinct domains of explanation afforded by multilevel selection's application to politics. Survival and propagation of an individual's genetic material are affected by the characteristics and cohesiveness of the groups of which they are perforce a part for reasons of security, genetic relatedness, proximity and population density, mutual advantage, and division of labor. Whether violently as in war or nonviolently as in politics, multilevel selection affords a biologically informed lens through which to view between-group competition in a way that explains why groups alternately cleave apart and cohere. Again, this first domain of explanation focuses on between-group competition, extending evolutionary lessons about violent conflict to nonviolent political competition. We are simply extending multilevel selection's explanation of between-group competition beyond its typical applications to warfare and religion to include politics. In this first domain of explanation, our point is merely that groups based on common descent, religious affiliation, or regional proximity, though they have competed violently, can also compete politically. We speak of these as political lobbies, factions, and interest groups.

However, there is a second domain of explanation, which we detail in this eighth chapter. This second domain is a nearly novel application of multilevel selection and one that uniquely explains within-group ideological differences that, as we have seen, arise from genetic, neurobiological, and temperamental traits. We explain that political ideologies are not just cobbled together but represent something more fundamentally connected in terms of evolutionary logic. Military spending, deference, hierarchy, stratification, authority, religion, and marriage, promoted by the right, augment group fitness; while abortion, tolerance, homosexuality, freedom, individualism, cohabitation, and equality, promoted by the left, advance individual freedom at the expense of group fitness. First, in abstract and without context, we simply develop this thesis, associating the political right with group-selection and the political left with individual-selection. After providing this summary statement about how different threats to survival and reproduction map onto different ideological poles we provide two more sections—one explaining the selective forces that maintain these ideological poles and the range of differences between them; and other describing how those selective pressures impart change on individual persons and populations. In reading each of the following sections, keep in mind their collective purpose: To explain the evolutionary process causing those genetic, neurobiological, and temperamental differences out of which political ideological differences arise.

2 Understanding Right and Left as Group and Individually Selected

Through a multilevel selection lens, we understand liberals and conservatives as oriented to different levels at which survival and reproduction can be threatened: conservatives fear external conquest; liberals fear domestic tyranny.[4] Conservative fear of external conquest is based on the reality that group survival is a precondition for individual survival. Therefore, conservatives express and reward patriotism, nationalism, and militarism, while attempting to punish, marginalize, and ostracize non-conformists that undermine group unity, and free riders externalizing fitness costs to the group at large. Basically Hobbesian in outlook, conservatives are innately oriented toward maintaining prevailing social stability which, as mentioned, marks mature states.[5] This orientation explains why conservatives have enlarged amygdalae and express, at higher levels than do liberals, moral emotions of *ingroup loyalty, authority/respect,* and *purity/ sanctity.*[6] Conservatives then attempt to order and organize society around these fears, protecting stable institutions, suppressing dissent, demanding loyalty, instituting hierarchies, and insisting on traditional familial and social arrangements. Lacking such an acute neurobiological orientation to fear, or these other-focused moral emotions, liberals innately take societal stability for granted. Focused, at higher levels than conservatives, on *harm/care* and *fairness/reciprocity,* liberals subordinate the state to the individual, emphasizing fairness, freedom, rights, and liberties, while opposing oppressive social norms, coercion, judgement, and the inertia of tradition. For liberals, the state is a social compact securing life, liberty and the pursuit of happiness to the individual; for conservatives, the individual is a subordinate vehicle, securing stability, order, and the guarantee of survival to the nation.

When conservative politicians and their constituents legislate and vote for capital punishment and police funding, they are activated by innate fears of internal disorder and disunity, just as when they vote for arms spending and military action they are activated by an innate fear of external conquest. The conservative concern with national stability also actuates anti-abortion legislation and traditional marriage, as both promote population growth, which counters the military and demographic competition of rival groups. Moreover, religion and cultural tradition, along with hierarchical order, feature into Right-leaning party platforms for their ability to augment unity, allowing a collection of individuals to function more like an integrated superorganism. Finally, anti-immigration stances reflect native xenophobic reactions against any erosion of biological and cultural homogeneity, which again may have negative fitness consequences for between-group competition, and furthermore blur boundaries between ingroups and outgroups. The political Left's opposite stance on these many issues is evidence of their fear of fascistic domination over and above external conquest. Conservatives often allow their risk-averse, anxiety-informed orientation to suppress fairness and freedom on the way to ensuring order and strength; while liberals often allow their enlightened humanitarianism to critique the existing order and pursue utopian fairness in such a way that ignores the existential realities of living in a competitive world of limited resources. Collectively, we then see that military spending, deference, hierarchy, stratification, authority, religion, and marriage, promoted by the Right, augment group fitness at the expense of individual freedom; while abortion, tolerance, intersectionality, freedom, individualism, cohabitation, and equality, promoted by the Left, augment individual freedom at the expense of group fitness.

With this in mind, let us unpack the overview provided above by more overtly developing this evolutionary logic by way of example. To do this we consider the post-Reagan Republican platform discussed in Chap. 2:

1. Traditional heterosexual marriage
2. Pro-life stance
3. Stratification and hierarchy
4. Authoritarian and executive dominated power arrangements
5. Tradition, deference, and custom
6. Religion
7. Controlled immigration
8. Group homogeneity
9. Strong local law-enforcement and capital punishment
10. Military spending and armament

The rationale of these intuitive preferences can be lost on either a Jeffersonian Liberal or a modern-day liberal Democrat. Nevertheless, this is not a random cacophony of positions cobbled together to no purpose. Instead, these ten positions collectively ensure the strength and unity of family, society, and country. They are preferences and positions ensuring that the family is stable and productive of progeny; that the social community is ordered and safe; that the tribe or state is competitive and defended.

Relating to one and two, traditional families can produce viable offspring, who arguably have a greater chance to survive and thrive. Beyond this, conservatives want this traditional family arrangement, not only for themselves, but for other group members. This is somewhat harder to see as functional. However, as societies compete against one another as in multi-level selection theory, demographics are as important as weapons. There is safety in numbers. More than this, a stable family is likely to thrive around other stable families.[7] Accordingly, mature males are reliably sequestered within the marriage bed, rather than remaining as roving competitors that may break up a happy home.

Relating to three and four, stratification and hierarchy within a society managed with strong top-down control can bring adaptive differentiation. As Sober and Wilson (1998, p. 179) observe, social stratification can potentiate group strength, or as they put it, "group-level success." This observation is made particularly in relation to larger societies, and so Sober and Wilson go on to note the following: "It is likely that human social groups must differentiate to remain functional as they increase in size." A strong measure of stratification and hierarchy encourages the concentration of capital and time in the hands of elites who can, but do not always, use this for the benefit of the group or strength of the nation. Temporal and monetary capital marshals resources to productive ends. Capital is what makes individuals act like units. The factory system is so productive because it represents a capital investment that provides the infrastructure to organize effort. Likewise, the group advantages of authoritarian, and executive-dominated power arrangements, can be seen through a brief examination of the writings of economic historian David Hackett Fischer, who finds the following pattern: In peacetime, most are apt to allow Jeffersonian principles to prevail. However, in wartime, active suppression often takes place. Fischer sees that almost all wars coincide with the suppression of liberty and freedom, but he also observes a patterned backlash wherein freedom and liberty rally at war's end.

Relating to six, seven, and eight, there is the issue of group composition. Religion specifically promotes ingroup loyalty and, less reliably, outgroup hostility. Religious members are more apt to trust one another, and religion can be harnessed by elites to encourage the flock to endure suffering and subordination, as it is part of the *Great Chain of Being* wherein all have their place. Also, as per the dictates of genetic similarity theory (Rushton, 1986), we befriend, marry, and mate with those that are most like us. While we have to avoid inbreeding depression, we can mitigate the genetic dilution of sexual reproduction by finding a partner that resembles ourselves across a broad array of physical and mental features, as this is a proxy for genetic relatedness (Rushton, 1989, 1998). Therefore, group homogeneity provides an ideal pool within which to find mates and reciprocators. Still further, religious, ethnic, and linguistic fractionalization, all markers of heterogeneity, relate directly and indirectly to internal instability.[8] Sectional struggles and civil war often occur when one state contains multiple nations. For instance, Woodard and Fischer separately describe the distinct peoples that settled in North America, and thereafter the factionalism, both political and military, that fell out on those same lines throughout post-colonial history.

Relating to nine and ten, there is the authoritarian use of power to enforce rules and laws within the nation, and security and borders without. These are psychological dispositions meant to prevent revolution and conquest. The conservative mind is more apt to see threats and anticipate danger. Collections of conservative minds at the policy table or voting booth are apt to give preference to military build-up. Note that this does not necessarily imply hawkish deployment of military arms; however, it does ensure readiness and offers an implicit defense against international aggression, as when John Adams spoke of wooden walls, which was his way of advocating for a strong navy that would protect trade and deter European aggression.

This evolutionary thinking is particularly able to resolve apparent paradoxes within the conservative platform. For example, it seems paradoxical to simultaneously advocate for capital punishment, but against abortion. One can charge inconsistency, or hypocrisy, but both dissolve at this deeper level of understanding. If the goal is to be humane, then yes, there is inconsistency; but as it stands the goal is group strength, and so it makes sense to oppose the restriction of births, while killing those who act against group interests. This grants, to the group, the largest possible demographic well from which those that undermine its fitness can be culled.[9]

For the sake of clarity, we articulated this theory in the above manner, but recognize that, were we to stop here, readers might come away with a number of misconceptions. For example, we are not contending that political platforms evolved; lower-level, genetically informed temperamental and cognitive dispositions evolved, which then give rise to predilections to political stances on these many issues. In other words, when we say there is coherence in this platform and that it is actuated by evolved ideology, this does not mean that the political platform is itself a product of evolution, or an intellectually coherent philosophy productive of logical outcomes. Instead, political platforms make psychological sense as reflections of broad temperamentally based dispositions, with that disposition as a whole having a reasonably reliable relationship with behavioral and social consequences. Also, we are not contending that political platforms are necessarily adaptive in a sociopolitical sense; it is just that they have an adaptive logic. Likewise, we have addressed each end of what of course is an ideological spectrum with all areas in between these poles occupied. In this chapter's remaining sections, we broadly contextualize the theory above in preference to envisioning and anticipating discrete critiques. To do this we specify pertinent evolutionary pressures and the way in which people and populations respond to those pressures—we specify the evolutionary mechanism and the subsequent response. More specifically, directly in the next section we pointedly discuss the nature of individual and group selective pressures, exploring how they sometimes dynamically push and pull in different directions; then in the subsequent section we consider how people and populations respond to the dynamics of individual and group selection. Jointly, these two sections are meant to explain how the exigencies of individual and group selection create a dynamic continuum of ideological difference.

3 Opposing Selective Pressures

There are otherwise excellent works on political psychology that misunderstand the evolutionary principles they use as explanatory tools. For example, political theorists Hatemi and McDermott (2011) relied on evolutionary theory to understand species-typical Paleolithic Era adaptations, without being more fully acquainted with the literature on those individual differences necessary to explain ideological variation within populations.[10] Ryan's (2020) *The Genetics of Political Behavior: How Evolutionary Psychology Explains Ideology*[11] argues that the varying requirements of survival at different points in history explain why liberals and conservatives have anatomically different brains as well as radically distinct behavioral traits. As per Ryan, Rightist traits, such as callousness and fearfulness, emerged early in evolution when violence was pervasive in human life and survival depended on the fearful anticipation of danger. Leftist traits, such as pro-sociality and empathy, emerged later as environmental adversity made it necessary for humans to live in larger social groups that required new adaptive behavior. However, this dichotomizes the fitness environment, and thereby dichotomizes the ideological continuum. At least by implication, conservatives are cast as evolutionary relics of a bygone age whose psychological equipment is rendered obsolete by modern culture. Furthermore, it disregards the continuing existence of warfare among modern states. Like Hatemi and McDermott, Ryan produced an insightful work that circled around the explanation of ideology, without ever seizing upon it. Both explain the evolution of ideology via *directional selection* without reference to *balancing selection*.

Directional selection occurs when there is an optimal adaptation to a prevailing selective pressure; the species at large moves in that direction, toward that single optimal solution. By contrast, balancing selection occurs in the absence of a single optimal solution, when there are instead opposing evolutionary pressures that pull in opposite directions, thereby selecting for a range of solutions across a continuum. Guppies illustrate the point. Experimental evolution on guppy populations demonstrates balancing selection across a continuum of *shyness* and *boldness*. There are opposing forces acting simultaneously on guppies: The need to eat, and the need to avoid being eaten. Shyness is a temperamental trait that predicts avoiding the predator filled depths of a pool or pond and is selected for by the possibility of predation. Boldness is a temperamental trait that predicts foraging within the depths of a pool or pond and is selected for by the possibility of nutrition. In consequence of this push and pull, we find, in nature, a range of guppy temperaments along the shy-bold continuum. The dynamics of the evolutionary pressures maintain this range of temperaments within the population. There is then a tradeoff between risk and reward that balances shyness and boldness, not within the individual but across the population. The ratio of shy and bold guppies, like the ratio of any trait evolving via balancing selection, will remain static so long as the risk and reward ratios remain static. It follows that the ratio of shy and bold guppies can change, as risk and reward ratios change. Reznick (1982) demonstrated this experimentally: removing predators alters the ratio, increasing proportions of bold guppies; introducing more predators alters the ratio in the opposite direction, increasing proportions of shy guppies.

Ryan most especially fails to recognize the underlying dynamics of balancing selection seemingly because the balance between these opposing forces changed over time. Ryan therefore mistakes the changing ratios within a balancing selection regime for directional selection. The time and place within which we and those authors are writing are peaceful, ordered, and stable, when compared to, for example, hunter-gatherer societies within the upper Paleolithic. Yet, this only changes the mean around which selection balances. So, there can be directional change in the mean across which a population is distributed, but there is still a range of values being balanced in response to opposing selective pressures. For clarity, consider an analogy to the stock market: There is both a long-term growth trend and cyclical ups and downs taking place amidst the growth trend. In Part II of this book, by viewing historical periods as different as Ancient Greece and Modern England, we are implicitly considering human populations with different ideological-relevant means; some of which are highly group selected and others of which are more individually selected. At present however, we focus on the nature of balancing selection itself, describing the opposing forces that maintain this balance.

Edward Wilson[12] (2012, p. 243) quite clearly sees that individual and group level selection vie within one another as opposing selective pressures within a balancing selection paradigm: "...the balance of selective pressures cannot move to either extreme. If individual selection were to dominate, societies would dissolve. If group selection were to dominate, human groups would come to resemble ant colonies." In the quote above, Wilson lays bare the implications of multilevel selection. In his *Social Conquest of the Earth*, from which the quote is taken, Wilson was capturing the idea of tradeoffs within a balancing selection regime. This logic ultimately explains a range of individual differences relevant to fitness, including the range of genetic, neurobiological, and temperamental differences reviewed across the first four chapters of this book. Wilson's quote captures the two exigencies inherent in multilevel selection: individual fitness and group fitness (understood as the aggregate individual level fitness). Again, individual fitness directly relates to the survival and reproduction of individuals, but as we have seen, individual survival and reproduction can hinge on the properties of the groups in which they find themselves. Evolutionary pressures are thus pulling down and up the ladder of aggregation at the same time; in other words, they are pulling down toward the individual level and up toward the group level. This tug of war creates dynamic tension wherein individuals are distributed along a range of genetic, neurobiological, and temperamental differences.[13] The need to survive and reproduce of course creates pressures for individuals to act in their own interest, to live free, marry who they will, pursue resources, resist control and coercion. Yet, we have seen that emergent properties of groups are also important to fitness, and so there is evolutionary pressure to act in the interest of higher-level groups, to organize, to create hierarchies, impose discipline, and punish. It is in consequence of this tension that we do not live either in anarchy or ant colonies.

To summarize, multilevel selection allows us to understand two evolutionary exigencies working in opposition to one another. Routinely there is not harmony between these levels, and they oppose one another and pull in opposite directions. This creates, instead of a single optimum, a range of adaptive solutions, which we can recognize as anchored on one end by individually selected types and at the other end by group selected types. And so when we henceforward speak of an individually selected person or population, we mean a person or population bearing, as a matter of degree and relative to others, the impress of individual selection, which imparts the imperative for the individual to survive and reproduce unhampered by being co-opted as a cell of state. Alternatively, when we henceforward speak of a group selected person or population, we mean a person or population bearing, again as a matter of degree and relative to others, the impress of group selection, which imparts the imperative for the group to avoid dissolution and conquest from being undermined by uncoordinated or selfish individual actors.

4 Developmental and Genetic Responses to Current Environments

We have taken pains to specify the two opposing selective pressures and the balancing selection regime that results from this dynamic. We have done so by providing descriptions of the selective regimes created by individual and group selection and then providing some examples of the evolutionary responses to those selective regimes.

In this section, we explain the mechanisms by which selective regimes can effect change across several different levels, amongst which there may be various causal transactions and interactions. The two major levels that we will discuss are developmental and genetic, but each of these also has some important sublevels within them. Each of these responses to current environmental contingencies happens on different time scales, across generations, between cohorts, or within minutes. As described in our primer on behavioral genetics in Chap. 2, selective regimes can impart evolutionary change at both the genetic and the cultural level, as well as through the interactive process known as gene-culture coevolution. These changes can take the form of either altering fixed (permanent and stable) traits or altering contingent (plastic or flexible) traits that are responsive to long-term or short-term environmental contingencies.

First, as we have already explored the genetic and evolutionary response, we will only add that selective regimes can alter the relative proportions of persons within the population that are more individually selected v. more group selected. To avoid confusion about these semantic labels, we reiterate: All people at all times experience both individual and group selection, and therefore bear the impress of both; neither the selective pressures nor their resultant impact on humans are dichotomous. Still, populations can differ in the relative degrees of group v. individual level selection experienced as evolutionary selective pressures, just as individuals can differ in the relative degrees of group v. individual level selection of which their genetically informed minds bear the impress. For example, Ancient Sparta, discussed subsequently in Chap. 9, was genetically shaped by extreme intragroup societal mores and extreme intergroup competition to evolve, individually and in aggregate, into a highly group selected society. Some modern societies, that imposed less martial discipline from within and experienced less existential threat from without, witnessed a relative increase in the proportions of individually selected members, and so their aggregate genetic profile drifted toward that of a more individually selected society. This is the aforementioned process of changes to the mean around which individual and group extremes are distributed.

Second, we can see the second form of change, cultural evolutionary change, by considering the work of Inglehart (2018). As per Inglehart, existential security leads to secularization. Then secularization brings reliable reductions in fertility; traditional "pro-fertility norms" erode, and there is no longer a penchant to "stigmatize any form of sexual behavior that is not linked with reproduction, such as homosexuality, abortion, divorce or masturbation" (Inglehart, 2018, p. 82). There are then universalizing tolerances of ingroup differences like homosexuality and outgroup sympathies wherein values are universalized globally. Inglehart claims that the diffusion of *individual choice norms* among high-income countries is one of the reasons for "adopting more peaceful, feminine orientations" (Inglehart, 2018, p. 104). Inglehart repeatedly links these developments with the population's reduced willingness to risk their lives through war. On page 109, Inglehart presents a graph showing that unwillingness to fight for one's country rises as a function of individual choice norms. Former Axis Powers have shifted greatly over time. Putting it all together, there is a shift toward secularization, and feminization, yielding a lower fertility rate and reduced willingness to "fight for one's country" (Inglehart, 2018, p. 3). The effect is to create societies more apt to succumb to rival societies, demographically and martially. In all these ways, Inglehart is effectively describing the process of European societies becoming more individually selected. His explanatory model is a process of cultural evolution; he is describing a sort of cohort effect, or a developmental model wherein being raised in the relative absence of existential threat causes a shift, within society and among its individual members, toward an individually selected worldview. This is like a process of developmental shunting or catalyzation. Among others, his most emphasized causal mechanism of all such change is increasing existential security of cohorts born after 1970; Inglehart emphasizes their security and satisfaction of material wants as imparting post-materialist values that prioritize free expression, autonomy, tolerance, and happiness, over and above their opposite materialist values. Existential security is a feature of the selective regime, and the shifts toward individual choice norms are nothing but shifts toward individual selection. Again, however, Inglehart is speaking of these processes taking place within a generation through cultural evolution, whereas we previously spoke of these processes taking place between generations through genetic evolution. Both processes can and do take place simultaneously. Selective pressures, in other words, can have the simultaneous effect of causing these cultural shifts Inglehart describes and the evolutionary shifts we previously described. As an aside, though without disagreeing with Inglehart in his capacity to represent cultural evolution, much of this change ascribed to culture may well be evolutionary. Post-WWII European societies are genetically different, having sloughed off millions of highly group selected males across two generations between 1914 and 1945. Then later generations experienced a selective regime with relatively stronger individual and weaker group selection. Thus, we can speak of a true process of gene-culture coevolution when contrasting present and past populations.

Third, individuals can facultatively shift in response to stress, manifesting higher or lower levels of authoritarianism, empathy, tolerance, and the like, as discussed implicitly by Stenner, the authoritarian personality researcher featured in Chap. 5. Stenner is certainly right about the shift in activation. We can see this in the wake of *9/11* where individuals and the nation at large assumed a more hawkish position supportive of interventionist national policy. Or consider that priming people about mortality augers "…support for charismatic leadership" (Cohen et al., 2004; Van Vugt & Schaller, 2008). Similarly, when under threat, people prefer taller, more dominant, masculine, and healthier looking leaders (Little, 2007). Facultative responses are responses to what Jost (2006) speaks of as *situational factors,* which are contrasted with *dispositional factors.* Instances of terrorism (Godefroidt, 2023), injustice, and mortality threats caused dispositional liberals to facultatively favor conservative stances on capital punishment, abortion, and homosexuality (Nail et al., 2009), which corroborate earlier findings that *system threat* and *mortality threat* reliably cause shifts toward conservatism, and toward endorsement of leaders and ingroup preferences (Jost, 2006). Perceptions of demographic replacement also induce facultative shifts toward the right (Craig & Richeson, 2014). This threat-induced shift toward conservatism is consistent with *Compensatory Control Theory's* post-threat cognitive changes (Landau et al., 2015). System and mortality threats dispose human populations toward a group-selected coherence, as morphogen differentiation inducing factors like ammonia depletion catalyze free living slime molds to form a siphonophoric-like stalk preparatory to dispersal (Van Es et al., 1994). These threat-induced facultative shifts temporarily reduce "ideological asymmetry in existential motivation" that exists temperamentally across the political spectrum (Jost et al., 2017). Here are the same threats to existential security discussed by Inglehart, but causing reactions within a discrete time frame; not across generations, but across months or even minutes.

Thus, these perspectives are all valid, they only focus on different modes of change and timescales in which that change occurs. In sum, Stenner emphasizes real time facultative shifts in the mind; Inglehart emphasizes slow cultural shifts in generational cohorts; whereas we emphasize even slower genetic shifts across generations.

5 Conclusions

Risking redundancy, we summarize our thesis: Partisans on each side of the ideological divide are temperamentally trained on different threats, or sets of potential selective events threatening survival and reproduction. The Right's fear of external conquest is based on the reality that group survival is a precondition for personal survival. As documented by the historical record *ad nauseam*, diaspora peoples, tribes, religious dissidents, ethnic minorities, and other groupings of people are subject to genocide, demographic displacement, and other forms of aggression and competition. Such threats can scale up to include whole peoples and nations. These amount to threats at the group level, and of course, undermine the survival and reproductive success of individuals belonging to those groups. These selective threats at the group level evoke and evolve dispositions toward order, strength, duty, tradition, patriotism, nationalism, and militarism; added to this are dispositions toward altruistic punishment meant to marginalize, ostracize, and eliminate nonconformists and nonreciprocators. From this underlying logic laid bare by multilevel selection theory, we can see the conceptual convergence of the conservative Right in contemporary US politics across the aforementioned ten stances. The rationale of these intuitive preferences can be lost on either a Jeffersonian Liberal or a modern-day liberal Democrat. Nevertheless, this is not a random cacophony of positions cobbled together to no common end. Instead, these ten positions are sociopolitical manifestations of more basic psychological dispositions selected to ensure the strength and unity of family, society, and country. To be sure they are shaped to some extent by elite messaging, but are also preferences and positions which ensure: (1) that the family is stable and productive of progeny; (2) that the social community is ordered and safe; and (3) that the tribe or state is competitive and defended. By contrast, those on the political Left are more temperamentally oriented to threats at the individual rather than the group level, and consequently emphasize fairness, equity, freedom, rights, and liberties, while opposing oppressive social norms, coercion, judgement, and altruistic punishment. In consequence, we speak of those on the ideological left as more *individually selected;* that is, to facultatively, culturally, and genetically bear the impress of evolutionary selection at the individual level; alternatively, we speak of those on the ideological right as more *group-selected*; that is, to facultatively, culturally, and genetically bear the impress of evolutionary selection at the group level.

Notes

1. Removing religion from the political sphere, balancing powers among three branches of government, creating a lower and upper house to balance plebian and patrician interests, delineating spheres of authority as in state and federal power, these are some of the girders of the constitutional cage.
2. This can be seen in the post-constitutional phase of US history wherein the founders, having formed a constitutional framework that controlled group-based factions, then rapidly succeeded to form into factions based on political ideology. This is the history of the contentious 1790s wherein emergent Federalists vied with emergent Republicans, which culminated in Jefferson's *Revolution of 1800.*
3. Some of the best minds in the Western pantheon have grappled with this question of tradeoffs. Madison hoped to pit lower level interest groups against one another, Adam Smith sanguinely hoped individual selfishness could promote group fitness, whereas Montesquieu sought to restrict population size and expanse sufficiently to ensure unity.

4. Anspach (2023) validates this dichotomy: “Those who are more sensitive to threat from others tend to hold socially conservative attitudes, while those who fear authority generally take more libertarian positions.”
5. A reasonable degree of stability is a precondition for ideological debate.
6. These moral emotions were articulated by Jonathan Haidt. Haidt, whose framework of moral emotions was referenced above, may well come closest to explicitly articulating this theory while, at the same time, presenting the necessary intellectual accoutrements to understand it. In *The Righteous Mind,* Haidt (2013) jointly discusses religion, morality, and political affiliation alongside an explanation of evolution and group selection. He outlines a newly expanded set of moral emotions, showing that conservatives identify with morals outside the *care/harm* value set, to concern themselves, for instance, with whom their neighbors are sleeping. Then, after an explanation and defense of group selection, Haidt goes on to identify group selection as promoting a *hivish* morality that is binding, and which instills trust, efficiency, and cooperation. Additionally, Haidt portrays how group selection can benefit the ingroup while harming the outgroup. Group selection causes others to rally around the flag, the leader, the nation, the cause. There is, too, a proper understanding of group selection taking place during a time of rapid, Neolithic evolution where the main selective pressures were other humans who punished non-reciprocators and non-cooperators in a process of self-domestication. Haidt’s review of group selection is thorough and the concept is applied to political affiliation alongside moral reasoning and religious persuasion. That conservatives are group selected and liberals are individually selected is implied, seemingly suggested, easily discerned, but not, it seems, overtly stated. We believe, from these moral emotions and a group-selected mind more generally can be extrapolated many fundamental beliefs of the conservative mind and tenets of the conservative platform.
7. Consider the relation between consanguinity and social instability, and the conflict caused by divorce, and the abuse perpetrated by stepparents.
8. This is a complex literature. Each category of heterogeneity has a different relationship with social stability in a complex and contested web of associations. The interested reader can pursue these topics in Wegenast and Basedau (2014), Alemu (2016), Karnane and Quinn (2019), or Arbatlı et al. (2020).
9. Similarly, those on the ideological right are more religious. Religion, especially Christianity, is often said to be hypocritical in that it preaches

tolerance but promotes bigotry. Religion erects ingroup-outgroup boundaries, promoting harmony and reciprocation among the faithful while differentiating them from infidels.

10. Hatemi and McDermott (2011), in their defense, were writing at a time when the literature explaining individual differences was less robust. For example, Buss and Hawley's volume on the Evolution of Personality and Individual Differences had just been published. Chapters in that book were of course consolidating and developing themes that had been previously published, but the evolutionary study of individual differences was somewhat immature then. And also to their credit, they understood that this is the exact variety of evolutionary theory necessary to explain political behavior.
11. The book also explores new evolutionary theories that emphasize the role of the environment in shaping not only human political behavior but also humans' genetic architecture. With implications for the future of politics, the book explores how the niche worlds we build for ourselves through political action can have consequences for the evolution of the species. Proposing a new way of understanding human politics, this is fascinating reading for students and academics in psychology, the social sciences, and humanities, as well as general readers.
12. His *Social Conquest of the Earth*, much like all else Wilson writes, contains brilliant insights scattered throughout the text. Among these is an explanation of political ideology, much like the explanation we are putting forth. Like ourselves, Wilson understands political ideology to be explicable only through multilevel selection, though his thoughts on the subject are not fully explained or highlighted.
13. We are here referring to a process regulated by negative frequency-dependent balancing selection as described specifically in papers on the *Coral Reef Model* (Figueredo & Jacobs, 2010; Figueredo et al., 2011; Figueredo et al., 2005). The coral reef model emphasizes the heterogeneity of the social environment (interactions among conspecifics), as opposed to the external physical ecology or the "community" ecology (biotic interactions among allospecifics), and proposes that sociality generates social microniches to which different personality and life history variants are differentially adapted. This model is generally supported by a review of the animal personality literature (reported in Figueredo et al., 2005) wherein it was observed that systematic individual differences in behavior were largely associated with sociality across a wide variety of nonhuman animal species (Sherman et al., 2013).

Integrative Metacommentary: Part III. Ultimate Causation

Aurelio José Figueredo, Steven Charles Hertler, and Mateo Peñaherrera-Aguirre

Ultimate Causation

According to multilevel selection theory, political behavior consists of a dynamic interplay between competition and cooperation. Chapters 6, 7, and 8 all dealt with the formation of social coalitions of increasing complexity and their deployment in the context of both within-group and between-group competition. Describing this progression partially addressed two of Tinbergen's Four Questions, those concerned with ultimate rather than proximate causation, tracing the evolved adaptive functions of such social coalitions through a brief (albeit incomplete) sketch of their evolutionary history. Throughout this continuum of sociopolitical complexity, we found that the level of lethal aggression produced by this competition generally decreased, although paradoxically peaking at the level of tribal horticultural societies.

A. J. Figueredo • M. Peñaherrera-Aguirre
University of Arizona, Tucson, AZ, USA
e-mail: ajf@arizona.edu; mpeaher@email.arizona.edu

S. C. Hertler (✉)
Saint Elizabeth University, Morristown, NJ, USA

Chapter 6 used within-group conflict as an indicator of sociopolitical behavior, including the formation of agonistic coalitions and alliances. This chapter explored relations between neurocognitive anatomy and within-group conflict in human and nonhuman primates and showed that higher endocranial volumes predicted higher levels of within-group conflict. Phylogenetic analyses revealed that, among primates, within-group conflict was highest in Australopithecines and Hominins, and higher in Hominins than Australopithecines.

Chapter 7 used Murdock's Standard Cross-Cultural Sample to examine the patterns of exclusively human lethal aggression both within and between groups ranging widely in sociopolitical complexity from hunter-gatherer bands to organized states, while concentrating mostly on non-state societies. The cascade model we constructed estimated the sequential effects of factors such as the use of different domesticated animals, parasite prevalence, Culture of Honor norms supporting revenge, rewards accorded warriors, frequencies of internal and external conflicts, and sociopolitical complexity. Contrary to much previous theory, we found that possession of herding livestock had no direct effect on Culture of Honor, but only an indirect effect that was fully mediated by possession of horses. While possession of horses had no statistically significant effect on parasite prevalence, possession of herding animals had the expected large effect on parasite burden. Curiously, Culture of Honor had no statistically significant direct effect on homicide, but only an indirect effect that was fully mediated by rewards accorded warriors. In addition, while parasite burden had the expected large effect on internal conflict, it had no direct or indirect effect on external warfare. Finally, external warfare, possession of horses, and reliance on farming all had substantial effects on sociopolitical complexity.

Chapter 8 described the characteristics of factions, interest groups, and political parties in the modern state, relying primarily on the United States of America as a convenient model for a contemporary constitutional republic. In contrast to the two previous chapters, the emphasis here is on nonviolent political competition. This chapter reviews the underlying values (such as binding and individualizing values) that presumably undergird conservative and liberal ideologies as *dispositional* influences, then goes on to review the extrinsic morbidity and mortality factors (such as the role of external threats to the survival of the group or the individual) as *situational* influences. Together, these provide the bases for a functional explanation of many of the biosocial influences on political ideology.

Part IV

Cycles of Integration and Disintegration in Multilevel Sociopolitical Coalitions

9

As Far as Our Spears can Reach: *Asabiyyah* and the *Dulotic Progression* in Sparta (743 BC–192 BC)

Mateo Peñaherrera-Aguirre, Aurelio José Figueredo, and Steven Charles Hertler

A people trained consummately but exclusively for warlike contact with its neighbors found itself suddenly compelled, by the outcome of one particular war, to enter into nonmilitary relations for which they were not only unprepared but were positively unfitted by their own peculiar institutions and habits and ethos. These peculiarities which the Spartans…had developed in order to grapple with a previous problem, and which had given them superhuman strength within the limits of the narrow environment within which their lines had previously been cast, now took their revenge upon this peculiar people by making them inhumanly or infra-humanly incompetent to live in the wider world into which the fortunes of war had eventually carried them. The very exactness of their adaptation to their previous environment made any re-adaptation to a new environment so

M. Peñaherrera-Aguirre • A. J. Figueredo
University of Arizona, Tucson, AZ, USA
e-mail: ajf@arizona.edu; mpeaher@email.arizona.edu

S. C. Hertler (✉)
Saint Elizabeth University, Morristown, NJ, USA

S. C. Hertler et al., *The Evolution of Political Ideology*,
https://doi.org/10.1007/978-3-031-83941-2_9

difficult for them as to be virtually impossible; and the very qualities which had been the secret of their success in the one situation became their worst enemies when they found themselves in the other.

—Arnold Joseph Toynbee[1]

1 Introduction

The above quoted historian, Arnold Toynbee, attributed the fate of Sparta to excesses of specialization. In making his point, Toynbee invoked the dystopian novels of the day, writing of Huxley's *Brave New World* and Wells' *First Men on the Moon*, both of which feature dehumanizing degrees of specialization.[2] He does not neglect to compare human excesses of specialization to complex insect societies, drawing parallels between insect castes and the human social engineering that constrains the specialist. At the time, it was believed to be the nature of social insect instincts to be exquisitely adapted, but perfectly inflexible.[3] In human affairs, Toynbee warns, "from such inflexibility it is a short step to folly" (Toynbee, 1951; volume III, p. 109). Policies, intelligently designed procedures, instructive mannerisms, and productive practices of all kinds can become instilled into a society long after their original object has disappeared. Inertia of rigid forms results.

In the remainder of this chapter, Sparta is discussed as an archetype of the extremely group-selected society and analyzed empirically so as to ascertain whether, as Toynbee asserted, martial specialization cut both ways, first creating Sparta's legendary power and then undermining Sparta's continued competitiveness. Although there might be some historical validity to Toynbee's position, this single explanation is incomplete. We will examine the analogy to social insect societies in greater detail and determine whether and to what extent it is predictive of the rise and fall of Spartan military success over the span of several centuries. Nevertheless, we will also examine certain confounding biodemographic factors as alternative, if not mutually exclusive, hypotheses.

2 *Asabiyyah* and Spartan Ultrasocial Institutions

"Undoubtedly the greatest work of its kind that has ever been created by any mind in any time or place" (Irwin, 2018, p. ix). This is what Toynbee said of the *Muqaddima,* a book written in fourteenth century Arabia by Ibn Khaldun (1377/1958). The *Muqaddima* is perhaps the first great installment in the genre of cyclical history later featuring the writings of Toynbee, Spengler, Vico, Gibbon, and Hegel. Within the *Muqaddima*, and central to its explanation of decline and fall, was Ibn Khaldun's construct of *Asabiyyah,* which referenced a group's cohesion, collective will, and tribalism, all of which predicted military conquest. *Asabiyyah* is an expression of group selection; both concepts emphasize group cohesion, subordination of the self to group ends, martial vigor, and hivishness. Consequently, evolutionary researchers and human behavioral ecologists have expanded the construct of *Asabiyyah* and examined its dynamics through mathematical modeling and macrohistorical empirical studies (Hertler et al., 2020; Figueredo et al., 2018; Turchin, 2003, 2006). These investigations undertook to explain how *Asabiyyah* evolved and how it was maintained in larger groups.

According to Hertler et al. (2020), multilevel selection theory predicts that a society's *Asabiyyah* emerges in evolutionary and ecological settings evincing intense between-group competition, promoting the rise and persistence of social mechanisms sustaining within-group cooperation *via* gene-culture coevolution. Moreover, mores, institutions, and related cultural variants modify selective pressures (cultural niche construction; Odling-Smee et al., 2003) and potentially reinforce a group's *Asabiyyah* (Hertler et al., 2020). In other words, *Asabiyyah* is wrought by conflict, territorial disputes, and other threats imposed by rival groups. These threats from those outside the group then have the secondary effect of changing dynamics between those inside the group. Groups so threatened by outsiders become less tolerant to internal dissent, cowardice, selfishness, and in turn reward conformity, bravery, self-sacrifice. So societies then move, culturally and evolutionarily, toward a more group-selected level of *Asabiyyah,* from both the external threat itself as well as their reaction to it.

It is relatively easy to envision how this group selected dynamic of *Asabiyyah* evolved in small-scale societies by means of kin selection, inclusive fitness, and reciprocity[4] (Hertler et al., 2020). However, explaining the maintenance of *Asabiyyah* and related prosocial and group selected behaviors in populous societies like Sparta requires social incentives and punishments (Hertler et al., 2020; Peñaherrera-Aguirre et al., 2022; Turchin, 2015). In the sections that follow, we review Sparta's ultrasocial institutions[5] (790 BC–210 BC) to the end of showing how external competition caused Sparta to: (1) incentivize cooperators and punish defectors; (2) increase *Asabiyyah*; (3) change structural-demographic dynamics in such a way that first augmented power and dominance; but (4) ultimately undermined power and dominance by restricting adaptability and demographic growth.

3 The Civic Structure of Sparta

The civic structure of Sparta entailed a three-tiered rigidly defined hierarchy: (1) Spartan citizens referred to as *Homoioi* or *Spartiates* (Bayliss, 2022; Campbell, 2012); (2) *Perioikoi,* the lower class of free people that usually inhabited the Spartan outskirts (Bayliss, 2022; Campbell, 2012); and (3) *Helots,* the state-owned serfs, generally former citizens of rival regions (e.g., Messenia; Campbell, 2012). The status of *Homoioi*, or *Spartiates*, was restricted for the sons of *Homoioi* who successfully completed their civic and military preparation at the educational and training program (the *Agoge*), allowing them to serve in the Spartan army (Hodkinson, 2002). Moreover, until the age of 60, *Spartiates* were required to attend a mess hall and dine every evening with their peers (Campbell, 2012; Hodkinson, 2002) and contribute monthly with barley, cheese, pork, olive oil, and wine (Bayliss, 2022). A modest dress style concealed wealth disparities among *Spartiates*, with vibrant colors restricted to military attires and colorful clothing viewed as a sign of vanity and a potential source of conflict among *Spartiates* (Bayliss, 2022).

According to Hodkinson (2002), this austere lifestyle rested on three principles: (1) *uniformity*, wherein wealthy individuals were encouraged to adopt a more modest way of life; (2) *collective interests*, wherein society's goals were prioritized over individual desires; and (3) *conformity*, wherein individuals were expected to follow customs and regulations. Besides the persistence of traditions and institutions designed to reduce public signs of wealth and social disparities, it was not rare for individuals to lose their full-citizen status if they failed to consistently contribute to the mess hall (e.g., *Hypomeiones*; Bayliss, 2022; Campbell, 2012; Cawkwell, 2002). It was not uncommon for wealthier Spartans to sponsor the sons of *Hypomeiones* (*Mothakes*), allowing them to participate in several spheres of Spartan education, training, and development, becoming adopted brothers (*Syntrophoi*) alongside the sons of *Spartiates* (Bayliss, 2022; Whitby, 2002). This patronage enabled lineages of disenfranchised Spartans to eventually regain their status as *Homoioi* (Bayliss, 2022). Inferior classes included the illegitimate sons of *Helot* women and *Spartiate* men (*Nothoi*), and "tremblers" (*Tresantes*), *Spartiates* accused of behaving cowardly during battle (Kennell, 2011; Whitby, 2002). *Tresantes* shaved half of their faces, wore patched clothing, occupied denigrating chorus positions, yielded to younger *Spartiates*, and were not considered as potential marriage partners (it was especially shameful to accept a *Tresante* as a son-in-law or father-in-law; Bayliss, 2022).[6]

All *Spartiates* older than 18 years of age participated in an assembly (*Apella*).[7] Out of these *Spartiates*, 28 elders were elected for life to participate, alongside the Spartan co-kings, to a council of elders in charge of the development and discussion of public policies (*Gerousia*; Carr, 2012) before presenting these proposals to the citizenry in a general assembly (*Ekklesia*; Kennell, 2011; Morris, 2009). Spartan citizens were not allowed to debate the policies or alter the agenda; instead, they could accept or reject the *Gerousia*'s proposals (Bayliss, 2022; Morris, 2009). In addition to convoking the general assembly, the *Gerousia* also had the faculty to veto a resolution and acted as direct intermediaries between foreign dignitaries and the *Ekklesia* (Bayliss, 2022). Five senior magistrates (*Ephoroi*), one per village out of the *Spartiate* body, were also elected for a single, year-long term (Morris, 2009). The *Ephoroi* supervised the actions of the Spartan diarchy (Andrewes, 2002; Morris, 2009). For example, at least one *Ephor* accompanied one of the kings during military campaigns (Morris, 2009).

During the Persian wars, *Spartiate* soldiers were mustered from five villages (*obai*) within Sparta (Lupi, 2006). Individuals (from 20 to 59 years old) were recruited from the same *obai* and consequently trained and fought in the same regiment (Hodkinson, 2002).[8] This parochial organization was subsequently replaced by a cross-village recruitment system, wherein each regiment comprised individuals from different villages (Hodkinson, 2002). Such an organizational shift suggests the emergence of additional ultrasocial institutions that enforced cooperation among *Spartiates* who probably did not interact frequently and were not closely related at a genealogical level.

According to Campbell (2012), the army was organized into five regiments (*lochoi*): (1) "Hell-bent" (*Arimas*); (2) "Devourer" (*Edolos*); (3) "Center-leader" (*Messoages*); (4) "Thundercloud" (*Ploas*); and (5) "Ravager" (*Sinis*). The *lochagos* acted as military officers and led each regiment (*lochos*), which in turn were organized into sworn bands (*enômotiai*) commanded by the *enômotarchês* (Bayliss, 2022; Campbell, 2012). Although Sparta relied on its regiment-based structure for multiple military campaigns, the army eventually replaced this organization with six military subunits of smaller bands (*morai*; Bayliss, 2022; Campbell, 2012). The stratified nature of the Spartan army also rested on the presence of efficient communication pathways connecting various military strata, starting at the level of the co-kings, passing through the high-ranking military officers (*polemarchoi*; Campbell, 2012; Hodkinson, 2002) and the regimental commanders, and finally reaching the leaders of the sworn bands who instructed soldiers under their command (Campbell, 2012; Hodkinson, 2002). The martial discipline of this system is captured in the following quote:

> I shall fight while I live, and I shall not put life before being free, and I shall not desert my *taxiarchos* [brigadiers] nor my *enomotarches*, whether he is alive or dead, and I shall not depart unless the leaders lead us away, and I shall do whatever the general command, and I shall bury on the spot the dead among my fellow-fighters, and I shall leave no-one unburied. (Campbell, 2012, p. 44)

Demographic reconstructions suggest that *Spartiates* were a minority relative to other social groups in Spartan society, with some estimations indicating no more than 10,000 *Homoioi,* with this number declining to fewer than 1,000 at the Battle of Leuctra (371 BC; Figueira, 1986). Several historians, including Aristotle, hypothesized that this demographic collapse significantly contributed to the decline of Sparta, as evidenced by the expansion of rival *poleis*, including Thebes and Macedon. Figueira's (1986) estimates further support the reported decline in *Spartiates* relative to the rest of the Spartan army (479 BC to 371 BC). According to the author, the Spartan military deployed at the battle of Platea entailed 50% *Spartiates* and 50% *Perioikoi*. By 425 BC, these proportions shifted, wherein 38% of *Spartiates* and 62% of *Perioikoi* participated in the battle of Pylos.[9] As per Figueira, the pivotal battle of Leuctra (371 BC) featured minimal participation from *Spartiates* (26%) relative to *Perioikoi* (65%).

4 Perioikoi and Helots

As Sparta expanded its sphere of sociopolitical dominion via conquest or diplomatic influence, the inhabitants of several nearby communities became incorporated into the Spartan civil structure (Kennell, 2011). Thus, the term *Periokioi* emerged when referring to non-*Spartiate* Lacedaemonians who remained second-class citizens, subordinate to the Spartan *Homoioi* (Bayliss, 2022), and who were required to participate in military campaigns, even as they were barred from contributing to the development of foreign policies (Bayliss, 2022). Although *Perioikoi* communities featured some degree of sociopolitical autonomy, often forming military leagues, these communities remained subservient to Sparta (Campbell, 2012; Kennell, 2011). [10]

According to recent estimations, *Perioikoi* owned less than a third of arable fields, whereas *Spartiates* maintained a monopoly over most agricultural lands (Bayliss, 2022). This disparity was not due to demographic differences but wealth disparities (Bayliss, 2022). The economic activities of *Perioikoi* included trade, manufacturing, and mining (Ducat, 2018; Kennell, 2011). *Spartiates* were barred from participating in these economic endeavors (Bayliss, 2022). Thus, *Perioikoi* provided an essential influx of goods, services, and liquidity to Spartan society (Kennell, 2011; Shipley, 2002). Moreover, *Perioikoi* were responsible for developing Spartan cultural epiphenomena, such as sculpture, painting, and ornamental metallurgy. Regarding their involvement in military activities, some researchers argue that as *Perioikoi* participated in several Spartan battles; they must have experienced at least a quasi-*Spartiate* military instruction (Bayliss, 2022). Moreover, the historical record indicates that it was not uncommon for *Spartiates* to consider foreign younglings who underwent the latter training as foster kin (*trophimoi*; Bayliss, 2022; Cawkwell, 2002; Kennell, 2011). Consequently, current research questions relate, not to whether *Perioikoi* received military training, but rather whether their martial education reached similar rigorous levels as that of *Spartiates*.[11]

Thus far, we have seen how Spartans kept the *Perioikoi* at a distance even in the backdrop of demographic decline. As demographic decline caused a shortfall of soldiers, so it caused a shortfall of farmers. In their need, Spartans turned to outgroups for assistance, but ensured that they remained outgroups. Specifically, after the conquest of Laconia and Messenia, *Spartiates* established a system of serfdom (*Helotry* or *Helotage*) wherein conquered peoples were forced to predominantly act as agricultural laborers, producing pork, cheese, wine, barley, and olive oil (Bayliss, 2022). *Spartiates* also demanded from *Helots* a portion (as a fixed amount) or proportion (a share or ratio) of crops produced on the land (Bayliss, 2022; Figueira, 2018). Other responsibilities included domestic tasks akin to those performed under chattel slavery (Bayliss, 2022; Ducat, 2002). Although several biodemographic reconstructions suggested that the *Helot* population fluctuated between 140,000 to 375,000 individuals (generating *Helot:Spartiate* ratios ranging from 7:1 to 20:1), current estimations based on available arable land (*Spartiates* owned between 115,000 ha to 145,000 ha) indicate a *Helot* population between 75,000 to 120,000 individuals (Bayliss, 2022; between 85,000 and 115,000 individuals, according to Figueira, 2018).

Some historians have called into question the extent to which *Spartiates* could control *Helots'* daily lives. (Bayliss, 2022). Although these accounts still acknowledge that *Spartiates* displayed open disdain for *Helots*, the fact that *Spartiates* preferred to remain in Sparta, as well as the sizeable population of *Helots* remaining in Messenia, precluded *Spartiates* from exerting total dominion over their serfs (Bayliss, 2022). According to Bayliss (2022), the presence of the interposing Taygetus mountain range precluded *Spartiates* from regularly visiting and overseeing their estates in Messenia. This logistical challenge often required the appointment of an overseer or manager in charge of maintaining order and meeting expected production quotas. Although additional research is needed, some historians argue that *Spartiates* either hired *Perioikioi* or assigned specific *Helots* to manage their estates, with *Spartiates* granting these *Helot* overseers certain "privileges" that otherwise were generally refused to their fellow laborers (Bayliss, 2022).

5 The Dulotic Progression

In a case of evolutionary irony, extremes of group selection undermined Sparta by, for example, rigidly distancing outgroups even when they could not muster numbers enough to man the lines and fields. Herein we specifically explore one manifestation of this broader trend, by considering the practice of enslaving all of the Messenian people. A frankly speculative analogy from the natural sciences can be found in the evolutionary dynamics of slavery (*dulosis*) in ants. Rather than a sole explanation, this speculative analogy is offered as working synergistically with other contributory factors to explain Sparta's decline.

The *dulotic progression,* a hypothesized evolutionary sequence, has been reconstructed using extant examples (so-called "living fossils") of the presumed successive stages of the evolution of slavery in ants. This sequence was first proposed by the eminent myrmecologist William Morton Wheeler in 1907 and elaborated in his seminal 1910 monograph, but the present retelling has been updated with subsequent findings. The most significant addition (Hölldobler, 1976) since the seminal work of Wheeler (1907, 1910) is the finding that some existing species, such as the Honeypot Ants (*Myrmecocystus mimicus*) of the Southwestern United States, practice slavery within the same species, wherein the winning colony in intergroup territorial confrontations (called a *tournament* due to the ritualized agonistic displays) may overrun the losing colony and enslave some of its members. This was the first instance of intraspecific (within species) slavery described in ants. Some researchers (e.g., Alloway, 1980; Pollock & Rissing, 1989) have proposed that slave-making evolved from the intraspecific brood raiding that may occur as a consequence of territorial conflicts, where larvae and pupae of the vanquished colony were taken back to the victorious colony to be cannibalistically consumed. However, some of them matured there and became functional slaves.[12] This reconstructed evolutionary history for the origin of slavery in ants was first suggested by Charles Darwin in 1859.

The next step in this progression was the evolution of *facultative* interspecific (between species) slavery, where one species (such as *Formica sanguinea*) enslaves members of what are the most closely related species (such as *Formica fusca*), according to Emery's (1909) Rule, but can survive without them. This idea was developed in the greatest detail by Wilson (1971) and then Holldobler and Wilson (1990), with more recent reviews provided by D'Ettore and Heinze (2001) and Buschinger (2009). The basics of the reconstructed story goes as follows: Allopatric speciation creates two closely related species of ants in geographical proximity, such that they often come into ecological competition due to similarity. As with the intraspecific case, territorial conflicts and brood raiding ensue, leading to the enslavement of the members of the sibling species that most frequently loses this competition. The primary function of interspecific brood raiding gradually changes to the habitual capture of immature members of the vanquished species for slaves, instead of for food. Consistent with this evolutionary ecological theory, a phylogenetic study of the genus *Formica* (Borowiec et al., 2021), which is the ant genus in which the highest number of socially parasitic species are found, has confirmed that the higher frequency of slave-making in this genus may be related to its higher rates of species diversification.

The step that followed in this same progression was the evolution of *obligate* interspecific slavery, where one species (such as *Polyergus rufescens*) enslaves members of what are typically the most closely related species (such as, once again, *Formica fusca*), according to Emery's Rule (1909). Some American "Amazon ant" slave-makers (such as *Polyergus lucidus*) may enslave members of any of several other related species, such as *Formica schaufussi, Formica nitidulus, Formica subsericea,* and *Formica neocinerea* (Dobrzanska & Dobrzanski, 1989; Wheeler, 1910). These obligate slave-makers, however, are *not* able to survive without their slaves, often being unable to feed either themselves or their offspring. As Wilson (1975a, p. 108) puts it, "But those of the morphologically most advanced forms are completely helpless, having been transformed into specialized fighting machines capable only of conducting efficient slave raids." Thus, although this may be considered the first step in the degeneration of the social parasites, such specialization boosts their martial effectiveness, at least in the short term.

The seemingly final step in this progression was the evolution of *degenerate* or *inquiline* interspecific slavery, in which the slave-making "warrior" caste is either mostly or entirely lost (as seen in various species of the genera *Temnothorax, Tetramorium,* and *Epimyrma*) and it is instead the slave-maker queen that murders the host queen (as by decapitation) and enslaves her colony by simply usurping it (Engels, 1990; Parmentier, 2020; Winter & Buschinger, 1983). This usurpation is often accomplished by chemical deception using pheromonal mimicry instead of violent slave raids (Guillem et al., 2014). Rabeling (2020, p. 12) has described the anatomical and behavioral degeneration known as *inquiline syndrome* to "include important modifications associated with the socially parasitic life history, such as reduced mouthparts, reduced body size, smooth and shiny cuticle, reduction or absence of the worker caste, intranidal mating with close relatives, and polygyny." Wilson (1975b, p. 32) once again puts it most eloquently: "The evolution of social parasitism in ants works like a ratchet, allowing a species to slip further down in parasitic dependence but not back up toward its original free-living existence." He later adds: "Together they display an evolutionary descent that begins with casual raiding by otherwise free-living colonies, passes through the development of full-blown warrior societies, and ends with a degeneration so advanced that the workers can no longer even conduct raids" (Wilson, 1975b, p. 35).

Another interesting aspect of this analogy to the history of Spartan-Messenian relations is the fact that enslaved ants have recently been found to engage in various kinds of effective subversion and resistance (Czechowski & Godzinska, 2015). These behaviors range from direct physical attacks on the slave-makers (which presumably reduces their ability to raid nearby nests of the slave species, who are potentially related genetically to the rebellious individuals) to slave reproduction within the slave-maker colony (as by workers laying male eggs), to "sabotage" of the slave-maker colony (as by surreptitiously killing their brood), to slaves fully emancipating themselves from their enslaved condition (as by either taking over the slave-maker colony or escaping, founding a separate colony, and recruiting conspecific queens). These behaviors are not just isolated incidents but have been found to have a widespread distribution (across at least four US states) in the host species *Temnothorax longispinosus* and the obligate slave-making parasite *Protomognathus americanus* (Pamminger et al., 2012). The ensuing mortality can reach as much as 60% of the slave-maker pupae being reared by the enslaved host species in mixed colonies, compared to only 20% of the host species pupae (also being reared by the host species) in unenslaved or "free" colonies, indicating that there is no lack of competence in the rearing of young by the host species, at least when caring for their own. These behaviors often include the disproportionate killing of slave-maker queen pupae by the rebellious slaves charged with caring for them, with the presumed adaptive function of reducing the local parasite pressure on the enslaved species (Achenbach & Foitzik, 2009).

6 Sparta's Decline in *Asabiyyah* and the Dulotic Progression

Inspired by the entomological research conducted with eusocial insects featuring slave-making dynamics (as a form of brood parasitism) and its corresponding phenotypic sequelae, we propose the presence of similar phenomena observed at the macrohistorical level in Sparta. We reconstructed Sparta's *dulotic progression* as divided into four main stages. The first phase corresponds to the Late Greek Dark Ages[13] in Sparta and culminates with the First Messenian War (Early Archaic period); although it is suspected that during this historical timeframe, Sparta had serfs and slaves, their social roles resembled that of other Greek city-states. Sparta also adopted a new style of warfare during the Archaic period, wherein individuals (*hoplites*) fought as part of a tight formation (*phalanx*), although, at this early stage, Sparta's military specialism did not differ considerably relative to other polities.

The second phase began with the annexation of Messenia and the rise of *Helotage* (slavery) and lasted until 650 BC. This phase was characterized by a greater dependence on serf-slave labor and an increasing level of military specialism on the part of the Spartans. During this period, Spartan sociopolitical and military control of Messenia was not absolute, as evidenced by the *Helot* revolt that eventually led to the Second Messenian War. The third phase of the Spartan *dulotic progression* entailed the end of the Archaic period and culminated in the Corinthian War (395–387 BC). Corresponding to the Early and Middle portions of the Classical period, this third phase evidences Sparta's complete reliance on the activities of *Helots* and *Perioikoi* to sustain its sociopolitical organization. Sparta's military specialism reached the historical zenith of its martial attributes, which contributed to its involvement and victory in the Persian and Peloponnesian Wars, two multi-faction conflicts that consolidated Sparta's hegemony. The last phase overlaps with the aftermath of the Corinthian War and the subsequent Boeotian War (378–371 BC), a conflict that ended Sparta's military and political supremacy in the region (as evidenced by Thebes' victory at the Battles of Tegyra, 375 BC, and Leuctra, 371 BC). During this phase, the contingent of *Perioikoi* in the Spartan army considerably outnumbered the *Spartiate* regiments, a military outcome attributable to *Spartiates* facing high war mortality rates without being able to replenish their ranks with new *Homoioi*. Thus, by this phase, Sparta had partially outsourced its military caste to the *Perioikoi*, soldiers who were not fully regulated by *Spartiate* ultrasocial institutions.

7 Methods: Sample & Analyses

The present study sampled 60 battles and sieges dated 743 BC to 192 BC.[14] Following Miller and Bakar's (2023) description of *Battle Outcomes*, each battle was also classified according to the enemy's and Sparta's presence of allies in the battlefield. Lastly, a *meta-ethnic* six-point scale was developed measuring the biocultural and geographical proximity between Sparta and each rival faction: (0.0) Rival faction is located in the Peloponnese; (0.5) Rival faction is located in the Peloponnese whereas allies of rival faction are located outside of the Peloponnese but within Greece; (1.0) Rival faction is located outside of the Peloponnese but within Greece; (1.5) Rival faction is located outside of the Peloponnese but within Greece, whereas allies of rival faction are located outside of Greece; (2.0) Rival faction is located outside Greece; and (2.5) Rival faction and its allies are located outside Greece. Information on Spartan's kings (Agiad and Eurypontid dynasties) was collected from Carr (2012). A Sequential Canonical Cascade Model was constructed analyzing time, time^2, the *meta-ethnic* scale, and the number of Spartan rulers per five years (*lustrum*) as main effects, while also analyzing the interaction between the *meta-ethnic* scale and the number of rulers in the model. The analyses then examined the contribution of the frequency of enemy coalitions per *lustrum*, and the frequency of Spartan coalitions per *lustrum*, on the frequency of Spartan victories per *lustrum*. This statistical procedure is computed following a theoretically specified sequence where a criterion variable in a step of the analyses operates as a predictor in an ensuing stage (Figueredo et al., 2017). All models were estimated using the statistical platform *UniMult 2.0* (Gorsuch, 1991).

8 Results & Measurement Models

The Sequential Canonical Cascade Model (SEQCA) reached statistical significance and explained 61% of the variance (Table 9.1).[15] The ominibus test for the first step of the model did not reach statistical significance; thus, no further interpretations are provided. Although the frequency of Spartan coalitions was not significantly predicted by that of enemy coalitions or the linear effect of time, the quadratic effect of time (time^2) had a negative and significant influence on Spartan coalitions, indicating a curvilinear function shaped like an inverted letter *U*. These results suggest that either opportunities or motivations to establish coalitions and alliances between Sparta and other Greek polities remained relatively low between the Eighth and the Sixth Centuries BC. A considerable increase

Table 9.1 Sequential Canonical Cascade Model examining the influence of time, time^2, a meta-ethnic scale, enemy coalitions, Spartan coalitions, on frequency of Spartan victories

Model	Effect size (E)	90% CI	*F*-value	df1, df2	*p*-value
Overall (*V* = 0.748)	0.61	0.00,1.00	2.95	15, 75	0.001
Variables	**Effect Size (sR)**	**90% CI**	***F*-value**	**df1, df2**	***p*-value**
Y variable: *Enemy coalitions*					
Time	-0.17	-0.51, 0.22	0.92	1, 25	0.35
Time^2	-0.08	-0.44, 0.30	0.24	1, 25	0.63
Meta-ethnic scale	0.37	0.00, 0.65	4.63	1, 25	0.04
Number of rulers	0.10	-0.28, 0.45	0.31	1, 25	0.58
*Meta-ethnic scale*Number of rulers*	0.25	-0.13, 0.57	2.12	1, 25	0.16
Multiple R (Xs only)	0.50	0.00, 1.00	1.64	5, 25	0.19
Residual: Mean = 0.00; SD = 0.96; Skew, Kurtosis = 1.03, 2.40; Range = -2.15–3.15					
Y variable: *Spartan coalitions*					
Prior Y variable					
Enemy coalitions	0.22	-0.17, 0.55	2.37	1, 24	0.14
X variables					
Time	-0.01	-0.38, 0.36	0.01	1, 24	0.94
Time^2	-0.38	-0.66, -0.01	7.04	1, 24	0.01
Meta-ethnic scale	0.54	0.21, 0.76	14.21	1, 24	0.0009
Number of rulers	0.05	-0.33, 0.42	0.14	1, 24	0.72
*Meta-ethnic scale*Number of rulers*	0.18	-0.20, 0.52	1.61	1, 24	0.22
Multiple R (Xs only)	0.68	0.37, 1.00	4.6	5, 24	0.004
Residual: Mean = 0.00; SD = 0.63; Skew, Kurtosis = -0.33, -0.09; Range = -1.32–1.03					
Y variable: *Frequency of victories*					
Prior Y variables					
Spartan coalitions	0.42	0.06, 0.69	11.58	1, 23	0.002
Enemy coalitions	0.26	-0.13, 0.58	4.32	1, 23	0.05
X variables					
Time	0.26	-0.12, 0.58	4.43	1, 23	0.05
Time^2	-0.18	-0.52, 0.20	2.16	1, 23	0.15
Meta-ethnic scale	0.20	-0.18, 0.54	2.70	1, 23	0.11
Number of rulers	-0.23	-0.55, 0.16	3.29	1, 23	0.08
*Meta-ethnic scale*Number of rulers*	-0.45	-0.71, -0.10	13.3	1, 23	0.001
Multiple R (Xs only)	0.63	0.38, 1.00	5.18	5, 23	0.003
Residual: Mean = 0.00; SD = 0.71; Skew, Kurtosis = 1.41, 3.81; Range = -1.20–2.60					

in Spartan coalitions occurred during the fifth and fourth centuries BC, declining after the third century BC. The model also revealed a positive and significant influence of ethnolinguistic distance (the meta-ethnic scale) on the frequency of Spartan coalitions. The analysis indicated that the frequency of Spartan victories was positively, significantly, and directly predicted by the frequency of Spartan coalitions, and thus only indirectly predicted by that of the ethnolinguistic distance. The model also revealed a negative and significant interaction between the ethnolinguistic distance and the number of rulers on Spartan victories, suggesting that lustrums featuring multiple Spartan leaders and frequent military interactions with bioculturally distant rivals reduced the frequency of Spartan military victories. A unit-weighted estimation also indicated the presence of a general Spartan sociopolitical factor that loaded positively onto the frequency of Spartan victories ($\lambda = 0.81, p < 0.0001$) and negatively onto the number of rulers ($\lambda = -0.81, p < 0.0001$). Figure 9.1 features the manner in which the Spartan sociopolitical factor scores varied as a function of time.

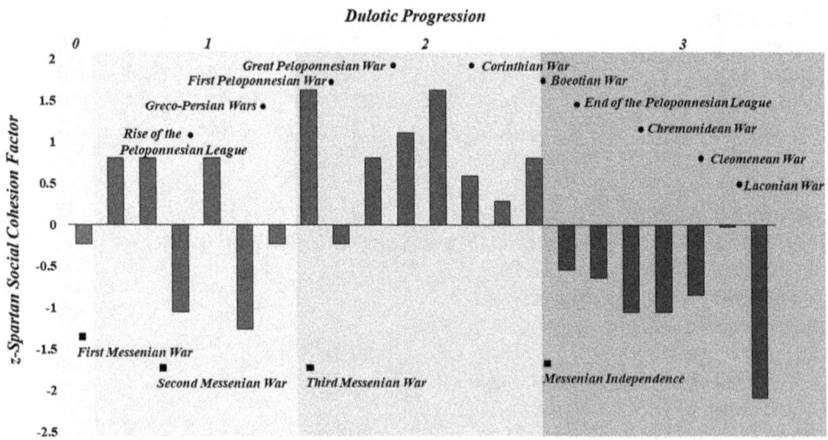

Fig. 9.1 Standardized Spartan Social Cohesion factor scores fluctuating as a function of time (years BC)

9 Discussion

Although chroniclers and historians have described the rise and decline of Sparta in great detail, few studies have examined these macrohistorical dynamics through an evolutionary and quantitative lens. Inspired by the *dulotic progression* model, we see that the same institutions contributing to Sparta's supremacy (the emergence of a professional army maintained in part due to the economic and social activities of *Helots* and *Perioikoi*) also ordained its eventual downfall (Sparta's outsourcing its military due to the declining number of *Spartiates*). The present chapter provided an historical review and empirical examination of Spartan ultrasociality, wherein high meta-ethnic scores, hence greater biocultural distance between Sparta and the corresponding rival factions, favored the establishment of Spartan coalitions and alliances, which in turn predicted Sparta's military victories. In contrast, the interaction between the number of Spartan rulers and the meta-ethnic scale reduced the frequency of Spartan victories. Although the sparsity of available data precludes estimating polynomial functions, Fig. 9.1 suggests an inverted U function spanning from the Messenian enslavement (ca 670 BC) to the Laconian War (195 BC), with the Spartan fortunes first peaking (with greater military specialization) and then declining (with degeneration). This is consistent with the *dulotic progression* model.

Although the data displayed in Fig. 9.1 might make it easy to suggest that the decline of Sparta was entirely due to the degenerative consequences on Spartans of the enslavement of Messenians, we also note that the period of peak Spartan military dominance, peak *Helotage,* coincided with the highest number of outgroup alliances as well as the highest number of *Spartiates*. Similarly, the period of declining Spartan military dominance and the emancipation of the *Helots* also coincides with a decreased number of outgroup alliances as well as the declining number of *Spartiates*. Thus, it is difficult to infer any direct or indirect causal relations between these contemporaneous processes. We may speculate that the specialization of the *Spartiate* population into military functions was made possible by consigning the other necessary social roles to the *Helots* and *Perioikoi* during the period of peak Spartan military dominance. This military specialization might have led to both greater mortality and lower reproductive rates (below replacement levels relative to mortality, perhaps due to decreased reproductive effort) among the *Spartiate* class, fostering even greater reliance on outside auxiliaries and promoting what might be characterized as a vicious feedback cycle. Whether there was any biological or cultural degeneration of the *Spartiate* population during this period of a few centuries (as predicted by the *dulotic progression* theory) is impossible to determine with the present data. Perhaps archaeogenetic studies in the future might be able to address this question. Either way, this extended analogy remains speculative.

10 Conclusions

Obviously, the biological details of slave-making in ants cannot be generalized too literally to human societies. However, increasing dependence and the consequent degeneration of the slavers parallel the loss of independent physiological functions by parasitic species; a very common, although not wholly universal, outcome of the parasitic lifestyle throughout the animal kingdom (e.g., Kern, 1915; Lewis et al., 2002). As Helanterä (2021) has pointed out, however, there are multiple routes to social parasitism in ants, and a single, oversimplified *dulotic progression* cannot capture the full complexity of the evolutionary dynamics of this phenomenon. The same is undoubtedly true of Spartan society. Thus, we cannot be entirely confident that the patterns described here are a complete and comprehensive parallel to the historically documented Spartan decline as it relates to Messenian enslavement. Still, we can merely offer this account as highly suggestive of the possibilities for an evolutionary ecological analogy.

The *dulotic progression* is a specific manifestation of specialization and militarism, which is in turn a manifestation of nearly unalloyed group selection. Competition and territoriality pushed Sparta toward group selected extremes, which was an ultimately unsuccessful evolutionary and cultural trajectory, creating a closed society with reduced biological and cultural plasticity. The same process could be said to have affected nineteenth century Prussia, a bygone European power that spent upwards of seventy percent of its gross domestic product on its military, such that

Count Mirabeau quipped, Prussia is "not a country with an army but an army with a country" (Showalter, 2004). Still, this is a rare fate. Most societies begin with relatively high levels of group selection, but end, not when group-selected traits are intensified, but when group-selected traits erode, as marked by hedonistic norms, corrupt leaders, and increasing evolutionary trends toward individually selected members. As later chapters will develop, societies such as eighteenth century Britain and the United States of America struck highly competitive balances between group and individually selected extremes, pairing group-selected powers with individually selected adaptability that is documented historically in the inclusion of liberal ideology into a truly political dialectic.

Notes

1. (1951; volume III; p. 72)
2. This militant and merciless pressure of Spartiate public opinion, to which the individual Spartiate was subject, was perhaps the most odious feature of Spartan life. A Study of History (1951; volume III; p. 64).
3. This of course ignores many entomological complexities, but refers to caste; and the point is taken. Nevertheless, many advanced social insects, such as honeybees, were subsequently found to be quite capable of both advanced learning and symbolic communication.
4. Reciprocal altruism and indirect reciprocity
5. Although numerous publications have described in detail the sociopolitical structure of Sparta, few studies have examined this historical evidence through an evolutionary lens.
6. Historians continue to debate whether *Tresantes* persistently experienced these humiliations. For example, even though more than a hundred *Spartiates* were dishonored for surrendering at the battle of Sphacteria (425 BC), their punishment was eventually revoked based on their civil and political status (Bayliss, 2022). Similarly, there is no mention of any public denigration of the hundreds of Spartans who surrendered the battle of Leuctra (371 BC; Bayliss, 2022; Cawkwell, 2002).
7. Some debate persists concerning the political and social function of this institution, or even whether it existed, since chroniclers often mention this assembly alongside religious rituals rather than political contexts (Andrewes, 2002; Carr, 2012; Kennell, 2011).

8. Although the evidence suggests that each regiment contained individuals from different age groups, little to no evidence exists on whether members of the same family were assigned to the same subunit (Hodkinson, 2002).
9. Other estimations suggest 35–36% of *Spartiates* and 64–65% of *Perioikoi*.
10. There is some discussion about how distance affected the ability of Sparta to control and integrate the *Perioikoi*: Shipley (2002) argued that the geographical distance between *Perioikoi* communities and Sparta limited the extent to which Sparta could frequently and directly act on *Perioikoi* societies. According to the author, communities closer to Sparta were less likely to feature a local culture than more distant settlements. Ducat (2018) provides a similar interpretation, wherein *Perioikoi* settlements in Laconia (Gytheion, Geronthrai, Las, Oinous, Pellana, and Sellasia) featured institutional and sociocultural similarities with Sparta. In contrast, *Perioikoi* societies located between the Southern portion of the Eastern Peloponnesian coast and the Parnon mountain range (including Anthana, Epidauros, Eua, Kyphanta, Prasiai, Thryrea, and Zarax) remained culturally distinct from Sparta. This process is attributable, in part, to Argos' political influence in the area (Ducat, 2018). In addition to Laconia, *Perioikoi* communities were also situated in Messenia (Asine, Aulon, Kardamyle, Korone, Kyparissia, Mothone, Pharai, and Thouria), albeit little to no information remains concerning their sociopolitical organization (Ducat, 2018). Although several authors often referred to *Perioikoi* organizations as city-states (*poleis*), other researchers argue that *Perioikoi* communities' lower level of sociopolitical complexity indicates that *Perioikoi* frequently inhabited villages and small towns but not *poleis* (Shipley, 2002). Thus, it is unclear whether these communities had any sovereignty over their surrounding territory (Shipley, 2002). Through this lens, the interaction between Sparta and the neighboring *Perioikoi* settlements is reframed from one between city-states to an asymmetrical association of Sparta with towns/villages.
11. Contributions of *Perioikoi* in war remain a subject of intense debate. For example, Lazenby (1985) stated that *Spartiates* adamantly avoided deploying the *Perioikoi* (at least against Peloponnesian rivals), a military doctrine that was abandoned by the time of the Battle of Leuctra (371 BC; Lazenby, 1985). The author's argument rests on the assumption that suspicion and force characterized most *Spartiates-Perioikoi* interactions. Consequently, per Lazenby (1985), *Spartiates* refrained from mustering and deploying *Perioikoi* in battle, and even when Sparta faced a consider-

able decline in *Spartiates* numbers (fifth to fourth centuries BC), inferiors rather than *Perioikoi* were recruited into Sparta's military ranks. Current empirical evidence does support Lazenby's perspective. According to Ducat (2018), the number of inferiors, as a demographic group in Sparta, was insufficient to compensate for the loss of *Spartiates*.

12. For a developmental theory on what behavioral mechanisms made this possible, see Le Moli, 1980.
13. The collapse of Mycenaean kingdoms during the Late Bronze Age (Late Helladic IIIB-IIIC; Cavanagh, 2012, 2018) reverberated over Messenia and Laconia (Cavanagh, 2012, 2018). Peloponnesian towns formerly under the administration of Mycenaean polities experience demographic declines, leading to the abandonment of vast land areas (Cavanagh, 2018). Archaeologists and classists describe this period (~1200 to 800 BC) as featuring severe sociopolitical disruption and anarchy (Cavanagh, 2018) as well as a sharp reduction in sociopolitical complexity. By the end of the ninth century BC, networks of relatively persistent communities emerged between Acarnania, Achaia, Elis, Messenia, and Laconia (Cavanagh, 2018). The historical transition between the post-palatial Late Bronze Age to the Early Iron Age evidenced an array of cultural and political shifts including: (1) The rise and spread of West Greek protogeometric pottery followed by geometric styles (Cavanagh, 2012, 2018); (2) Mortuary practices, such as individual cremation, replacing beehive and chamber tombs (Cline, 2024); (3) the substitution of Linear B script for Phoenician-type inscriptions (Cline, 2024); (4) origin and diffusion of Dorian dialects (Cline, 2024); and (5) metallurgical iron-based innovations (Cavanagh, 2018). The latter archaeological evidence has led researchers to vehemently debate whether these cultural modifications suggest that Dorian groups dispersed (migration or invasion) across the Peloponnese or local communities resettled in these regions during this period of sociopolitical instability (Cavanagh, 2018; Cline, 2024). According to Bayliss (2022), several contemporary archaeologists and historians prefer the term Dorian "*ethnogenesis*" to describe the rise of a relatively homogeneous material culture among Archaic and Classic Peloponnesian communities, circumventing the aforementioned debate. Future archaeogenetic studies are required to determine whether the origins of Dorian groups are to be found within the Peloponnese.
14. Data on Spartan battles were collected from the Historical Conflict Event Dataset (HCED; Miller & Bakar, 2023) and complemented with additional historical sources (Bayliss, 2022; Campbell, 2012; Carr,

2012). The HCED relied on Jacques' the Dictionary of Battle and Sieges (2007) as its primary source. Miller and Bakar (2023) also reviewed several Guides and Encyclopedias on military history, including Ashby et al. (2014), Bennett (2005), Eggenberger (2012), Harbottle (1981), Rogers et al. (2010), and Tucker (2010). Miller and Bakar's final database contained information on 8800 conflicts (1468 BC to 2003 AD). Following previous military datasets (such as the Correlates of War), Miller and Bakar provided information on the various factions involved in each event and the conflicts' outcomes. The database also features information concerning the battles' scale (following Lehmann and Zhukov, 2019), whether one of the factions perpetrated a massacre after the siege or battle, and the *theater* in which the confrontation occurred (land, sea, air, or a combination of the latter three).

15. An alternative SEQCA model with the abovementioned predictors using standardized residuals after controlling for temporal autocorrelations (extracted from a Linear Mixed Model with century as a random effect) reached statistical significance (p = 0.002) and explained 51% of the variance. Residualized enemy coalitions were not significantly predicted by any of the variables in the model. The residualized ethnolinguistic distance positively predicted residualized Spartan coalitions (sR = 0.53, p = 0.003). Lastly, the residualized Spartan victories were positively predicted by residualized Spartan coalitions (sR = 0.32, p = 0.03), enemy coalitions (sR = 0.36, p = 0.01), and ethnolinguistic distance (sR = 0.30, p = 0.04). In contrast, the residualized number of rulers negatively and significantly predicted the residualized Spartan victories (sR = -0.42, p = 0.006).

10

Latin and Turco-Saracen Conflict during the High Middle Ages (AD 1099–1187)

Mateo Peñaherrera-Aguirre, Aurelio José Figueredo, and Steven Charles Hertler

> Among motives for negative association, fear provides the strongest link because it speaks to another bottom line, the fundamental concern with self-preservation. When this concern is heightened, and threats to survival and security loom large, individuals and small social groups find a bond in their common fear that enables them to set their differences aside and unite in the pursuit of goals that are otherwise unattainable.
>
> —Ioannis Evrigenis[1]

1 Introduction

Albeit Latin pilgrimages and military expeditions to the Levant during the High Middle Ages have often been portrayed as archetypal clashes between the putatively monolithic blocks of Christianity and Islam, the

M. Peñaherrera-Aguirre • A. J. Figueredo
University of Arizona, Tucson, AZ, USA
e-mail: ajf@arizona.edu; mpeaher@email.arizona.edu

S. C. Hertler (✉)
Saint Elizabeth University, Morristown, NJ, USA

S. C. Hertler et al., *The Evolution of Political Ideology*,
https://doi.org/10.1007/978-3-031-83941-2_10

historical record describes a dynamic geopolitical landscape wherein political factions cooperated and competed across a *meta-ethnic frontier* (a biocultural boundary between two or more human groups). The present chapter qualitatively reviews these complexities, thereafter quantitatively demonstrating a cyclical oscillation wherein victory and unity are inversely correlated. This chapter's macrohistorical analyses offer empirical evidence that: (1) Turco-Saracen within-group competition facilitated Latin victories; (2) Latin victories over Turco-Saracen factions facilitated Latin within-group competition; and (3) Latin within-group competition facilitated Turco-Saracen victories. This historical period then perfectly illustrates Evrigenis' *negative association* as described in the quote above. A broadly shared cultural heritage and common religion were not always sufficient to keep these groups bound together and clearly demarcated from one another. Triumph over a common foe relaxed bonds; defeat by such a foe strengthened them. When on their heels and behind their battlements, either group assumed the proportions of a monolith; like a sedimentary rock whose layers are bound by pressure. However, victory slowly undermined solidarity, relaxing these pressures, exposing underlying layers of association.

2 Demographic-Structural Theory

The presence of ultrasocial institutions (promoting cooperation among genetically distantly related individuals inhabiting large-scale societies), such as ecclesiastical organizations, provided a necessary, yet not sufficient, sociocultural scaffolding for ingroups to cooperate in competition with outgroups. Were ultrasocial institutions both necessary and sufficient in creating group unity, a neat binary would result. Christendom and Islam would then form impregnable wholes, fighting one another like rival ant colonies. In reality, we have a sort of Punnett Square scenario that plays out over time. The frequency of these dynamics varies as a function of each biocultural group's *Integrative-Disintegrative* cycle.[2] Figure 10.1 visually represents two biocultural hierarchical structures: Latins (L; Principality of Antioch, County of Edessa, County of Tripoli, and the Kingdom of Jerusalem) and Turco-Saracens (TS; Seljuk, Burid, Artuqid, and Zengid Turks; Fatimid Egyptians; and Ayyubid Kurds)

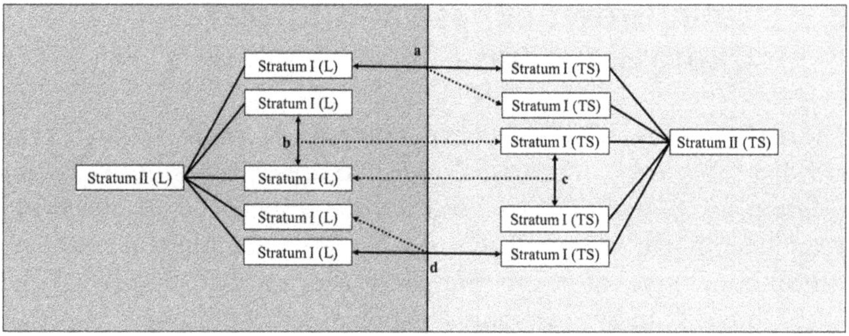

Fig. 10.1 Hierarchical biocultural structures of Latin and Turco-Saracen polities and their corresponding cooperative and agonistic interactions within and beyond a meta-ethnic frontier. Whereas double-headed solid arrows indicate coalitions and alliances between factions, single-headed dotted arrows represent a faction's agonist actions toward a target. See the text for further detail concerning the various dynamics depicted in this image

cooperating and competing within and across a meta-ethnic frontier. The model represents four types of dynamics: (a) Turco-Saracen and Latin coalition against a Turco-Saracen rival; (b) Latin and Latin coalition against a Turco-Saracen rival; (c) Turco-Saracen and Turco-Saracen coalition against a Latin rival; and (d) Latin and Turco-Saracen coalition against a Latin rival. Even though some historical accounts have restricted their review of Latin and Turco-Saracen interactions in the Levant to between-group agonistic encounters (hence, b and c dynamics in the model), these descriptions ignored the presence of coalitions and alliances beyond a meta-ethnic frontier (a and d dynamics). Turco-Saracen and Latin coalitions against Turco-Saracen rivals (a), and Latin and Latin coalitions against Turco-Saracen rivals (b), are expected to occur during Latin integration and Turco-Saracen disintegration stages. In contrast, Turco-Saracen and Turco-Saracen coalitions against Latin rivals, and Latin and Turco-Saracen coalitions against Latin rivals, are predicted to occur during Turco-Saracen integration and Latin disintegration stages.

As we descend into the subsequent section's historical narrative, with its many names and tortuous happenings, it is important to keep an eye on how these four alliance combinations change as a function of victory and defeat.

3 Latin Growth Phase and Turco-Saracen Depression Phase

The decline of the Abbasid caliphate in the tenth century facilitated the growing influence of numerous Kurdish and Arab tribes in the Levant (Morton, 2020). A century later, Turkic tribes expanded across the region *via* military might and conflict instigation, as by reigniting further internal confrontations out of past grievances (Morton, 2020). Although the Turkic presence in Syria was initially limited to tribal organizations that were socioeconomically dependent on practicing transhumant pastoralism, raiding local settlements, and serving as mercenaries for intra-Syrian conflicts, subsequent migratory waves established a more formal and persistent occupation in the region (Morton, 2020). Turkic leaders initially demanded the diplomatic submission of Arab-ruled cities such as Mosul, Aleppo, and Damascus; this practice was later abandoned in favor of conquest and direct Turkic control over these polities. In response to this expansion, Fatimid rulers in Egypt launched several campaigns against the Turks, often recruiting Syrian and Iraqi allies. However, these failed Fatimid expeditions encouraged a Turkic invasion of Egypt (Morton, 2020).

In addition to this northern threat, Egypt faced a combination of drought, famine, and economic disruption that interfered with the state's continuous financing of Turkic mercenaries, culminating in a Turkic revolt (Morton, 2020). Under these dire circumstances, the Fatimid commander, Badr al-Jumali, defeated the northern Turkic force and subdued the rebelling Turkic mercenaries in AD 1077 (Morton, 2020). Seljuk Turks and Fatimid Egypt reached a military and political impasse, leading to the division of the Levant, with the former occupying the northern region and the latter the southern territories (Morton, 2020). A brief period of geopolitical stability characterized the 1080s, wherein Fatimid Egypt successfully suppressed its internal strife and the Seljuk Turks, under the command of Sultan Malik Shah, consolidated dominion over northern Syria (AD 1084–1085; Morton, 2020). Multiple factions conquered and reclaimed Jerusalem during these hostilities (between the late 1070s and 1080s; Lock, 2009). Malik Shah died in 1092, reigniting intense rivalries among the Seljuk nobility, which resulted in several civil wars (Jotischky, 2013; Lock, 2009; Morton, 2020). The death of Malik Shah's brother, Tutush, further aggravated this regional schism. Tutush's sons (Duqaq and Ridwan) inherited the cities of Damascus and Aleppo and proceeded to wage war against each other (Morton, 2020). Although Seljuk military campaigns continued in Anatolia from 1090–1095, Turkic hegemony in Syria and parts of Palestine was noticeably diminished.

The Battle of Manzikert marked a decisive victory of the Seljuk faction over the Eastern Roman Empire in AD 1071 (Jotischky, 2013). In its aftermath and the subsequent expansion of Turkic forces across Asia, Eastern Roman Emperor Alexios I's envoys requested Pope Urban II's assistance (Holt, 2004; Jotischky, 2013). In November of 1095, Pope Urban's pronouncement at the Council of Clermont persuaded European nobles[3] to participate in a military campaign that would reclaim Jerusalem (Asbridge, 2010; Venning, 2015). Godfrey of Bouillon conquered Jerusalem on July 15, 1099, three years after the expedition was launched (Jotischky, 2013; Murray & Nicholson, 2006; Venning, 2015). By July 22, Godfrey was elected ruler of the newly formed Kingdom of Jerusalem.[4] Latin forces annexed other polities in the vicinity. For example, from AD 1100 to 1101, Frankish armies conquered Haifa, Arsuf, and Caesarea (Jotischky, 2013; Morton, 2020). Although Kilij Arslan, the Seljuk sultan, briefly interrupted this series of victories, Baldwin of Boulogne (Godfrey's successor to the throne of Jerusalem; Edgington, 2019) and Raymond of Toulouse led several successful military campaigns against Egypt, Damascus, and Homs (AD 1101–1102; Venning, 2015).

Fatimid Egypt and Damascus established a political and military coalition destined to vanquish the newly formed Kingdom of Jerusalem (AD 1104–1105; Jotischky, 2013), launching several campaigns from Ascalon to the southern borders of the crusader state (AD 1105–1113). However, the conquest of Sidon, the lack of significant Latin incursions into Fatimid Egypt, and the fact that the Crusader states provided a geopolitical buffer between Fatimid territories and Seljuk polities contributed to the downfall of the Fatimid-Damascene coalition (Jotischky, 2013). During the first decade of the twelfth century, the prevalent animosity among Turkic leaders interfered with the success of military expeditions against the Latin states. According to Jotischky (2013), rulers such as Ridwan of Aleppo, Kerboga of Mosul, and Tughtegin of Damascus opted in favor of a tolerant coexistence and potential cooperation with the emerging Latin states in the Levant instead of banding together in supralocal sociopolitical organization. Responses of Turkic rulers to the recently constituted Latin states in Syria and Palestine ranged from open antagonism to cautious interest in potential coalitional partners against other Turkic polities (Barber, 2012; Jotischky, 2013; Morton, 2020).

Furthermore, during the first quarter of the twelfth century, several Turkic leaders preferred not to continue their military campaigns against Latin forces, even after severely crippling their Frankish enemies in battle. For example, instead of laying siege to Antioch as it remained relatively defenseless following the Battle of the Field of Blood in 1119 (Morton, 2018), il-Ghazi, the first Artuqid ruler of Aleppo, focused on looting and taking captives from smaller settlements (Barber, 2012). In a similar vein, Balak, il-Ghazi's nephew defeated and seized Joscelin of Courtenay, Count of Edessa, in the vicinity of Saruj in 1122, but did not launch a major military expedition against the county in subsequent years (Barber, 2012). Balak's capture of King Baldwin II[5] at Shenchrig in 1123 followed a comparable outcome, wherein the captors were more interested in negotiating Joscelin and Baldwin's terms of release (Barber, 2012) rather than conquering the Kingdom of Jerusalem. In 1124, Timurtash, il-Ghazi's son, released Baldwin after the king relinquished the fortress of Azaz, paid a hefty ransom, and promised to support Aleppo against his enemies (Barber, 2012). The deaths of il-Ghazi and Balak (respectively, in 1122 and 1124), along with the release of Baldwin II, initiated a brief period of geopolitical stability in the region (Barber, 2012). Even as il-Ghazi and Balak's actions did not severely compromise the integrity of the Latin polities in the Levant, they evidenced the states' political and military vulnerability and provided a sociopolitical foundation for the eventual unification of Turkic territories.

4 Latin Stagflation Phase[6] and Turco-Saracen Depression-Intercycle Phase

Two significant events transformed the balance of power in the region: the seizure of Aleppo[7] and the death of Tughtigin of Damascus. However, rather than moving against Aleppo, Baldwin II focused his efforts on Damascus (Venning, 2015). Ultimately, Baldwin's failed Damascene campaign became one of his last military expeditions, with the king dying in 1131 (Tibble, 2018; Venning, 2015). His successor and son-in-law, Fulk of Anjou, faced the increasing threat of Turkic domination of Aleppo during his rule (Tibble, 2018). In two years, Zengi of Mosul conquered Montferrand and Homs. Although the Latins regained Baynas in 1140 (Tibble, 2018), the death of Fulk in 1143 produced one of the most severe sociopolitical crises experienced by any Latin states since their inception. Melisende, Fulk's wife, and Baldwin III, their son, were crowned in December of the same year (Tibble, 2018; Venning, 2015). Fulk's death and the absence of Joscelin II, Count of Edessa, motivated Zengi to assemble a sizeable Turcoman force and attack his County (Barber, 2012). Joscelin requested immediate aid to the Principality of Antioch and the Kingdom of Jerusalem. Raymond of Antioch delayed his assistance due to a lingering enmity between Antioch and Edessa, and instead waited for Melisende to assemble Jerusalem's army and march it north (Barber, 2012).

Even though the city of Edessa was well fortified, it lacked an organized defense, allowing Zengi to breach its walls (Barber, 2012; Conder, 2013). The Turcoman army sacked the city, with thousands of Edessans falling in the onslaught (Barber, 2012; Tibble, 2018; Venning, 2015). Raymond's and Melisende's forces could not reach the city in time (Barber, 2012). Zengi's intentions of seizing Damascus were arrested when a slave murdered the Atebeg in 1146 (Venning, 2015). News of the fall of Edessa reached Europe, with Pope Eugene III proclaiming a military expedition destined to reclaim the city. Kings Conrad III of Germany and Louis VII of France, alongside other European nobles, heeded the Pope's call and launched the Second Crusade (Jotischky, 2013). Nur al-Din, Zengi's son, decided to continue with his father's expansion plan, which included capturing Damascus. The leaders of the Second Crusade and the High Court of Jerusalem met near Acre, at Palmarea, in 1148 (Tibble, 2018). The war council decided to besiege Damascus instead of reclaiming Edessa or attacking Ascalon (Barber, 2012; Morton, 2020). As the Latins advanced in the city, Mu'in ad-Din Unur of Damascus requested aid to Aleppo and Mosul (Barber, 2012). Although medievalists continue to debate the contributing factors behind the failed siege of Damascus (i.e., intra-Latin treachery or Zangid military might; Barber, 2012), the current historical evidence strongly suggests that the Second Crusade revealed increasing factionalism among Latin leaders and rising cohesion among Turkic forces (Barber, 2012; Jotischky, 2013; Morton, 2020). Even after the disastrous Damascene campaign, Conrad III attempted to persuade the Latin nobility to counterattack (Barber, 2012). The Latin's refractory response forced Conrad to return to Europe and publicly denounced the Latin nobility's apathy (Morton, 2020).

5 Latin Crisis Phase and Turco-Saracen Growth Phase

In AD 1152, Baldwin III insisted on a second coronation, which would legitimize him as the sole ruler of the Kingdom of Jerusalem (Lock, 2009; Venning, 2015). Melisende and her supporters opposed Baldwin III's ascension and rebelled against the crown (Barber, 2012; Tyerman, 2006). Consequently, the Latin states entered into a period of intense civil strife. As Frankish polities struggled to maintain internal cohesion, Nur al-Din continued his expansion and unification campaigns (Barber, 2012; Jotischky, 2013; Morton, 2020). Nur al-Din acknowledged that a protracted siege of Damascus often resulted in a recalcitrant response from the population. Instead, he convinced the Damascenes to rebel against their leaders and surrender the city (Morton, 2020; Riley-Smith, 1999). Following Nur al-Din's unification of most of the Turkic Syrian enclaves and Jerusalem's defeat at the Battle of Jacob's Ford in 1157, Latin leaders opted for reallocating their military efforts toward Fatimid Egypt instead of Turkic Syria (Barber, 2012; Lock, 2009). At the time, the Egyptian caliph had limited influence over governmental affairs; instead, Shawar, ruler of Upper Egypt, and Dirgham, a veteran military leader, were embroiled in direct competition (Jotischky, 2013; Morton, 2020).

In 1163, Amalric succeeded King Baldwin III and decided to attack Fatimid Egypt (Barber, 2012; Jotischky, 2013). Although the newly enthroned Latin king defeated Dirgham, he preferred not to conquer Cairo, allowing the Fatimid commander to remain in power (Barber, 2012). Shawar requested and Nur al-Din supplied military support, sending his Kurdish general Asad al-Din Shirkuh to overthrow Dirgham (Barber, 2012; Jotischky, 2013; Morton, 2020). Dirgham died during the conflict, which allowed Shawar to obtain complete control of Egypt (Barber, 2012). Shawar's suspicion increased as Shirkuh remained in the South and attacked the city of Bilbais instead of returning to Syria (Morton, 2020). Hence, when Amalric marched on Egypt for a second time, Shawar promised more extensive tribute, and pledged to cover Amalric's military expenses if Shirkuh's forces were expelled from Egypt (Barber, 2012).

Although Amalric defeated Shirkuh, the King's triumph was short-lived (Barber, 2012). In 1164, Nur al-Din routed a sizeable Latin army at the Battle of Artah (Lock, 2009), conquered Harim and Banyas, and seized Bohemond III, Raymond III, Joscelin III, and other critical political figures (Barber, 2012; Venning, 2015) (Lock, 2009; Tibble, 2018). Thus, the King of Jerusalem was forced to temporarily abandon his southern campaign to deal with the northern crisis (Barber, 2012). Shawar was assassinated in 1169, possibly on Shirkuh's order (Barber, 2012; Jotischky, 2013; Morton, 2020). Shirkuh, himself, died two months later, allowing his nephew, Salah ad-Din, to succeed him (Venning, 2015). In 1171 Salah ad-Din proceeded to abolish the Fatimid caliphate, establishing the Ayyubid dynasty (Barber, 2012; Jotischky, 2013; Venning, 2015). Although Turco-Saracen polities at last surrounded the Latin states, Salah ad-Din refused to support Nur al-Din (Barber, 2012; Riley-Smith, 1999). Though thwarted by his death in 1174, Nur al-Din planned a military expedition against Egypt in response to Salah ad-Din (Venning, 2015; Lock, 2009).[8] Thus, at the ebb of Christian power came Turco-Saracen disunity.

6 Latin Depression Phase and Turco-Saracen Stagflation Phase

Contrary to past succession crises, the Latin court readily accepted Baldwin IV, Amalric's leprous son, as their king (Barber, 2012; Holt, 2004; Lock, 2009). Among the many subjects requiring the new king's attention (such as implementing several military operations against Muslim territories), none became as controversial as finding Sibylla, the king's sister, a suitable husband (Barber, 2012). Baldwin relied on the advice of Raymond III, Count of Tripoli, to address this marital conundrum (Barber, 2012). Following Raymond's recommendation, Sibylla married William Longsword, the son of William V Montferrat, a noble with close ties to the Holy Roman Emperor Frederic Barbarossa (Barber, 2012; Lock, 2009). Although the marriage was short-lived, as Longsword died less than a year after arriving in the Holy Land, Sibylla and William did conceive a son, Baldwin V (Lock, 2009).

For a time, tense relations between Christians and Muslim states diminished the latent dissent among Latin nobles, allowing Baldwin IV to allocate his military efforts north toward Damascus (Lock, 2009). In response, Salah ad-Din, the Ayyubid sultan, invaded the kingdom's southern borders and marched toward Jerusalem (Lock, 2009). Despite the sultan's superior force, the Latin king defeated Salah ad-Din at the Battle of Montgisard (1177 AD; Hamilton, 2000; Tibble, 2018). In the absence of an immediate external threat, increasing tensions within Jerusalem's court resurfaced. Once again, court scheming ensued, centering on Sibylla's second marriage. This precipitated a political impasse, which culminated in a military expedition organized by Raymond III and Bohemond III of Antioch, demanding Sibylla be wed to Baldwin of Ibelin (Lock, 2009). Facing the Latin incursion, Baldwin IV hastily arranged his sister's marriage to Guy de Lusignan (Riley-Smith, 1973), forcing Raymond and Bohemond to relinquish their attempted coup (Barber, 2012). Although the king's decision brought temporary stability, it also produced the political fault lines that would eventually lead to the downfall of the Kingdom of Jerusalem.

Fearing potential strife within Jerusalem's court, Baldwin proposed a two-year truce to which Salah ad-Din agreed (Holt, 2004; Lock, 2009;

Venning, 2015). However, before the proposed truce was ratified, Reynald of Châtillon (Lord of Kerak) raided a merchant caravan, reigniting the conflict between the crusader states and the Ayyubid sultanate (Barber, 2012; Tibble, 2018; Venning, 2015). Due to his deteriorating health, Baldwin briefly appointed Guy as regent. Yet when Guy inadequately responded to Salah ad-Din's incursions, Baldwin removed Guy and crowned his five-year-old nephew, Baldwin V, as co-king (Barber, 2012). Guy responded with open insubordination, from refusing to open the gates of Ascalon to the king, to massacring Bedouin caravans who were under the king's protection (Barber, 2012). Baldwin IV died at the age of 24 in 1185 (Venning, 2015).

Baldwin V's rule was brief, with the death of the nine-year-old king plunging the Kingdom of Jerusalem into its most severe succession crisis (Holt, 2004; Venning, 2015). Joscelin III, Sibylla's uncle, advised Raymond to return to the city of Tiberias to appease the nobility (Barber, 2012). Raymond abandoned Jerusalem and summoned most Latin nobles to Nablus to discuss the subject of succession (Barber, 2012). Hence, a political schism materialized at the burial of Baldwin V. Several Latin nobles (Joscelin III of Courtney, Reynald of Châtillon, William V of Montferrat), the patriarch of Jerusalem (Eraclius), and the leaders of the military orders (including Gerard of Ridefort, Grand Master of the Knights Templar, and Roger des Moulin, Grand Master of the Knights Hospitaller) were present at the boy's funeral (Barber, 2012). In contrast, most barons attended Raymond's council (Barber, 2012). After the burial, Sibylla claimed she should succeed her son, and the nobility agreed if she elected any co-ruler other than Guy (Edbury, 1991; Venning, 2015). Following her coronation as Queen of Jerusalem, Sybilla defiantly designated Guy as monarch (Lock, 2009; Venning, 2015). In response to Sibylla's and Guy's coronation, Raymond and the barons argued that Isabella's husband, Humphrey IV of Toron, should seize the throne of Jerusalem by force (Lock, 2009; Venning, 2015). Humphrey abandoned Raymond's council in the face of civil war and swore fealty to Sibylla and Guy, forcing the remaining barons to follow suit (Baldwin, 1969; Barber, 2012; Lock, 2009).

As the count of Tripoli refused to acknowledge the new rulers of Jerusalem (Lewis, 2017), Gerard of Ridefort recommended that Guy lay siege to Raymond's keep at Tiberias (Barber, 2012). Facing an imminent

Latin incursion, Raymond warned Salah ad-Din of Guy's military actions. In response, Salah ad-Din indicated his willingness to provide military assistance to Tiberias (Baldwin, 1969; Barber, 2012). Moreover, as a gesture of good faith, Salah ad-Din released several captive knights and promised to support Raymond as ruler of the Franks (Barber, 2012). Balian of Ibelin persuaded Guy to abandon his military expedition against Raymond and volunteered to mediate the conflict between the king and the count (Venning, 2015). Alas, Raymond's conditions (including having the county of Tripoli take control over Beirut) brought the negotiations to an impasse (Barber, 2012). Guy also feared losing the tenuous support of several barons who had expressed resentment toward him as a king (Barber, 2012). Moreover, loyal nobles threatened the crusader states' stability and security. Reynald of Châtillon raided a wealthy caravan marching toward Damascus, breaking the four-year truce between the Latins and Salah ad-Din (Baldwin, 1969; Venning, 2015). As Reynald refused to return the spoils and free the captives, Salah ad-Din assembled his army and promised to execute the warmongering Lord of Kerak (Barber, 2012).

Simultaneously, Gerard of Ridefort, Roger des Moulins, Balian of Ibelin, Reginald Grenier (Lord of Sidon), and the Archbishop of Tyre journeyed to Tiberias to settle with Raymond (Barber, 2012; Venning, 2015). Under Raymond's permission, an Ayubbid contingent (~700–7000 men) passed through Galilee en route to Nazareth (Barber, 2012). The Latin force eventually fissured, with Gerard and Roger's band facing the Ayyubid force at the Battle of Cresson (Lock, 2009; Philips, 2002; Tibble, 2018; Venning, 2015). The battle was disastrous; Gerard and three knights were the sole survivors (Barber, 2012). Raymond witnessed a macabre parade of Ayyubid soldiers marching through his lands with Latin heads mounted on their lances (Barber, 2012; Venning, 2015). Eventually, Balian, Reginald, the Archbishop of Tyre, met with Raymond, who remained emotionally distressed after Cresson. The count agreed to immediately reconcile with Guy (Barber, 2012).

Meanwhile, Salah ad-Din consolidated his position and reinforced his 30,000-strong army with additional volunteers (Barber, 2012). The threat of a full-fledged Ayyubid campaign against the crusader states hastily drove Guy and the Latin barons to muster all available forces (Barber,

2012). Messengers arrived with alarming news that Salah ad-Din laid siege to Tiberias (Baldwin, 1969). In response, Guy summoned a war council to discuss immediate strategies against the Ayyubid incursion. Raymond warned the king against facing the sultan in open battle, even if Tiberias and the count's wife were under siege (Barber, 2012; Holt, 2004; Murray & Nicholson, 2006; Venning, 2015). The count instead insisted that Salah ad-Din's army would withdraw due to severe heat or would eventually fracture, allowing the Latins to launch a counter-offensive (Barber, 2012). Although the council agreed with Raymond's strategy, Gerard of Ridefort and Reynald of Châtillon privately met with Guy, persuading him to pursue Salah ad-Din (Baldwin, 1969; Barber, 2012; Venning, 2015).

On July 3rd, 1187 AD, Guy ordered the Latin army (~18,000 to 20,000 men) to abandon the water-springs of Saffuriya and advance toward Mount Turan (Barber, 2012). Scorching temperatures, the absence of adequate water wells, and the continuous harassment of Muslim mounted archers forced Guy to make camp close to Maskana, a midpoint between Saffuriya and Tiberias (Barber, 2012). The following day, the bulk of Salah ad-Din's army intercepted the Christian forces and hindered their access to the village of Hattin (Holt, 2004; Jotischky, 2013; Lock, 2009; Smail, 1995; Venning, 2015). The Latins' attempt to reach Lake Tiberias devolved into a rout (Baldwin, 1969; Barber, 2012). Although the military contingents of Raymond III, Count of Tripoli, Balian of Ibelin, and Joscelin III managed to break through the Muslim lines, the rest of the army remained surrounded (Baldwin, 1969; Barber, 2012; Morton, 2020; Venning, 2015). In the battle's aftermath, Salah ad-Din captured key Latin leaders, including Guy de Lusignan, Gerard of Ridefort, Humphrey IV of Toron, and William V of Montferrat; and following on his promise, the sultan personally executed Reynald of Châtillon (Baldwin, 1969; Barber, 2012; Lock, 2009; Venning, 20155). In a single day, Salah ad-Din successfully removed the principal military protection of Latin states in the Levant. Numerous Frankish rural settlements were raided and destroyed after the Battle of Hattin, with captives being executed or sold into slavery (Barber, 2012). Christian settlements and fortresses were seized with relative ease, albeit in some instances, minor opposition occurred. In the upcoming months, Salah ad-Din conquered Acre, Ascalon, Beirut, Jaffa, Nablus, Sidon, and Toron, with only Tyre and Jerusalem remaining under Latin rule (Baldwin, 1969; Barber, 2012; Venning, 2015). Salah ad-Din eventually laid siege to Jerusalem, with Balian of Ibelin surrendering the city on 2 October 1187 (Venning, 2015).[9]

7 Statistical Patterns

To expose trends within the foregoing narrative, we turn to a series of statistical analyses.[10] Several Linear Mixed models (with Maximum Likelihood estimation and random intercept) were computed to empirically examine the following sequence after controlling for the effects of time: (1) *Turco-Saracen Within-Group Competition* positively predicts *Latin Between-Group Victorious Competition*; (2) *Latin Between-Group Victorious Competition* positively predicts *Latin Within-Group Competition*; (3) *Latin Within-Group Competition* positively predicts *Turco-Saracen Between-Group Victorious Competition*. A Sequential Canonical Cascade model also examined the previous sequence and incorporated the effects of time as linear and quadratic functions.

The present study also subtracted the *Latin Within-Group Competition* from the *Latin Between-Group Victorious Competition* (*Latin Difference* vector) and the *Turco-Saracen Within-Group Competition* from the *Turco-Saracen Between-Group Victorious Competition* (*Turco-Saracen Difference* vector). These vectors provided information on the frequency of internal strife relative to the number of conflicts (victories) with foreign biocultural groups. As exemplified in Fig. 10.2, the standardized vector for Latin Differences remains predominantly positive (higher frequency of victories during external conflicts compared to internal strife) for the first quarter of a century.

Although Latin states experienced occasional instances of greater frequency of *Between-Group Victorious Competition* relative to *Within-Group Competition*, the predominant pattern between the second and fourth quarter of a century was of prevalent internal instability. Three principal *Latin Within-Group* crises feature values below −1 standard deviation: (1) 1130, corresponding to the ascendency and initial rule of Fulk and Melisende of Jerusalem; (2) 1150, corresponding to internal strife between Baldwin III, and his mother, Melisende of Jerusalem; and (3)

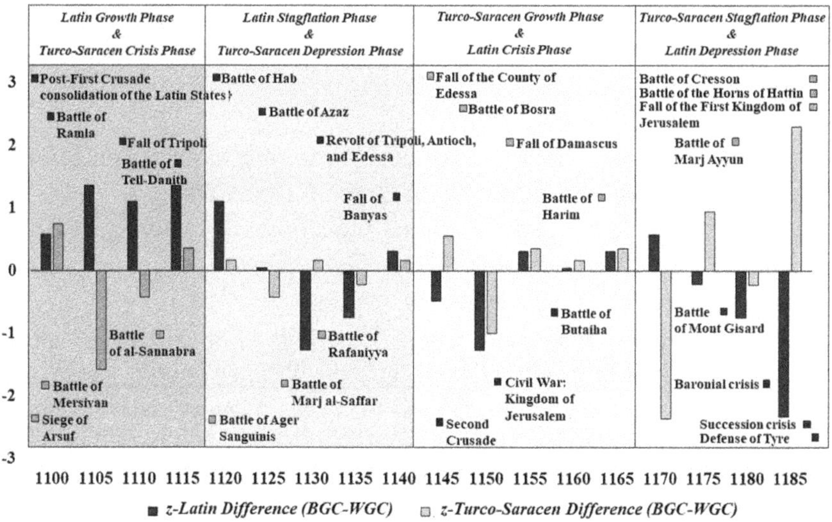

Fig. 10.2 Temporal trend of a standardized Latin Difference vector (*Latin Between-Group Victorious Competition* minus *Latin Within-Group Competition*) and a standardized Turco-Saracen Difference vector (*Turco-Saracen Between-Group Victorious Competition* minus *Turco-Saracen Within-Group Competition*). Latin States include: The Kingdom of Jerusalem, the County of Edessa, and the Principality of Antioch. The County of Tripoli was founded in AD 1102. The color of the squares, respectively, represents critical historical events in Latin and Turco-Saracen polities

1185, corresponding to the succession crisis after the death of Baldwin V further fracturing the political and military integrity within Latin states. In a similar vein, the macrohistorical reconstruction revealed three principal *Turco-Saracen Within-Group* crises: (1) 1105, internecine events involving Ilgazi, Kilij Arslan, Jawali, and Ridwan; (2) Nur-ed-Din and Shirkuh expansion and consolidation campaigns; and (3) Salah ad-Din expansion and consolidation campaigns.

7.1 Linear Mixed Models

Several Linear Mixed models (Maximum Likelihood estimations) were computed to determine the sequential influence of *Turco-Saracen Within-Group Competition*, *Latin Between-Group Competition*, and *Latin*

Within-Group Competition on *Turco-Saracen Between-Group Competition.* The models included Time as a random effect (operationalized in terms of generations).

Step 1: *Latin Between-Group Victorious Competition* was not significantly predicted by *Turco-Saracen Within-Group Competition.*
Step 2: *Latin Within-Group Competition* was positively and significantly predicted by *Latin Between-Group Competition* but not by *Turco-Saracen Within-Group Competition.*
Step 3: *Turco-Saracen Between-Group Victorious Competition* was positively and significantly predicted by *Latin Within-Group Competition*, but not by *Latin Between-Group Victorious Competition* or *Turco-Saracen Within-Group Competition*. These results are further described in Table 10.1.

7.2 Macrohistorical Sequential Canonical Cascade Model

A Sequential Canonical Cascade Analysis (SEQCA) reached statistical significance and explained 52% of the variance. Table 10.2 and Fig. 10.2 describe these results in further detail.

Step 1: *Turco-Saracen Within-Group Competition* was not significantly predicted by either *Time* or $Time^2$.
Step 2: *Latin Between-Group Victorious Competition* was positively and significantly predicted by *Turco-Saracen Within-Group Competition.* The model also detected a significant negative effect of Time, indicating a decrease in Latin victories over *Time*. $Time^2$ positively and significantly influenced *Latin Between-Group Victorious Competition*, above and beyond the linear temporal effects.
Step 3: *Latin Within-Group Competition* was negatively and significantly predicted by *Latin Between-Group Victorious Competition*. Neither *Turco-Saracen Within-Group Competition*, *Time,* nor $Time^2$ contributed significantly to this model step.

Table 10.1 Linear Mixed Models examining the sequential influence of *Turco-Saracen Within-Group Competition, Latin Between-Group Victorious Competition, Latin Within-Group Competition*, on *Turco-Saracen Between-Group Victorious Competition*

Variables	*b*	*Std. error*	*Df*	*t-value*	*p-value*
Y variable: Turco-Saracen Within-Group Competition					
Intercept	6.111	0.870	17.00	7.02	<0.0001
Y variable: Latin Between-Group Victorious Competition					
Intercept	1.630	1.698	8.33	0.96	0.3640
Turco-Saracen Within-Group Competition	0.323	0.185	13.94	1.75	0.1020
Y variable: Latin Within-Group Competition					
Intercept	2.389	2.097	4.44	1.14	0.3123
Latin Between-Group Victorious Competition	0.779	0.218	13.42	3.57	0.0033
Turco-Saracen Within-Group Competition	-0.089	0.167	12.15	-0.53	0.6046
Y variable: Turco-Saracen Between-Group Victorious Competition					
Intercept	3.748	1.343	14.00	2.79	0.0145
Latin Within-Group Competition	0.486	0.197	14.00	2.47	0.0272
Latin Between-Group Victorious Competition	-0.030	0.217	14.00	-0.14	0.8915
Turco-Saracen Within-Group Competition	-0.200	0.192	14.00	-1.04	0.3159

Analyses computed using Maximum Likelihood estimations and including generational Time as a random factor

Step 4: *Turco-Saracen Between-Group Victorious Competition* was positively and significantly predicted by *Latin Within-Group Competition*. Neither *Latin Between-Group Victorious Competition*, *Turco-Saracen Within-Group Competition*, *Time*, nor *Time*^{2} contributed significantly in this final analysis stage.

The main results of this sequential canonical model are simplified and depicted in Fig. 10.3.

In addition to the previous analysis, the present study also examined a rival sequence.[11] The comparison of these sequences strongly indicates that *Latin Within-Group Competition* did not operate as a predictor but as an outcome of *Latin Between-Group Victorious Competition*.

Table 10.2 Sequential Canonical Model examining the influence of *Turco-Saracen Within-Group Competition, Latin Between-Group Victorious Competition, Latin Within-Group Competition,* on *Turco-Saracen Between-Group Victorious Competition*

Variables	*Effect size (E)*	*90 CI, 90% CI 90 CI 90 CI*	*F-value*	*df1/ df2*	*p-value*
Overall (V = 1.084)	0.52	0.00, 1.00	3.85	8/26	0.004
Variables	*Effect size (sR)*	*90 CI*	*F-value*	*df1/ df2*	*p-value*
Y variable: Turco-Saracen Within-Group Competition					
Time	-0.28	-0.68, 0.26	1.26	1/15	0.28
Time²	0.09	-0.43, 0.56	0.12	1/15	0.73
Multiple (Xs only)	0.29	0.00, 1.00	0.69	2/15	0.52
Residual: Mean = 0.00; SD = 3.49; Skew/Kurtosis = 1.07/0.38; Range = -5.06–8.90					
Y variable: Latin Between-Group Victorious Competition					
Prior Y variable					
Turco-Saracen Within-Group Competition	0.41	-0.12, 0.76	23.46	1/14	0.0003
X variables					
Time	-0.50	-0.80, 0.01	34.39	1/14	<0.0001
Time²	0.69	0.29, 0.89	66.49	1/14	<0.0001
Multiple (Xs only)	0.85	0.77, 1.00	5.44	2/14	<0.0001
Residual: Mean = 0.00; SD = 1.65; Skew/Kurtosis = -0.24/-0.55; Range = -2.84–3.15					
Y variable: Latin Within-Group Competition					
Prior Y variables					
Latin Between-Group Victorious Competition	0.47	-0.05, 0.79	5.13	1/13	0.04
Turco-Saracen Within-Group Competition	-0.05	-0.54, 0.47	0.06	1/13	0.81
X variables					
Time	0.35	-0.19, 0.73	2.94	1/13	0.11
Time²	0.32	-0.22, 0.71	2.40	1/13	0.15
Multiple (Xs only)	0.48	0.00, 1.00	2.67	2/13	0.11
Residual: Mean = 0.00; SD = 2.76; Skew/Kurtosis = 0.09/-0.92; Range = -4.45–5.13					
Y variable: Turco-Saracen Between-Group Victorious Competition					
Prior Y variables					

(*continued*)

Table 10.2 (continued)

Variables	*Effect size (E)*	*90 CI, 90% CI 90 CI 90 CI*	*F-value*	*df1/ df2*	*p-value*
Latin Within-Group Competition	0.54	0.05, 0.82	5.96	1/12	0.03
Latin Between-Group Victorious Competition	-0.13	-0.60, 0.41	0.33	1/12	0.58
Turco-Saracen Within-Group Competition	-0.22	-0.66, 0.32	1	1/12	0.34
X variables					
Time	-0.11	-0.59, 0.42	0.24	1/12	0.63
Time²	0.18	-0.36, 0.63	0.64	1/12	0.44
Multiple (Xs only)	0.21	0.00, 1.00	0.44	2/12	0.65
Residual: Mean = 0.00; SD = 2.67; Skew/Kurtosis = 0.55/-0.29; Range = -4.56–6.03					

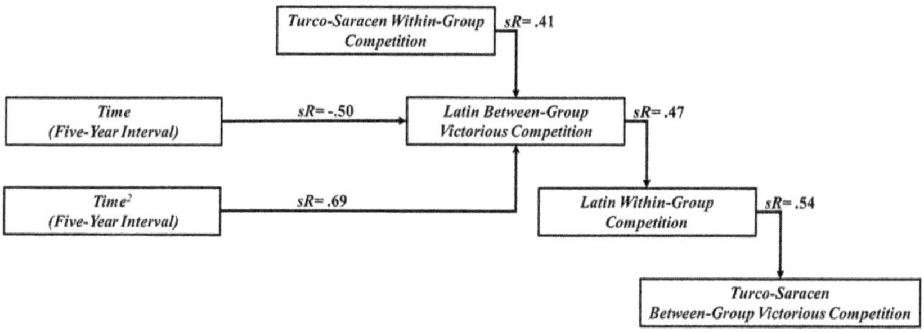

Fig. 10.3 Abridged Sequential Canonical Analysis examining the influence of *Time²*, *Time*, *Turco-Saracen Within-Group Competition*, *Latin Between-Group Victorious Competition*, *Latin Within-Group Competition*, on *Turco-Saracen Between-Group Victorious Competition*

8 Discussion

The current *cliodynamic*[12] examinations revealed that Latin and Turco-Saracen factions experienced asynchronous *Integrative-Disintegrative* dynamics. Waning Seljuk hegemony, the rise of autonomous Turkic polities, and increasing internecine conflict in Fatimid Egypt contributed to the rise and expansion of Latin states in the Levant. At the time, it was not uncommon for Turkic and Fatimid leaders to establish diplomatic and military agreements with the new lords of Outremer against other Turco-Saracen rulers. Hence, coalitions and alliances emerged across the religious and biocultural divide. The present statistical models indicated that: (1) greater *Turco-Saracen Within-Group Competition* increased *Latin Between-Group Victorious Competition*; (2) Latin successes raised the frequency of *Latin Within-Group Competition*; and (3) Latin internecine conflict contributed to *Turco-Saracen Between-Group Victorious Competition*. In other words, unity predicted victory, and victory predicted disunity. Furthermore, examining the statistical fit of alternative models suggested that *Latin Between-Group Victorious Competition* was a predictor, rather than an outcome, of Latin Within-Group Competition. In other words, these juxtapositions between unity in defeat and disunity in victory were causal rather than coincidental.

9 Conclusions

The history of the Latin Crusades in the Holy Land and North Africa only conforms to its typecast role as a classic clash of civilizations when between-group conflict intensified. By contrast, the relaxation of between-group conflict exposed nested levels of allegiance within groups. This is a feature common to all groups. It is not to be forgotten when considering, as will subsequent chapters, wars within civilizations, civil wars, revolutions, party competition, and third-party movements. Also not to be forgotten as we move on to consider modernity with its separation of church and state, is that axial age religions functioned much like political ideologies, for example, in their prefiguration of reality and in their tendency to differentiate ingroups from outgroups. As Azar Gat's *Ideological Fixation* (2022) demonstrates at length, like political ideologies, religious ideologies are nevertheless features of mind expressed in a material reality in which they compete with other bases of group formation in the pursuit of power.

Notes

1. Quoted in *Fear of Enemies and Collective Action* (Evrigenis, 2007, p. xiii).
2. This *Integrative-Disintegrative* cycle references the work of Turchin and Nefedov. According to Turchin and Nefedov (2009), Secular Cycles are classified into *Integrative Secular Trends* (featuring an *Expansion Phase* followed by a *Stagflation Phase*) and *Disintegrative Secular Trends* (evidencing a *Crisis Phase* and a *Depression/Intercycle Phase*). Although Turchin and Nefedov (2009) provided a detailed account of demographic, economic, and sociopolitical variables and the manner through which these change across phases, the present chapter focuses on four principal domains: *Elite Dynamics, State Strength and Collective Solidarity, Sociopolitical Instability*, and *Internal Peace and Order*.

   Per Turchin and Nefedov (2009), the *Expansion Phase* is characterized by: (1) a limited number of nobles or aristocrats; (2) low levels of elite resource consumption; (3) a small elite/commoner ratio; (4) increasing social unity and collective solidarity; and (5) internal peace and increasing social order. In the *Stagflation Phase*, a "Golden age" is reached wherein: (1) elites increase in number, leading to rising intra-elite competition for available positions; (2) some portions of the population display conspicuous consumption; (3) rise of counter-elites; (4) social unity and collective solidarity decline slightly; and (5) slight reduction in internal peace and social order. The *Crisis Phase* exhibits: (1) increasing elite factionalism escalating to open conflict; (2) significant corruption; (3) rising income inequality; (4) social fragmentation; and (5) rebellions, commoner-peasant uprisings, and conflict among the elite. The *Depression/Intercycle Phase* evidences: (1) civil wars, magnicides, and downward-social mobility reduce the number of the elite; (2) reduction of the elite resource consumption; (3) persistent attempts to re-establish sociopolitical stability, followed by disintegration; (4) frequent civil wars; and (5) foreign military incursions.
3. The leaders of this First Crusade included Raymond II of Flanders, Raymond IV of Toulouse, Stephen of Blois, Robert of Normandy, Hugh of Vermandois, Eustace and Baldwin of Boulogne, Godfrey of Bouillon, and Bohemond and Tancred of Taranto (Barber, 2012; Jotischky, 2013; Morton, 2020; Venning, 2015).
4. Albeit the Frank noble refused the designation of King of Jerusalem and instead adopted the title *Defender of the Holy Sepulcher* (Venning, 2015).

5. As the Principality of Antioch requested Baldwin II for military assistance in response to the latter Turkic incursions on numerous occasions, the king remained less than 40% of his time (1119 to 1126) in Jerusalem. Such absence generated open discontent among the Latin nobility.
6. Turchin and Nefedov (2009) proposed the construct of stagflation, describing a rapid increase in prices and rents, allowing the aristocracy to accumulate wealth. The state is then forced to sell elite ranks in response to fiscal challenges, further increasing social mobility and increasing a society's number of elites. Although the present chapter will not focus on the fiscal or monetary policies of Latin and Turco-Saracen states (and consequently, the inflationary dimension associated with the construct of stagflation), this entry will continue to use the term stagflation to maintain consistency with the previous cliodynamic literature.
7. This was accomplished by Zengi, the governor of Mosul.
8. However, as it happened, a month later, Amalric succumbed to a sudden illness (Riley-Smith, 1999), setting the stage for the collapse of the Kingdom of Jerusalem.
9. Even though debate persists among Medievalists concerning the fate of Raymond III of Tripoli, some researchers argue that the annihilation of the Latin army at Hattin severely impacted his health (Lewis, 2017). The count died before the end of AD 1187 (Venning, 2015).
10. Macrohistorical data on within-group competition of Latin (County of Edessa, County of Tripoli, Principality of Antioch, and the Kingdom of Jerusalem) and Turco-Saracen polities (Seljuk Turks of Anatolia, Turkic rulers of Aleppo and Damascus, Egyptian Fatimids, and Ayyubids), and between-group competition of Latin and Turco-Saracen polities, were collected from Venning (2015).
11. *Model 2*

    (1) *Time* and *Time*^{2}
    (2) *Turco-Saracen Within-Group Competition*
    (3) *Latin Within-Group Competition*
    (4) *Latin Between-Group Victorious Competition*
    (5) *Turco-Saracen Between-Group Victorious Competition*

This alternative model considers the influence of *Latin Within-Group Competition* on *Latin Between-Group Victorious Competition*, wherein a lower frequency of competition amongst Latin factions is predicted to increase the likelihood of success against Turco-Saracens. The macrohistorical model comparison revealed that relative to *Model 1* (nine non-significant paths), *Model 2* featured fewer non-significant paths (eight). Following the latter comparison (analogous to the hyperplane count in factor analysis; Gorsuch, 1983), *Model 1* is considered more parsimonious than *Model 2*. These results strongly indicate that *Latin Within-Group Competition* did not operate as a predictor but as an outcome of *Latin Between-Group Victorious Competition*.

12. Cliodynamics is a subdiscipline statistically analyzing historic events to unearth cyclical patterns.

11

The Inquisitions

Aurelio José Figueredo, Mateo Peñaherrera-Aguirre,
and Steven Charles Hertler

For there must be also heresies among you, that they which are approved may be made manifest among you.
—1 Corinthians 11:19 KJV

1 Introduction

The present chapter is about the adaptive functions of Inquisitions as ultrasocial institutions. As reflected by the quote above from one of the New Testament Epistles of the Apostle Paul, we advance the proposition that so-called "heretics" are investigated and persecuted for the purpose of reinforcing the political, social, and cultural bonds among the so-called "faithful," thereby delineating *nonconformists* from *conformists*. Often,

A. J. Figueredo • M. Peñaherrera-Aguirre
University of Arizona, Tucson, AZ, USA
e-mail: ajf@arizona.edu; mpeaher@email.arizona.edu

S. C. Hertler (✉)
Saint Elizabeth University, Morristown, NJ, USA

S. C. Hertler et al., *The Evolution of Political Ideology*,
https://doi.org/10.1007/978-3-031-83941-2_11

these persecutions can be shown to mask ulterior motives, such as appropriating the wealth and property of those so targeted, while also serving as instruments of interethnic strife among different biocultural groups competing for sociopolitical supremacy.

Because there have been too many historical Inquisitions to cover in a single chapter, we focus on two specific test cases for these hypotheses: (1) the *Medieval Episcopal* and *Papal Inquisitions* against the heterodox Christian sectarians in the south of France; and (2) the *Royal Spanish Inquisition* of the Early Modern Era against the *Conversos* or "New Christians" of Spain. We show how an examination of these particulars supports the expectations of Multilevel Selection Theory. In both cases, we delve somewhat into the ideological bases of the perceived threats to orthodoxy to better understand what specific characteristics of the targeted groups triggered the sociological equivalent of a strong immunological reaction on the part of the persecuting regimes.

2 Historiographical Polemics Regarding the Albigensian Heresy

The Albigensians (or so-called "Cathars"[1] from the Greek καθαρός, meaning purified, as by fire) were a Christian sect that was deemed heretical[2] by the Roman Catholic Church (Pegg, 2001, 2008). Some scholars (e.g., Given, 2018) have proposed that the Gregorian Reforms (~AD 1050–1080), which centralized the Roman Catholic Church and established it as an institution of much greater autonomy from the secular powers and authority over matters of "belief, practice, and discipline" (p. 6), provoked widespread adverse reactions throughout Europe, leading to the many dissenting popular religious movements of the twelfth century that expressed resistance to this perceived usurpation of power and influence.

Before we delve any deeper into these matters, however, we must address certain polemics in the historiography of this phenomenon that might otherwise derail our entire analysis. Some radically revisionist historians (e.g., Moore, 2012) have denied the existence of any organized or coherent heretical movement, systematic theology, or established alternative ecclesiastical hierarchy in the Languedoc early in this historical period. These skeptics claim that it was instead a social construct created by the forcefully leading questions of the Inquisitors,[3] which were presumably informed by outdated stereotypes of heretical theologies drawn from the Patristic writings of up to a millennium before, and later conjured into being by the understandably hostile reactions on the part of the dissident Christians in the region to these Inquisitions and Crusades.

As evolutionary scientists, we do not subscribe to the *historical particularism* (e.g., Boas, 1896) evidently advocated by many of the radically revisionist critics of the traditional view of the Albigensians. To the extent that we find it consistent with the data, we therefore view this dissident Christian movement in the evolutionary context of previous heterodoxies which show clear homologies that are probably phylogenetic according to both medieval accounts and modern reconstructions (e.g., see Wakefield & Evans, 1991).[4] Thus, we maintain that adhering slavishly to only the bare bones of the existing data is unrealistic.[5] In what follows, we therefore made some educated inferences, as others have done before, regarding certain aspects of the Albigensian religious ideology that might not be directly in the surviving documentary evidence. Reflecting this reconstructed cultural phylogeny, Wakefield and Evans (1991) provide a sketch of the ideological thread they called "*The Dualist Tradition*" (pp. 9–19) that ran through several successive Christian sects.

In support of the traditional view, Wakefield and Evans (1991) also provide the full text of several authenticated works of the self-professed dualists themselves. These include works such as *The Book of the Two Principles* (~AD 1240), which is the longest work of this kind and lays out the philosophical basis of this heterodox theology,[6] as well as *The Book of the Secret Supper* (~AD 1190), also known as *The Book of John the Evangelist* (*Interrogatio Iohannis*), of which we have only a single surviving manuscript. On the other hand, this same book engages in various polemics, not only against the Inquisitions that persecuted them, but also against divergent heretical groups. These polemics provide a glimpse into the heterogeneity and disunity within this heterodox movement.[7]

Arnold (2016), while acknowledging many of the points made by the skeptics, rejects the idea that the entire movement was a collective figment of the fevered imaginations of Inquisitors and Crusaders. He upholds the traditional view that a heterodox religious movement of some kind must have existed to provoke such a reaction, even if somewhat distorted by its persecutors. He also observes that many of the skeptics appear to be motivated by an ideologically leftist animus against what these critics characterize as the overwhelming hegemonic power and domination putatively wielded by both the Medieval Church and State.

3 Origins of the Albigensian Heresy

The *Inquisitor's Manual* of Bishop Bernard Gui (~AD 1324, translated by Shirley, 2006; also in Robinson, 1905) condemned the Albigensians as "Manichaean heretics." In what follows, we examine the historical and evidentiary basis for this claim in light of modern scholarship.

According to conventional wisdom (e.g., Arendzen, 1910; Hopkins, 2001; Widengren, 1946), the third century syncretistic religion of the Parthian prophet Mani merged various elements of Judaism, Mandaeism, Chaldean Astrology, Parthian Magianism, Persian Mazdaism, Zoroastrianism, Zurvanism, Indian Jainism, and Buddhism with Christianity. Other scholars (e.g., Rudolph, 1987), however, have attributed the derivation of most of his doctrines to the *Gnostic*[8] baptist sect in which he was raised (Baumgarten, 1986). Consistent with this view, Stroumsa (2010, p. 286) has stated, "Recent research, however, has tended to emphasize that Judaism, rather than Persia, was a major origin of Gnosticism." Drower (1960, p. xv) has also commented that, "Heterodox Judaism in Galilee and Samaria appears to have taken shape in the form we now call Gnostic, and it may well have existed some time before the Christian era."

Drower (1960) attributes the common origin of both Gnosticism and what is now mainstream Christianity to the survival of adherents to the pre-exilic Israelite cult of the First Jerusalem Temple (purportedly commissioned by Biblical King Solomon) that resisted the centralizing reforms of the legendary King Josiah of Judah (~640–609 BC). These reforms, whether actually mandated by the Biblical King Josiah or instead implemented by others in the post-exilic (Second Temple) period, prohibited sacrifices outside of the Jerusalem Temple (such as in the previously used "High Places") and instituted a strict monotheistic worship of Yahweh.[9] These reforms were directed at ending many features of traditional Israelite worship, including that of various other gods, and burning, and scattering the ashes of all articles associated with their worship in the Kidron Valley. These other gods included the deities such as Asherah (originally *Ashratah*), also known as the Queen of Heaven.[10] The reforms also included banning child sacrifice, removing statues of the horses and the chariot of the sun that were standing at the entrance of the Jerusalem Temple, and destroying the altars of various other divinities located in the Mount of Olives.

The Israelite religion of the First Temple Period was thus polytheistic, with the national god Yahweh being one of the many sons of the El Elyon (the Most High God). According to this ancient Israelite belief, El Elyon had 72 sons,[11] to each of whom was assigned a different nation, and Yahweh's allotted portion was the nation of Israel (Hess, 2007; Smith, 2003). According to Drower (1960), the dissenting remnants of the pre-exilic Israelite temple cult that gave rise to both Gnosticism and Christianity continued to adhere to these common Western Semitic beliefs and thus frequently referred to "the Lord" Yahweh as the *Son of God* and to "the Most High" El Elyon as *The Father* (Barker, 1992).[12] In contrast, Yahweh and El Elyon had been syncretized into a single God by the aforementioned Deuteronomistic reforms of King Josiah that gave rise to mainstream Second Temple Judaism.

Among the dissenters from mainstream Second Temple Judaism were the so-called *Essenes*, described in the *Judean War* by Flavius Josephus (Mason & Honora, 2008). Lightfoot (1875) cites the *Panarion* of Church Father Epiphanius of Salamis as describing two distinct groups within the Essenes, the Osseans and the Nasaraeans, both of which he says predated Christianity. Drower (1960) and others have pointed out that Nasaraeans (*Νασαραίοι*) or Nazoraeans (*Ναζωραῖοι*) cannot be philologically derived from meaning citizens of Nazareth.[13] Instead, Nasoraean (*naṣuraiyī*) is an Aramaic word used to this day by the Mandaeans to mean a "guardian" or "possessor" of secret doctrines who observes all laws of ritual purity (Drower, 1960). The word Mandaean (*mandayye*) actually means Gnostic in the Mandaic dialect of Aramaic (Thomas, 2007), with *manda* meaning

Gnosis. The Mandaeans still claim John the Baptist as a member of their sect and attribute various Gnostic sermons to him as one of their prophets, claiming that they originated in the Jordan Valley of Israel (Thomas, 2007) but had to flee to Southern Mesopotamia to escape persecution by the Jews during the *First Judaean War*, or revolt against the Roman Empire. In fact, the Gnostic baptist sect in which Mani himself was brought up, the *Elkesaites*,[14] was an offshoot of the Mandaeans.

Furthermore, the New Testament strongly implies that Jesus the Nasoraean started his religious career as a follower of John the Baptist and recruited some of his disciples from that sectarian group, so the designation of Nasoraean for Jesus makes more sense than it would otherwise. In addition, many scholars have concluded, on independent evidence, that John the Baptist was probably an Essene (e.g., Anderson, 2011; Betz, 1990), so that also makes sense given that the Nasoraeans were one of two groups described by Epiphanius as constituting the Essenes (Lightfoot, 1875). In light of the chain of syllogisms being constructed within the present argument, it would follow that Jesus the Nasoraean was also probably a Proto-Gnostic (at least), and possibly an Essene as well, and that the very earliest Christianity was therefore Proto-Gnostic (see Barker, 1995). This conclusion is also supported by several Gospel passages[15] wherein Jesus imparted to his disciples secret doctrines not disseminated to the "multitude," which is a characteristically esoteric Gnostic practice. As Drower (1960, p. 101) speculates, "It might be in Galilee that our original Nasoraeans are to be sought…."[16]

Whatever its origin, the heterodox dissenting movement deriving from pre-exilic, or First Temple Israelite polytheism, soon split into several different factions, including at least the following three major ones:

1. Proto-Orthodox Christianity, which eventually became the state religion of the Roman Empire, evolving and adaptively radiating into most contemporary branches of Christianity (e.g., Catholic, Orthodox, and Protestant)
2. Non-Christian Gnosticism, such as Mandaeism, which rejected the Old Testament and Yahweh, considering Jesus a False Prophet
3. Christian Gnosticism, such as Marcionism, which also rejected the Old Testament and Yahweh, but instead considered Jesus a true messenger of the Most High God[17]

Barker (1992) maintains that in the early days of Proto-Orthodox Christianity, characterizing Jesus as the "Son of God" was interpreted as meaning that he was literally an incarnation of Yahweh. Because Yahweh had traditionally been called both "God" (although not the "Most High God") and "Lord," as well as "Son of God" (meaning son of El Elyon), then Jesus was also bestowed with these same designations. Of course, later Proto-Orthodox Christianity developed more involved Christologies, such as Trinitarian Doctrine, after the First Council of Nicaea in AD 325.

Meanwhile, Non-Christian Gnosticism rejected and reviled Jesus as the False Messiah[18] sent by the evil Creator of the material world, Yahweh. In contrast, Christian Gnosticism interpreted Jesus as an angelic messenger of the benevolent Most High God, who sent this other Son of God to save us from the domination of Yahweh, who was viewed as a malevolent Deity in this religious tradition as well. Nevertheless, because the Albigensians of the eleventh and twelfth century AD clearly fell within the Christian Gnostic camp,[19] we restrict ourselves in what follows to the historical development of those heterodoxies for the purpose of this chapter.

Within the tradition of Christian Gnosticism, it is evident that most of the heterodox doctrines of the Albigensians that were styled heretical appear to have been derived from those of the Marcionites of the second century AD. By most accounts, it was Marcion of Sinope (~AD 100–170) that produced the first edition of the New Testament[20] in the mid-second century AD, which the Proto-Orthodox Church claimed to be plagiarized and redacted from the Gospel of Luke and the Epistles of Paul. Marcion, on the other hand, asserted that the Canonical Gospels were plagiarized and modified from his "true" and authentic version of the "Gospel of the Lord" (Marcion of Sinope, p. 144, translated by Hill, 1891). For his heterodox teachings, Marcion of Sinope was excommunicated from the Proto-Orthodox Christian Church and he and his followers were branded as heretics (Ehrman, 2003, 2005). Nevertheless, at one point in the second century, Marcionites actually outnumbered the Proto-Orthodox Christians across much of the Mediterranean world, especially in Asia Minor (Ehrman, 2005).

Although Marcionism clearly influenced Manichaeism, some have speculated that Manichaeism then served as the missing link between Marcionism and later Christian Gnostic sects, such as the eleventh and twelfth century AD Albigensians (O'Grady, 1994). This indirect progression was presumably mediated by way of the seventh century AD Paulicians of Armenia and Anatolia and the tenth century AD Bogomils of Bulgaria and Macedonia over the major overland trade routes (see Curta, 2019). This reconstructed cultural phylogeny posits that Manichaeism had spread out from Babylonia through most of the Roman Empire, as far West as *Hispania*, and that most Marcionites had then converted to Manichaeism. After the Vandal invasion virtually extinguished Manichaeism in North Africa, the remnants of the Manichaeans in the Eastern Roman Empire then survived as the Paulicians and later Bogomils,[21] and these Bogomils afterwards influenced the Albigensians (O'Grady, 1994; Obolensky, 1972). This reconstructed cultural phylogeny, although somewhat oversimplified,[22] can thus be summarized as follows:[23]

Nasoreans → Marcionites → Manichaeans → Paulicians
→ Bogomils → Albigensians

Like the Marcionites, the Albigensians believed their version of Christianity to be the original, authentic teachings of Jesus, and that of the Proto-Orthodox Church to instead be the heretical version (Ehrman, 2005). Also like the Marcionites, the Albigensians rejected the Old Testament entirely, asserting that the Creator depicted in Genesis was not the Supreme God, and furthermore that this Supreme and Most High God[24] had sent Jesus to save humankind from that Old Testament Deity and his flawed and impossible-to-follow laws (Ehrman, 2005). According to both Marcionites and Albigensians, the Creator depicted in Genesis[25] was a lesser and imperfect divinity equivalent to the Demiurge[26] in Platonism and Neoplatonism[27] and was sometimes even called just that. This Creator was sometimes even characterized as evil, but most often derogated as wrathful and vengeful when his defective and unreasonable commandments were not followed (Ehrman, 2005). In contrast, the Supreme and Most High God was omnibenevolent but more distant and less directly involved with the material world,[28] and it was purportedly this Supreme God who was "the Father" to whom Jesus was referring. Like the Marcionites, Albigensians believed that humans were originally angelic spirits that had been wrongfully trapped by the Old Testament

Creator or Demiurge[29] in material bodies in his flawed creation of the material world. Thus, it was the goal of individual humans, with the salvific aid of Jesus, to combat and reverse this travesty by returning to the spiritual plane from which they ultimately came. Jesus himself was purportedly *docetic* in that he only *appeared* to be mortal, but was instead an immaterial and hence immortal spirit who had no physical body and only *appeared* to suffer and die on the cross. This was purportedly done to deceive the vengeful Old Testament Deity, who maliciously demanded death as punishment for sin, into releasing the spirits of Christ-believers (Ehrman, 2005). In this view, Jesus was thus *not* to be interpreted as the promised Jewish Messiah, which was a future messenger to be sent to the Jews by Yahweh (Ehrman, 2005).

Bishop Bernard Gui (translated by Robinson, 1905) described the so-called Manichaean heretics as rejecting all the sacraments of the Roman Catholic Church, including that of the Eucharist (because Jesus did not have a material body, whereas the Host is a material object and hence unclean) and Baptism with water (which is a material and corruptible substance and hence unclean), while also denying the efficacy of Catholic priests to hear Confessions and absolve sins (given that they themselves were often sinful, avaricious, prideful, and hence unclean and thus unable to make others clean), and not venerating the Cross (which was a grisly instrument of torture and execution). One can easily infer how these various principled objections would constitute a grave threat to the *modus operandi* (some would even say the business model) of the Roman Catholic Church. We might note that although this source is an obviously biased and hostile one, this account is completely consistent with the beliefs explicitly professed by the surviving documents produced by this heterodox religious movement, "Cathar" skepticism and outright denialism notwithstanding.

4 The Medieval Inquisitions and the Albigensian Crusade

Pope Lucius III approved an Episcopal Inquisition against the Albigensian ("Cathar") and Waldensian heretics in AD 1184 and excommunicated them, issuing instructions to his local bishops to search out these heretics two times a year. Pope Innocent III declared heresy to be high treason in AD 1199 and then called for a "Crusade" (then called a *military pilgrimage*) in AD 1209 against the Albigensians in Southern France, offering the lands and other possessions of the so-called heretics to any French nobleman willing to fight against them. As stated in that period by Peter of Vaux de Cernay (translated by Sibly & Sibly, 1998, pp. 37–38), "Catholic inhabitants will take over from the displaced heretics and will serve God in holiness and righteousness according to the tenets of the true faith which you follow."

McGlynn (2015) asserts that "the war was not primarily a confessional one but instead one about the land and its people" (p. 10) and that "this was a war of conquest and regional survival. It was the struggle for the control of land that made the conflict so bitter" (p. 11). The nobles of Southern France were held responsible and condemned for not stamping out heresy in their domains and were thus deemed by their opponents to have thereby forfeited their feudal rights through this perceived negligence. Thus, they were consequently subject to legitimate expropriation. For example, Peter of Vaux de Cernay (translated by Sibly & Sibly, 1998) denounced Count Raymond of Toulouse as "a limb of the Devil, a son of perdition, an enemy of the cross, a persecutor of the Church, the defender of heretics, the oppressor of the Catholic faithful, the servant of treachery" (pp. 24–25) for refusing to punish a heretic.

It has been pointed out that the main leaders of the Southerners, the counts of Toulouse and Trencavel, were Catholic and not (at least openly) heretical, so that their fierce resistance over several decades can only be explained as a sustained defense of their local identity, culture, and folkways instead of any distinct set of religious beliefs (McGlynn, 2015). The path of least resistance (as well as personal survival) for these Catholic counts would easily have been to either persecute their heretical subjects themselves or instead afterwards surrender them to the crusaders and inquisitors. However, even though we now call this region "Southern France," some consider it to have been in essence a separate *nation* at that time, meaning a distinct biocultural group, even though it was not a politically unified *state* (McGlynn, 2015). Various names for this region that are partially synonymous and overlapping over historical time are *Le Pays de Lange d'oc*, *L'Occitanie*, and *La Provence.*

This crusade of the nobles of Northern France against the nobles of Southern France conforms quite well to Turchin's (2003a, 2003b, 2006, 2013) idea of elite overproduction and consequent elite competition. Not only the nobles of Northern France, but also many mercenaries and martial pilgrim commoners (*ribalds*) joined the Crusaders' so-called "Army of God" in its march against the cities of Southern France, looting and pillaging those that would not surrender (and even some that *did* surrender), as well as slaughtering all or most of their populations (Pegg, 2008). Pegg (2008, p. xi) thus interprets the Albigensian Crusade as "a bitter war of national unity" and "a war of religious persecution, colonization, and racial extermination." Although there had been occasional massacres before within the Levantine Crusades, such as in the capture of Jerusalem in AD 1099, this general animus was purportedly something new in the crusading movement (Pegg, 2008). Some historians go so far as to have considered the Albigensian Crusade to have been an act of genocide (Pegg, 2008; Pinker, 2011; Lemkin, 2012), others called it a democide (Rummel, 1994). The total number of Albigensian "heretics" killed has been estimated at anywhere from 200,000 to 1,000,000 (Brookmyre, 2011; Dimont, 1994; Gus, 2001; Ellerbe, 1995; Newton, 1998; Rummel, 1994; Tatz & Higgins, 2016).[30]

5 The Papal Inquisitions

In AD 1215, at the Fourth Lateran Council, Pope Innocent III required all faithful Catholics to testify against any "who differ in their life and habits from the normal way of living of the faithful" (Pegg, 2008, p. 145). Distinctive badges and attire were to be required of heretics, Jews, and Muslims to prevent inadvertent social contact or intermarriage with the faithful, and their possible contamination with the metaphorical pestilence of heresy. Such mandates represented unprecedented requirements for an extreme degree of sociocultural conformism. We therefore interpret this phenomenon as reflecting more of an interethnic conflict than a mere doctrinal dispute.

Pope Gregory IX appointed the recently established Dominican Order ("Order of Preachers") to carry out a Papal Inquisition in AD 1233 against the Albigensians ("Cathars") in Southern France and the Waldensians in Northern Italy. Pope Innocent IV approved the use of torture by the Dominicans to extract confessions from the heretics in the Papal Inquisition in AD 1252. Confessed heretics were required to wear yellow crosses on their clothing as badges of infamy. A second conviction for heresy after a confession carried the death penalty and half of the heretics' property was confiscated by the Church.

The Waldensians[31] were another Christian sect that was considered heretical by the Catholic Church. Like the Albigensians, they rejected the authority of the Church to dictate the content of their preaching, but the Waldensians did not dispute Church doctrines and actually opposed those of the Albigensians (Pegg, 2008). In fact, the Waldensian author Durandus of Osca composed a *Liber antiheresis* in AD 1207 or 1208 accusing the Albigensians of many of the charges of heterodoxy leveled by the Inquisition (D'Avray, 2016). The Waldensians later aligned themselves with Calvinist Protestantism in AD 1532 (Pegg, 2008).

Rummel (1994, p. 61) provides the following assessment of mortality due to these Inquisitions, including 140 at Minerva, 400 at Lavaur, 60 at Casse, 183 at Montwimer, 210–15 at Montségur, and 80 at Barleiges. Even a cursory comparison of the mortality statistics, however, reveals that the casualties of the Albigensian Crusade were orders of magnitude higher than those of the Medieval and the Papal Inquisitions combined.

Both the Albigensians and Waldensians nevertheless challenged the hegemony of the Church, albeit in somewhat different ways, and thus became an obstacle to the creation of a culturally homogeneous nation state by the imposition of religious uniformity. The Dominican Inquisitor Bernard Gui (translated by Shirley, 2006) considered his role to be that of a physician seeking out society's disease. The Papal Inquisitions that followed closely upon the Albigensian Crusade were designed to root out any remnant of doctrinal dissent[32] and impose cultural and doctrinal conformity upon the recalcitrant population of the unruly provinces. Whatever the specific content of these religious beliefs, this requirement for complete conformity led to the cultural and political dominance of Northern France over Southern France. Needless to say, the mass transfer of lands and other properties from the former Southern nobles to the Northern nobles and to the Church reinforced this effect.

6 Expecting the Spanish Inquisition

Limiting ourselves to the case of the Spanish Inquisition, we first review the evolutionary history of antisemitism. MacDonald (1994) made the argument that Judaism could be seen to represent a "group evolutionary strategy" for collective survival and intergroup competition. MacDonald (1998) made the corollary argument that antisemitism therefore represented a counterstrategy implemented by Gentiles to neutralize any competitive advantages gained by Jews by means of implementing their presumed adaptive strategies. Nevertheless, MacDonald (1994, 1998) does not attribute any unusual motivation to Jews in pursuing their adaptive strategies, but merely attributes to them the same drives for survival and reproduction commonly found in other human groups (Figueredo, 1999). The negative (antisemitic) responses by other groups are ostensibly attributable to the relative success that Jews achieve by means of these reproductively relevant cultural practices, which rival groups view as a threat to their own relative success.

Whatever his personal views might or might not be, MacDonald (1998) does not explicitly *justify* anti-Semitism in the book, at least in the moral sense of the word, but merely *explains* why it can be expected to occur and recur over historical time. MacDonald also reviews the diversity of Jewish responses to antisemitism, including the observation that persecution by external agencies may promote group cohesion by increasing the need for collective self-defense (see also Dawidowicz, 1986). MacDonald reviews how persecution may increase the incentives for defection and create divisions regarding how best to deal with such threats, as well as also pointing out how a sudden release from persecution[33] might eliminate any previous incentives for group cohesion (see also Wertheimer, 1993). These observations are all completely consistent with the expectations of Multilevel Selection Theory (Hertler et al., 2020).

There are several major references on the subject that generally agree with the basic dynamics of MacDonald's theory (Cervantes, 1994; Gerber, 1992; Gitlitz, 1996; Kamen, 1998; Netanyahu, 1995). The Spanish Inquisition was a Royal one, unlike the Papal Inquisition in the 4th or Albigensian Crusade. Furthermore, the Spanish Inquisition was not directed at overtly practicing Jews, since they had been expelled from Spain in AD 1492. Instead, several mass conversions of Spanish Jews, including the forced conversions of AD 1391, created a distinct class of *New Christians*, or *Conversos*. The Spanish *Conversos* attracted the attention of the Inquisition because many were suspected of being *Crypto-Jews*, or practicing Judaism in secret while publicly pretending to be Catholic.

Whereas MacDonald evidently treats this Crypto-Judaism as widespread, other scholars (e.g., Netanyahu, 1995) attribute a very low prevalence to this practice. Yet others take a moderate position (e.g., Gerber, 1992; Gitlitz, 1996; Kamen, 1998), attributing the perception of widespread Crypto-Judaism to a lack of adequate instruction in the Catholic religion after forced conversion episodes (such as that of AD 1391). Regardless of their religious beliefs, whether Catholic or heretical, the Spanish *Conversos* continued to be considered as a separate ethnicity or biocultural group by both themselves and others.

The scholarly consensus in this area of research (e.g., Gitlitz, 1996; MacDonald, 1998; Netanyahu, 1995) is that the defining characteristics of this group was not their religious affiliation, but their essentially Jewish culture, customs, folkways, and social ecology. For example, a seemingly simple thing like declining to eat pork continued to bring them under suspicion of practicing Crypto-Judaism, even if they were otherwise sincere, practicing Christians. As nominal converts to Catholicism, the *Conversos* were now permitted to eat pork, but were not accustomed to consuming it in the course of their daily lives. This is strikingly reminiscent of the Medieval Inquisition's super-conformist requirement to inform on any "who differ in their life and habits from the normal way of living of the faithful" (Pegg, 2008, p. 145). Furthermore, some professions from which Jews had been legally excluded (e.g., Medicine and Law) became available to them again after conversion to Christianity. As *Conversos* came to dominate those socioeconomic sectors, they became perceived as a barrier to upward mobility by the emerging Gentile middle classes.

Although these two scholars evidently harbor quite discrepant value judgments in the matter, Netanyahu (1995) and MacDonald (1998) have proposed virtually identical theories of antisemitism as well as of their extension to the later derived anti-*Converso* sentiment. This indicates a striking degree of convergent validity in the fact-based claims that they assert. Although MacDonald takes an explicitly evolutionary approach, Netanyahu takes a sociological and historical approach, deriving the relevant Jewish social, political, and economic strategies from as far back as the Jewish collaboration in the fifth century BC Achaemenid Persian domination of Egypt.

Either way, the use of the Spanish Inquisition against *Conversos* is strongly suggestive of an interethnic conflict, given that by most accounts the vast majority of them were by that time practicing Christianity and not practicing Judaism. As with the Medieval Inquisitions and Crusades against the Albigensians, more than just religious doctrine was at stake. Although only a small minority of those tried by the Spanish Inquisition over three centuries were actually executed for heresy (~1% according to Borromeo, cited in Arie, 2004; ~2.7% according to Levack, 1987), *all* of them were arguably terrorized into cultural conformity to mainstream Spanish customs and folkways (cf., Borromeo, 2003).

7 Comparative Mortality Rates in the Spanish Inquisition

Contrary to popular misconceptions, the Spanish Inquisition infrequently killed those accused of committing heresy. During its earlier stages (1478 to 1530), the Valencian Holy Office accused 2350 individuals of recanting their conversion to Christianity and practicing another faith (90% of these cases concerned *Conversos*; Rawlings, 2008), with 750 people sentenced to death (31.91%). From 1530 to 1609, this regional tribunal prosecuted 4250 individuals (66% being *Moriscos*, or former Muslims that had been converted to Christianity; Rawlings, 2008). A similar pattern occurred during these early stages in other regions. From 1480 to 1488, The Sevillian Office sentenced to death 700 *Conversos* and punished ~5000 accused (Rawlings, 2008). In a similar vein, historical records suggest that the use of torture varied as a function of region and accusation. The Ciudad Real tribunal tried 400 *Conversos* and tortured two accused from 1483 to 1485 (Kamen, 2014). Although the Valencian court initially used torture infrequently, more than 66% of accused *Conversos* suffered torture after the first quarter of the sixteenth century (Kamen, 2014). In Granada, ~7% experienced some form of torture (Kamen, 2014). As the majority of trials were classified as minor offenses, Inquisitorial tribunals refrained from using torture in their examinations (Kamen, 2014).

Contreras and Henningsen (1986) examined close to 50,000 records of trials concerning more than 20 regional Inquisitorial tribunals from 1540 to 1700. This Inquisitorial Database revealed that torture and capital punishment occurred at a considerably lower frequency than previous historical claims (Rawlings, 2008). Following Contreras and Henningsen's work (1986), Rawlings (2008) provided a detailed account of the percentage of individuals accused of major and minor heresies and their respective death sentences between 1540 and 1700. According to Rawlings, a sizeable proportion of individuals in Aragon were accused of minor heresies (n = 15131; 58.4%): (22.7%) *propositions*, entailing instances of public denunciations against the Christian faith; (9.9%) superstition; (8.6%) miscellaneous; (8.3%) acts against the inquisition; (6.1%), bigamy; as well as (2.7%) *solicitations*, including indecency, obscenity, and provocative actions or statements. Accusations of major heresies occurred at a slightly lower frequency (n = 10,759; 41.5%) converted individuals practicing: (28.8%) Islam; (8.8%); Lutheranism; (3.6%) Judaism; and (0.2%) Illuminism. Out of a total of 25,890 trials, 520 individuals were sentenced to death in person (2.0%) and 291 in effigy (1.1%).

Rawlings (2008) describes a similar pattern in Castille wherein a sizeable number of cases were classified as minor heresies (n = 10,683; 56.9%): (33.2%) propositions; (6.6%) acts against the inquisition; (5.6%) bigamy; (5.1%) superstition; (4.1%) miscellaneous; (2.3%) and solicitation. A slightly lower number of cases were categorized as major heresies (n = 8101; 43.1%), wherein converted individuals were suspected of practicing: (17.8%) Islam; (18.4%) Judaism; (6.5%) Lutheranism; and (0.4%) Illuminism. Out of the 18,784 cases, 306 individuals were sentenced to death in person (1.6%) and 487 (2.6%) in effigy. These estimates strongly suggest a considerable decline in capital punishment between the early and later phases of the Spanish Inquisition.

Consequently, the Spanish Inquisition virtually extinguished Protestantism in Spain by the mid-sixteenth century (Kamen, 1998) with only about 100 executions. This relatively low mortality stands in stark contrast to the bloodbath that transpired within several other European nations, such as Germany, France, and Britain, during internal religious conflicts between Protestants and Catholics (White, 2012): Between AD 1618–1648, the Thirty Years' War in the Holy Roman Empire took between 4 and 12 million lives; between AD 1562–1598, the Wars of Religion in France took between 2 and 4 million lives; and between AD 1639–1651, the Wars of the Three Kingdoms in the British Isles took between 315 thousand and 1480 thousand lives. If we instead compare just Inquisitions to Inquisitions, there were between 25 and 50 thousand executions on allegations of witchcraft in Protestant Germany during the Early Modern Era (Robisheaux, 2013). In stark contrast to this, with a few early exceptions in the Basque Country, there were almost none at all in Catholic Spain during this same period (Llewellyn, 1994).[34]

As adherence to the Albigensian Heresy had been previously, Protestantism was treated by the Spanish Inquisition as an indicator of being politically disloyal, if not outright treasonous, as well as possibly being in the employ of a foreign power (Rodríguez-Sala, 2009). As with the *Conversos*, Protestants or their potential sympathizers might have been deterred from straying from the dominant social norms by witnessing showy *auto da fé* spectacles (however infrequent) involving *Conversos* and *Moriscos* as well as Protestants *per se*. Similarly to the Medieval Inquisitions, the adaptive function of the Spanish Inquisition according to multilevel selection theory might also have been the attainment of national unity.

8 Conclusions

The purpose of this chapter is not to morally justify these various Inquisitions, but to investigate their possible adaptive functions from the perspective of multilevel selection theory. To equate adaptive function with moral justification would be to commit the *naturalistic fallacy*. Nevertheless, although both alternative outcomes might be considered morally deplorable, it is evident from these data that the mortality attributable to the Inquisitions was much lower than that attributable to internal religious warfare. Consistent with the expectations of multilevel selection theory, we therefore conclude that the Inquisitions had the effect of suppressing otherwise devastating internal warfare by the application of various, although admittedly unpleasant, means of strict social control. A similar tradeoff applies to political dissent; its suppression is historically less sanguinary than its expression.

Although represented as interethnic conflicts, the Inquisitions covered in this chapter highlight the sometimes subtle difference dividing *within*-group from *between*-group competition. We have characterized these conflicts as being *between* different biocultural groups in both Northern versus Southern French; and Old Christian versus New Christian Spaniards. On the other hand, these two conflicts could also be said to have occurred *within* the French and *within* the Spanish nation-states, respectively. This is because human societies are organized within multiple hierarchically nested levels,[35] so that discriminable subdivisions almost always exist and may sometimes come into conflict. Taking the case of United Kingdom as an illustration, it is composed of England, Scotland, Wales, Cornwall, and Northern Ireland, all of which have experienced wars between them prior to entering into their current union while still retaining some individual autonomy.[36]

Lastly, and most broadly, one should neither allow the elapse of centuries nor the substitution of religious for political content to obscure this chapter's significance. Religious ideology and political ideology are close cousins. Inquisitions are premodern expressions of group selected traits, binding moralities, and strong reciprocity before the age of constitutions and social compacts when religion, perhaps even more than state sovereignty and ethnicity, signaled group identity and loyalty. Just as we have tried to expose latent political content within religious inquisitions, one should not fail to see latent religious content within modern politics. Contemporary political causes can similarly press moral orthodoxies and police secular heresies which can in turn be resisted, recapitulating those familiar tensions between the right leaning, group selected, and the left learning individually selected.

Notes

1. The name *Cathar* did not become widely used to describe them until the late nineteenth century and they instead called themselves *good men (Bons Hommes), good women (Bons Femmes),* and *good Christians (bons Chrétians).* Technically, the term *Cathari* was only applied to the *Perfecti,* who were the spiritual elite that had achieved the highest levels of accomplishment within the movement.
2. From the Greek αἵρεσις, originally meaning a strong and distinctive opinion.
3. Nearly all historians are in agreement that the records kept by most extant sources, such as those of the Inquisition, are at least partially tainted by ideological bias. Nevertheless, this does not mean that no useful information can be extracted from them if the accounts are carefully winnowed through and sorted for relative credibility. This is because many of those sources are quite heterogeneous in content and self-evidently reflect many different points of view, including those of the deponents themselves (Arnold, 2001; Arnold & Biller, 2016).
4. Of course, one must always be mindful of the *caveat* that one cannot expect any cultural tradition to remain in a state of unchanging stasis for nearly a thousand years. Instead, one can reasonably expect it to evolve and adapt somewhat over that long stretch of time.
5. In fact, this "bare bones" approach is not even practiced in most palaeontological museums, wherein the application of the well-validated principles of comparative vertebrate morphology permits the reconstruction of the bones that remain missing from fossilized dinosaur skeletons. Although one should always remain cautious regarding which parts were original and which were reconstructed, often using different colors in museums for the plaster or composite casts, it has proven to be scientifically valuable to infer the overall structure of the skeleton from the existing parts.
6. Which purported solved the so-called "problem of evil" as a logical conundrum.
7. *Sennis (*2016*) has written a very fair and balanced review of the multiplicity of views on the coherence of this movement, ranging from traditional to skeptical.*
8. From the Greek word Gnosis (γνῶσις) or "knowledge." F*or a more complete description of Gnosticism, see Hoeller (*2002*), Martin (*2006*), Pagels (*1979*), and Rudolph (*1987*).*

9. Yahweh was previously worshiped only *henotheistically* (or *monolatrously*) meaning while acknowledging the existence of other gods.
10. The loss of this beloved Israelite female deity is also remembered as the divine figure of "Wisdom" in the books of *1 Baruch* and of *1 Enoch*, as well as in the much later *Pistis Sophia* (Barker, 1992). When the Prophet Jeremiah blames the destruction of Jerusalem on its past worship of the Queen of Heaven, the people make the counterclaim that the calamity instead represented the consequences of having *abandoned* her worship (*Jeremiah* 44: 17–19).
11. These were called the *Bene Elohim* in the Hebrew Bible or Old Testament.
12. As noted above, the Western Semitic beliefs were also shared by the Canaanite and Ugaritic cultures (see Hess, 2007; Smith, 2003).
13. It has been suggested by some (e.g., *Salm,* 2008) that City of Nazareth mentioned in the New Testament never even existed in the first century and was instead an invention by the authors of the Gospels. There was a city named *Sepphoris* near the traditional location of Nazareth, although the Synoptic Gospels never mention it (Casey, 2010). Furthermore, the first extrascriptural evidence of a town called *Nazara* was attested by the third century AD (see the *Church History* of Eusebius, in *Schaff & Wace,* 1890) and an Israeli Arab city called *an-Nāṣira* (Hebrew *Naẕerat*) exists there today, suggesting that the area might indeed have been inhabited. Nevertheless, it is quite plausible that the philologically unlikely derivation of Nasorean from Nazareth might have represented an intentional misdirection by the author of the *Gospel of Matthew* to obfuscate the real (Gnostic) meaning of that word, and that the historically later town of Nazareth was retrojected into the first century AD to fit the narrative.
14. Elkesai (Ἐλκεσσαΐ in Epiphanius) was an Ossaean Essene prophet, who was probably an Aramaic-speaking Mesopotamian Jew (Luomanen, 2011).
15. For example, *Mark* 4: 10–12, *Matthew* 13: 10–13, and *Luke* 8: 9–10.
16. Drower always designates the Mandaean meaning of Nasoraeans by an (M).
17. El Elyon, not Yahweh.
18. *Mšiha* in Mandaean.
19. Especially given that they self-identified as such.
20. Consisting of the Evangelion (*Ευαγγελίων*) and the Apostolikon (*Αποστολικών*).

21. These were the so-called *Neo-Manichaeans*, because their beliefs had evolved somewhat.
22. This is because it is not feasible to attempt to comprehensively enumerate the full evolutionary radiation of Gnostic sects (including Simonians, Dositheans, Menandrians, Sethians, Basilideans, Valentinians, Carpocra tians, etc.) within the present chapter.
23. It must be kept in mind that this is an *evolutionary* cultural progression, and that these various dualist sects were never completely identical.
24. This Most High God was sometimes called "the Source" or "the Stranger" in their literature.
25. This Creator was sometimes identified with Yahweh, other times identified with Satan or the Devil (the latter especially by Mani).
26. From the Greek *δημιουργός*.
27. Some have argued that the conception of the imperfect Creator Deity within the dualistic Christian tradition is more likely derived from Gnostic interpretations of Graeco-Roman philosophy than from Sassanid Persian Zoroastrianism.
28. Like the original El Elyon of the old Canaanite and Ugaritic pantheon, from which the pre-exilic Israelite religion was descended.
29. The *Rex Mundi*, or King of the World, whether called Yahweh or Satan.
30. According to Peter of Vaux de Cernay (translated by Sibly & Sibly, 1998), Count Raymond VI of Toulouse died suddenly of an unexpected illness in AD 1222.
31. Originally called the "Poor of Lyon."
32. Such as rejecting the Old Testament, even when fully accepting much of the New Testament.
33. Such as in the eighteenth century AD emancipation of European Jews.
34. This was because the Spanish Inquisition did not believe that witches actually existed, and treated it as simply a superstition.
35. See Chap. 7, The Collapse and Regeneration of Complex Societies, in Hertler et al. (2020).
36. As will be seen in Chapter Thirteen, even within England, the confluence of between-group competition and religion in post reformation scarcely allowed nonviolent political processes to emerge.

Integrative Metacommentary: Part IV. Cycles of Integration and Disintegration in Multilevel Sociopolitical Coalitions

Aurelio José Figueredo, Mateo Peñaherrera-Aguirre, and Steven Charles Hertler

Cycles of Integration and Disintegration in Multilevel Sociopolitical Coalitions

Part IV presents three historical instances highlighting the dynamics of within- and between-group cooperation and competition. Discussions of multilevel selection often oversimplify the problem as one of *individual* versus *group* selection, as if those were the only two levels of organization involved within hierarchically nested biosocial systems. These three chapters all illustrate how higher- and lower-level groupings, or sociopolitical *coalitions*, can transition between states of cooperation and competition and how those processes codetermine evolutionary outcomes.

Chapter 9 traces the history of Spartan military hegemony in relation to various factors, including the possible role of the enslavement of Messenians in first enhancing military efficacy by permitting martial specialization but afterwards leading to the weakening and diminution of

A. J. Figueredo • M. Peñaherrera-Aguirre
University of Arizona, Tucson, AZ, USA
e-mail: ajf@arizona.edu; mpeaher@email.arizona.edu

S. C. Hertler (✉)
Saint Elizabeth University, Morristown, NJ, USA

the *Spartiate* population by permitting biocultural degeneration. This reconstructed evolutionary history was done based on an extended analogy with the *dulotic progression* in slavemaking ants. A diachronic statistical model, however, also separately estimated the influence of enemy coalitions, Spartan coalitions, as well as the ethnolinguistic distances between the Spartans and their opponents. These results indicate that the rise and fall of Spartan military hegemony was not exclusively attributable to dependence on and then sudden dispossession of Messenian slaves.

Similarly, Chapter 10 traces the history of the political and military confrontations purportedly occurring between the Latin and the Turco-Saracen coalitions of the High Middle Ages. We refer to these higher-order coalitions as Stratum II because they were composed of lower-order Stratum I coalitions, such as the Latin or "Crusader" City States of the Levant. As with the Spartan and anti-Spartan coalitions reviewed in Chap. 9, these two Stratum II coalitions experienced alternating cyclical periods of integration and disintegration. We showed that following the victory of one Stratum II coalition over the other, there followed a period of disintegration of the victorious one, which was consequently followed by the victory of the opposing coalition and then its own disintegration. This pattern was repeated for most of the historical period in question, with Stratum I factions of one Stratum II coalition sometimes allying with Stratum I factions of the purportedly opposing Stratum II coalition against another member of its own. This is a far cry from the traditional "Clash of Civilizations" model so often promoted for the Crusading Era.

Finally, although starting with yet another "Crusade," Chap. 11 delves into the multilevel dynamics of the Medieval and Papal Inquisitions of the twelfth and thirteenth centuries, as well as the Spanish Royal Inquisition of the fifteenth century. Again, we see that supposedly internal conflicts can be reconceived as conflicts between higher- and lower-order sociopolitical coalitions, given that the current "nation-states" of France and Spain were at that time, lower-order groups within them still separated by nontrivial ethnolinguistic distances. These conflicts were all precipitated by the waxing power of higher-order sociopolitical strata asserting unprecedented hegemony over lower-order ones. Furthermore, religious differences were deployed as rationalizations for the establishment of such authority and "national" unity.

Part V

The Formation of Faction

12

Early English Reactions to Organized Opposition

Steven Charles Hertler, Aurelio José Figueredo, and Mateo Peñaherrera-Aguirre

By 1688 conspiracy and rebellion, treason and plot, were a part of the history and experience of at least three generations of Englishmen…By 1700 England seemed to have escaped the danger of arbitrary government only to succumb to political anarchy…by the middle 1720s the English political system had begun to assume the air not only of stability but also of historical inevitability; it had become a child of Time and of Providence, an object of veneration, the Burkean fantasy, and a halo of glory was forming about those muddled, incoherent events of 1688, events that had so very nearly spelt anarchy and ruin to the English nation.

—J. H. Plumb[1]

S. C. Hertler (✉)
Saint Elizabeth University, Morristown, NJ, USA

A. J. Figueredo • M. Peñaherrera-Aguirre
University of Arizona, Tucson, AZ, USA
e-mail: ajf@arizona.edu; mpeaher@email.arizona.edu

S. C. Hertler et al., *The Evolution of Political Ideology*,
https://doi.org/10.1007/978-3-031-83941-2_12

1 Introduction

During the Early Modern Era, political structures within the ecological patchwork of the West erred consistently on the side of despotism from struggling so mightily to subjugate their citizenry and defend their borders. Threats of external conquest were compounded by contested successions, attempted coups, princely rivalries, rebellious mobs, and ecclesiastical disputations that were respectively met with executions, imprisonment, silken cords, and repudiated papal bulls.

England's sovereign, King Charles I, asserted the traditional prerogatives of a divine-right-ruler. He had the will, but not the means. Charles I was then embroiled in a Civil War that, for him, would end by regicide in AD1649. That Civil War would then continue for another two years. Looking back on the dislocating hellscape that remained, Thomas Hobbes famously penned his *Leviathan*, invoking an iron-fisted ruler capable of suppressing faction. But Hobbes was looking to the past for a solution. England's actual future would struggle forward toward a monarchical republic, even as it first vacillated between a failed republic, with Cromwell serving as Lord Protector, and a reactionary return to traditional monarchy, with the accession of Charles II who was succeeded by James II. In the struggle over succession, some attempted to prevent the Catholic James II from acceding to the throne, while others advanced his claim. These vying factions became respectively known as *Whigs* and *Tories*. These were not yet parties, and they did not yet contend civilly. The term *Whig* was an epithet—a Gaelic term hurled at horse thieves. The term *Tory* was likewise an epithet—an Irish label for a papal outlaw.[2] After the so-called Glorious Revolution of AD1688, wherein William of Orange displaced James II, Whiggish constitutional constraints and parliamentary participation were legitimated over and above divine right absolutism. An inflection point was reached. Thenceforth, England had no more revolutions, even as it had eschewed despotic controls and began to tolerate opposition. England entered its metaphorical chrysalis in the 1640s and emerged in the 1680s something different in kind. It is this phase of Albion's history that Chap. 12 scrutinizes; for in it, we see the proverbial missing link—the transitional society standing between early modern absolutism and late modern constitutionalism. The transition to comparatively peaceable ideological debate confined within emergent political institutions only arose following a sequence of climatic, economic, demographic, religious, and ideological change—change which is reviewed for its stage setting functions within the second and third sections of this chapter.

2 Socioecological Preconditions for Political Reform and Revolution

During the Early Middle Ages, slavery in Northern Europe was established as an intergenerational station wherein the offspring of enslaved families contributed to the persistence of this self-sustaining institution (Wyatt, 2021). Slavery sometimes resulted when individuals sold their kin when facing starvation, destitution, or as punishment for a criminal offense (Wyatt, 2021). The rise and spread of penal and hereditary slavery in northern Europe eventually replaced raiding and warfare as the main slavemaking activities (Wyatt, 2021). Similar to those in classical antiquity, activities performed by slaves included many skilled occupations, agricultural tasks, concubinage, and domestic labor (Wyatt, 2021).

As the Early Middle Ages gave way to the High Middle Ages, slavery declined throughout Northern and Central Europe. According to the *Domesday Book* (a survey of livestock, people, and manorial assets commissioned by William I, King of England and Duke of Normandy), slaves constituted approximately 20% of the English population in AD1066. Two decades later, the number of slaves declined to 10% of the population (Wyatt, 2021). King William also set a legal precedent by proscribing slave trading beyond English borders (Hudson, 2012). Albeit their ecclesiastical edict had no legislative authority, the Council of London (AD1102) also denounced slavery within England (Pijper, 1909). Consequently, by the thirteenth century, slavery in the British Isles had decreased considerably in some regions and disappeared entirely in others. Historians and medievalists continue to debate whether medieval slavery persisted as serfdom (in particular when considering indebted servitude or debt bondage) or, due to the nature of manorial and feudal relationships, serfdom ought to be considered as sufficiently different from slave-related institutions (Rio, 2017). The rise of serfdom is often associated with the emergence of feudalism, wherein decentralized authorities managed local lands (Brenner, 1996). Demographic reconstructions indicate that approximately half of the English agricultural territories were classified as servile during the fourteenth century (Bailey, 2014). Due to common law prescriptions, the frequency of serfs and servile land did not increase during the 1300s. In contrast to free tenants, serfs were not allowed to abandon their lord's estate without a formal note of manumission (Bailey, 2014). It was not uncommon for the lord to request monetary compensation for the temporary or permanent alienation of his servile chattels (Bailey, 2014).[3]

Moreover, major climatic disruption (probably caused by increased volcanic activity) in the spring of 1315, and extending to the summer of 1317, led to widespread crop failures, culminating in the *Great Famine* (Alfani & Ó Gráda, 2017; Fagan, 2000).[4] The repercussions of the Great Famine extended across Europe, with disease, theft, infanticide, homicide, and cannibalism persisting until 1321 (Fagan, 2000). The biodemographic impact of these famines also increased European populations' vulnerability to subsequent catastrophes, such as the first outbreak of the Second Plague Pandemic (*Yersenia pestis*, AD 1348; Alfani & Ó Gráda, 2017; Fagan, 2000). Consequently, following the Black Death, England lost 40% of its population, reducing the number of serfs in the British Isles during the Late Middle Ages (Bailey, 2014). Bailey (2014) identified four main factors for the eventual decline of serfdom between the fourteenth and the fifteenth centuries: (1) migration; (2) peasant revolts; (3) economic and demographic dynamics; and (4) manumission.

As mentioned, these demographic, sociopolitical, and economic disturbances did not occur in an ecological vacuum; on the contrary, climatic fluctuations (as evidenced by the early stages of the Little Ice Age, AD1300–1850; Fagan, 2000) were also fundamental contributors to the *Crisis of the Late Middle Ages* (DeWitte, 2015; Fagan, 2000; Jordan, 1997; Tuchman, 1987). Concerning the first factor, historical demographers have identified that the geographical mobility of British populations (such as in Essex) increased before the first half of the fourteenth century

(Bailey, 2014). The drastic decline in population size and the lingering effects of sociopolitical disarray following the Black Death allowed former serfs to settle in novel geographical locations or redefine their terms of employment and landholding privileges (Bailey, 2014). In addition to the latter incentives, migration offered serfs a potential escape from raids, revolts, and other forms of local aggression and oppression (Bailey, 2014). Similarly, as labor became scarcer relative to land, several landlords agreed to reduce their rents as a means to attract and retain potential tenants (Bailey, 2014). In tandem, laborers agitated for higher wages and better working conditions (Bailey, 2014). In England, the Peasants' Revolt of 1381 (Ormrod, 1990) followed from the wage and price controls imposed by the aristocracy in response to peasant demands for higher wages related to the decimation of farm workers caused by plague-related mortality. Generally, famine, disease, and violence also contributed to serfs and peasants' discontent with their landlords; the fourteenth and fifteenth centuries AD witnessed numerous peasant revolts, with some of them severely crippling the sociopolitical stability of English feudal estates.

Another historical development that contributed to the destabilization of the feudal system was the growth of towns and cities and the burgeoning urban middle class, or *bourgeoisie*. We can classify two factors influencing peasant migration to cities: (1) *pull* factors, or those that served to attract individuals toward urban living; and (2) *push* factors, or those that served to drive rural individuals from an agricultural subsistence in the countryside. The pull factors had to do with the growing prospects for a better quality of life in the medieval towns and cities for any peasants that could somehow leave the land to which they had been previously bound. Postan (1975, p. 239) characterized late medieval cities as "non-feudal islands in the feudal seas." Some historians attributed the revitalization, and even wholesale generation, of towns and cities to the financial and mercantile networks established during the Levantine Crusades described in Chap. 10 of this volume. Blaydes and Paik (2015) provide compelling historical evidence concerning the invigoration of commercial activities in Western Europe due to the religious and military pilgrimages to the Holy Land. According to the authors, various Eastern products, such as silk, porcelain, and spices, were mainly demanded in Western European markets. Urbanization in Northwestern European regions increased during the thirteenth century, partly due to a rising mercantile class maintaining commercial ties to the East. This century also evidenced a transition from modest Mediterranean settlements to prominent geopolitical city-states, wherein merchant guilds influenced the political trajectories of these polities (Blaydes & Paik, 2015). The push factors are mostly obvious from our previous discussion of the peasant dislocations and unrest. The climate change events that led to the Great Famine of 1315 mentioned above marked the end of the Medieval Warm Period and the beginning of the Little Ice Age and thus had longer term effects on the destabilization of European agriculture, and not just on the roles of the peasant subsistence farmers within the feudal system. It is important to recognize that the Little Ice Age was not just a stable and persistent cooling of the climate, but a period of great climatic volatility that fluctuated in precipitation as well as temperature, dubbed by some a *chattering climate*. Given that fact, it is easy to see how the lack of predictability in the weather caused repeated crop failures throughout this entire period of time, leading to intermittently rising food prices and food insecurity (Fagan, 2000). In addition, the depopulation of the countryside by the Black Death pandemic favored land-intensive uses such as pastoral animal husbandry over plow agriculture, and this also required less labor to operate[5] (Fox, 1995; Kowaleski, 2014; Voigtländer & Voth, 2013). Both of these influences led to a large-scale loss of employment in the late medieval farming sector in comparison to the commercial sector that was growing in towns and cities. In summary, a combination of push and pull factors contributed jointly to driving erstwhile subsistence farmers out of the countryside and toward expanding urban centers, thereby swelling the ranks of the growing middle class.

As related in Woodley and colleagues (2017), what climatologists call the *Little Ice Age* was largely contemporaneous with what historians call the Early Modern Era. Illustrating the significance of this correspondence, Zhang et al. (2011) reported substantial temporal correlations ($\sim r = 0.50$) between global cooling and societal crises from AD1500–1800, including the prevalence of wars and war fatalities. Serving as mediators, catastrophic agricultural and economic failures produced by the climatic disturbances between AD1560–1660 were probably causal to the *General Crisis of the Seventeenth Century*, which was a period of major sociopolitical upheaval and biodemographic decline (Parker, 2008; Trevor-Roper, 1959). Woodley and colleagues (2017) proposed that food insecurity and consequent financial instability augmented between-groups competition for dwindling resources, and that this, in turn, led to greater group-selective pressure during most of that historical period.

Since their first recorded use in fourteenth century Europe (DeVries, 2017; Needham, 1986), another exacerbating factor has been the increasing availability, decreasing expense, and heightening lethality of portable firearms. These ranged from the unwieldy harquebus to the matchlock, wheel-lock, and then flintlock muskets. The first time firearms were deployed as a main military force was during the Italian Wars of AD1494–1559 (Shaw & Mallett, 2018). With such weapons, a properly armed and trained peasant could bring down a plate-armored and mounted aristocrat ("knight") with a single well-aimed shot. The main barrier to such use was initially the expense of the firearms, but in many so-called "peasant revolts" (including the eponymous one of AD1381 mentioned above, as well as the later *Cade's Rebellion* of AD1450), it has been strongly suspected that the peasants and townsfolk were secretly financed and armed by various trade guilds, merchants, and local aldermen (e.g., Robinson, 2009). Firearms also featured prominently in the Hussite Protestant Rebellion in Bohemia against the Holy Roman Empire and Catholic Church from AD1419–1436 (McLachlan, 2010).

Either way, there was a *proletarianization* of warfare (Leeds, 1979) starting in the Early Modern Era, possibly as a partial result of more commoners being recruited into armies to support the many wars that were breaking out throughout the period. This escalation of recruitment was made necessary by the increasing size of armies during the period. For example, the size of Spain's armed forces increased from a few tens of thousands to many hundreds of thousands during the late fifteenth century (Hume, 1907). The same phenomenon was also occurring during the seventeenth century in states like Britain and Holland, perhaps in competition with the prior Spanish increases in military force. These changes in the conduct of warfare have been referred to cumulatively as constituting the *military revolution* (Parker, 1988). According to Parker's version of this theory, there were two innovations that kicked off this revolution: (1) portable firearms; and (2) improved fortifications against bombardment by artillery. On the part of the attackers, these enhanced fortifications made them more difficult to capture, purportedly necessitating the increases in army size described above. On the part of the defenders, the construction and maintenance of artillery fortifications also required greater manpower. On the part of both attackers and defenders, these developments all required more extensive and sophisticated logistical supports and those in turn necessitated even more added manpower. Thus, the sizes of armed forces increased by an order of magnitude in several European nation-states between AD1500–1700.[6]

It is therefore likely that once a large enough number of commoners were trained in the use of firearms and had increasing access to them, it became impossible to put the proverbial cat back in the metaphorical bag. It is therefore tempting to speculate that the democratization of politics during this historical period followed indirectly from the *democratization of violence*, rephrased thus because not all the increased conflict was to be found in the context of warfare between nation-states, but much of it was instead to be found in the context of internal rebellions and civil wars. The *military* revolution, along with all foregoing factors within this section, therefore set the stage for the various *political* revolutions that defined the era. This section is our analysis of the broader climactic and cultural factors that ushered in Liberal Constitutions during the dawn of the so-called *Age of Revolutions*.

3 Religion & Reformation: The Midwives of Political Parties

Religion predates civilization, originating among small-scale societies. Religion initially aligned with kinship and other markers of group identity in nodal, small-scale societies, but, as groups grew, so did the importance of religion. Given an inevitable inverse relationship between group size and genetic relatedness, religion became a salient cultural marker of ingroup membership (Dunbar, 2013) that increasingly welded together ethnically and culturally distinct subgroups.[7] But religion welds groups together by collecting them into ingroups; and where there are ingroups, it follows that there are outgroups. We therefore have the myth of the *chosen people*, belonging to the god of this or that population. So the psychology of religion's ingroup altruism can, in turn, spur outgroup hostility.[8] Those inside the fold are trusted, favored, accepted, and succored; those outside the fold are shunned or converted if not conquered and subjugated (Norenzayan & Shariff, 2008; Norenzayan, 2013). In Chap. 10, we saw religion intermittently reinforcing bonds among Christians and differentiating them from Muslims, although internecine strife within both camps indicated that this glue did not always hold very tightly in the face of disruptive factional self-interests. In Chap. 11, we saw various Inquisitions' attempts to enforce orthodoxy, purportedly rooting out heretical views, but more often than not favoring one ethnolinguistic group by disfavoring others. In essence, these are the two functions of religion from a multilevel selection perspective—boosting conformity and cooperation within groups; and enabling them to better compete with rival outgroups (Wilson, 2015, 2019).

In this Twelfth Chapter, we view political proto-parties as they formed subsequent to the breakdown of Catholic orthodoxy and Papal authority, with many among the genetically and culturally different European North cleaving from their Southern European counterparts. This period would be of less interest if the *Reformation*[9] had been arrested, stopping with a clean break from Rome. Reformation era rebelliousness and freethinking factionalism were less controlled than medieval Christian heresies. Nevertheless, this different outcome was not from a want of repressive effort, as Foxe's *Book of Martyrs* demonstrates. After the end of the Thirty Years War, facilitated by the Treaty of Westphalia, Catholic nation-states could exist alongside Protestant nation-states; although within certain nation-states, staunch Catholics continued to exist alongside multitudinous Protestant sects.[10] The throne, the only central authority capable of controlling religious factionalism, was occupied by queens and kings holding different religious sympathies and varying levels of tolerance, to the end of creating a vacillating century of religious, regional, and class conflict. During the Elizabethan Era, threat from Rome and continental powers constrained internal debate among protestant England. At that time, England was at war with Catholic Spain, even while Queen Elizabeth did not have a named successor. Once external pressures were relaxed, internecine conflict augmented. Post reformation religious diversity reverberated across an Early Modern England in transition from medieval imposition of orthodoxy to modern tolerance of heterodoxy. With modern notions of separation of church and state years distant, this reality of religious pluralism existing in the absence of tolerance was unleashed into a society stressed by population increase, fiscal strain, and elite competition (Goldstone, 1991); but more than that, religious pluralism festered in the cleavages between England, Scotland, and Ireland, and became interwoven in several succession crises and struggles between kings and nobles. Many within the Stuart line were Catholic sympathizers, which created conflict with Protestant aristocrats. Charles I married a Roman Catholic to the ire of Puritans in Parliament. There followed the English Civil Wars,[11] the decapitation of the King, and the sanguinary instability of the interregnum, which was only ended when Charles II, eldest son of Charles I, was enthroned. Charles II was succeeded by James II, his younger brother. In large measure because of his Catholicism, James II was chased from the throne three years later by William and Mary in the *Glorious Revolution* of 1688. The instability was contained, but continued as the *Jacobites*, partisans of Catholic James II, pressed for the restoration of James II's son and, more famously, his grandson, referred to as *Bonnie Prince Charlie*. This combined religious and regency crisis did not really end until 1746 when the Duke of Cumberland defeated Jacobites in the Scottish Highlands in the *Battle of Culloden*. Few epochs are more absorbing to biohistorians or more relevant from a multilevel selection perspective; yet, even more relevant to this particular book is how the emergence of political parties was bound up in this larger process.[12] From this cauldron of religiously and regionally charged conflict among elites, political rivalry replaced physical violence.

4 Constitutionalism & Contractualism: The Preconditions for Peaceful Political Competition

The battle against absolutism began with Magna Carta, and so predates the Reformation by approximately three centuries. Yet, in narrating the progression from Magna Carta to constitutional monarchy, Whig histories elide over the indispensable role of monarchical self-restraint, of which the empowerment of sheriffs against baronial power is an example. Using the Exchequer, Common Pleas, and King's Bench,[13] English Kings slowly replaced provincial, feudal custom with generalized, rational law. These common law precedents and courts, though advanced by English kings to break down local feudal hegemonies, applied to them, limiting kingly prerogative powers. So, by multiple methods, absolutism was restrained; and this process accelerated between 1450 and 1642 during a period of English history described by Alan Cromartie (2006) as a *constitutionalist revolution*. As narrated by Cromartie, through this time period, evolution in thought had ended up with the revolutionary "…idea that arbitrary power was something of its very nature illegitimate" (Cromartie, 2006; p. 203).[14] Laws and procedures were built on a foundation of precedent and custom, with both jointly confining monarchical power. These are the constitutional fetters Charles I resisted by, for example, disbanding Parliament and utilizing *ship money*, an extralegal taxation. Such a narrative is, however, partly propagandistic, as is the term *absolutism* itself. Ansell and Samuels (2014; p. 2009) describe a generalized process of revolutions being led by a newly enlarged elite (merging a narrower landed aristocracy to an emergent commercial, high bourgeoisie) attempting to gain "control over the expropriative authority of the state." Some among the bourgeoisie were able to capitalize on changes collectively imparted by demographic increase, ecology, and agricultural advances, to the end of leaving others behind and creating additional levels of stratification. Revolutions grew out of attempts of these high bourgeoisie to consolidate their gains, protecting capital from expropriation. This process was clearly in play. Additionally and most importantly, whatever gains were conserved thereafter, the constitutional revolution discussed by Cromartie ended in civil war and the military dictatorship of Oliver Cromwell.

This trend toward constitutional constraints that would have England become a monarchical republic was most famously resisted by Robert Filmer, a royalist born in sixteenth century England. Informed by Lutheran convictions of human depravity, Filmer wrote against limited monarchy in support of his sovereign, Charles I. The title of Filmer's tract, *The Anarchy of a Limited Monarchy*, illustrates the Hobbesian fears underpinning his conservatism.[15] On the other side, arguing for legal limitations and natural rights in opposition to divine right rule was Puritan divine Phillip Hunton,[16] who enumerated five restrictions on royal prerogative (Ware, 2004):

1. Laws shall not be made except with Parliamentary consent.
2. Fundamental common laws and constitutional laws shall not be violated.
3. Appointment powers shall be limited.
4. The succession and transfer of power is not wholly the king's prerogative.
5. Subjects shall not be taxed without parliamentary consent.

These five restrictions exchange absolutism for separated powers, constitutionalism, and social contractualism (Ware, 2004). Filmer would have none of this. Absolute power followed from divine right. Though admiring him in certain respects, Filmer attacked even Hobbes' fundamental contractualism that made allegiance conditional. Hobbes argued that the right of self-preservation is inalienable, which seems a plain enough assertion; but for Filmer, it is precisely this right of self-preservation that must be surrendered to maintain any real distinction between civil order and the state of nature in which all are at war against all.[17] Filmerian fears came to fruition in Locke who, despite remonstrating about the restraints of reason and temperance, technically sanctioned the war of all against all in his *Second Treatise on Government* wherein he states, "he who attempts to get another man into his absolute power does thereby put himself into a state of war with him" (Cromartie, 2006; p. 278). As Ward (2004, p. 94) states, for Filmer, "auto-interpretation of obligation is the very problem of the state of nature." Inalienable rights, whether restricted to the Hobbesian right to *life*, or expanded to the broader Lockean rights to *life, liberty*, and *property*, implicitly undermine absolute sovereignty. From a Filmerian perspective, a subject's ability to negate any law, order, or directive deemed a violation of natural rights hamstrings a sovereign's ability to "execute war or maintain national defense" (Ward, 2004, p. 93). Is the king *the law*, or subject to *the law*; is law arbitrary or rational? These were stage setting battles. Obviously, the opposing sides for which Filmer and Hunton stood were not respectively conservatives and progressives contending within a delimited political framework over recognizably political issues. Yet, in the end, and to the chagrin of Filmer, absolutism was undermined, which, whatever it might mean for long-term social stability, was the most basic and first liberal success, and one that arguably created reliable, nonviolent rules of engagement within which political debate could later unfold.[18]

Similar to the summary above, Francios Guizot, in his, *The History of Civilization in Europe*, provides a synthetic interpretation of early modern events, all tending toward the freedom of thought that is a precondition for expressing the full range ideological thought. This interpretation connects the reformation with the English Revolution of 1688, which is in turn more broadly connected with the fate of absolute monarchy on the continent. As the interpretation goes, freedom of conscience expressed in Northern Reformation Europe resisted papal authority, with this same spirit of resistance thereafter devolving upon absolute monarchy. William of Orange is cast as the central actor. Guizot frames William of Orange's usurpation of the English throne as merely a strategic move in a broader war against absolutism being waged against Louis XIV. Guizot highlights these circumstances to remind us that progression from absolutism to modern freedoms played out intranationally in revolutionary events, but also internationally in geopolitical events.[19] Laski adds important material explanations to this narrative of post-revolution stabilization. Political stability was partially a product of economic diversification and population growth. It seems that these factors jointly functioned to break down regional isolation and wove the English gentry into the economy. Advances in administrative efficiency between AD1660–1715 are another nonpolitical factor nonetheless adding to political stability. Laski understands these material factors as drivers of broader social change. The philosophy of liberalism is manifest for Laski in religious pluralism, national sovereignty, movable wealth of bankers and merchants and manufacturers, scientific thought, progress, and perfectibility—it is from assimilating such social changes that the philosophy of liberalism was called forth. More than most intellectual historians then, Laski reverses the causal arrow, finding philosophical treatises to be reactions and explanations for material conditions, rather than the primary drivers of such change. As Laski asserts (1936, p. 19) liberalism was "a new ideology to fit the needs of a new world."

5 Coming Back to a New Form of Political Stability after AD1688

The French Revolution was a revolution of the *Third Estate*, encompassing petty *bourgeoisie* and commoners arrayed against clergy and nobility, which respectively encompassed the *First* and *Second Estates*. The English Civil Wars, by contrast, were begun and concluded by the propertied upper middle classes as they contended with the King.[20] It was a war between the *one* and the *few*, during which the *many* were courted and to some extent empowered. To this point, Bertrand De Jouvenel's *On Power* (1949) has a chapter entitled *Liberty's Aristocratic Roots*. Ineluctably, though inadvertently, plebeians were courted as allies, and so became sharers in power (Evans, 2006; Vallance, 2015). The general disorder and active recruitment of the lower orders into the fray granted room for maneuver to those seeking broader social change. Brailsford (1961) writes about social agitators and dissidents, referred to as *levelers*, who would lay low societal distinctions and advocate for freedoms. These were men like William Walwyn who would read pagan classics on the Sabbath, who expressed a "readiness to stand alone and defy the majority" (Brailsford, 1961; p. 61). Walwyn possessed the art, new in those days Brailsford (1961; p. 62) reminds us, of "…rousing public opinion and organizing men into parties…." Radicals among rebels, levelers sought nearly universal suffrage, an almost communistic economic equality, unrestrained freedom of conscience, and other innovations that would have made the English Civil Wars a true social revolution on the order of the French Revolution approximately 150 years hence. The *many* never ran away with the revolution. These levelers, after serving as a forceful faction that were perforce used and placated, were then surprised and subdued *en masse*, cracking the back of this movement, ensuring that a third civil war[21] would not ensue, and establishing a protectorate before descending further down the revolutionary road. Nevertheless, progress was made. Though journeymen, servants, and hired laborers, and those receiving alms and aid under the poor laws could not vote, all others were enfranchised. As Brailsford (1961; p. 528) puts it, "distressing though these exceptions seem to us, on the balance the gain to democracy from this advance was revolutionary."

Plumb explains how England found a new form of order at the conclusion of its many revolutions. An oligarchic peerage manufactured under William III in the wake of the Glorious Revolution went far in creating newfound stability. Grants of land and title created a *spoils system*, an exaggeration and corruption of patronage that secured government support by appealing to the greed of its recipients.[22] The creation of this oligarchic faction coincided with indirect executive influence through the creation of *placemen* and *pensioners* (Plumb, 1967). Political stability was generated ultimately by a "sense of possession." Executive institutions were proliferating, and their powers and emoluments were being distributed, "at a rate that was capable of absorbing or obliging, and thereby satisfying, an effective segment of the political nation" (Plumb, 1967; p. 112). Thereafter, a state of stability was furthered by an expanding administrative bureaucracy inclusive of tax collectors, mushrooming at a rate ahead even of tax collection itself. At once, this provided government with patronage power and economic power, both of which fostered stability. A lion's share of these newfound revenues was lavished on the Navy, which in turn became another expression of patronage power, wedding still more to the fate of the state. The army, too, nearly quadrupled its regiments, containing more than 70,000 soldiers by AD1711. In doing so, it became a reservoir for upper class sons whose families purchased commissions (Plumb, 1967). Well-born families could look to foreign service in the growing ranks of a diplomatic core that increasingly aped continental practices, routinizing pay and rationalizing promotion. This simultaneously expanded patronage still further while stitching together parts of the emergent empire. In these many ways, social stability came about because a newly expanded elite strata of society, largely from the ranks of the then swelling high bourgeoise that had provoked the English Civil wars, were now thoroughly insinuated into a bureaucratic state.[23]

6 Contemporaries on Emerging Political Parties

In the years after AD1688 it became increasingly easy to separate out members of parliament by party affiliation.[24] Parties divided on questions of war and peace, toleration of dissent, and treatment of colonies. Tories became associated with independent squires of the countryside advocating free, frequent, and fair elections, taxation that was modest and justifiable, and punishment for corruption and bribery (Plumb, 1967, p. 151). Whigs, by contrast, "had become much more closely identified with both aristocracy and government." They were a smaller, more cohesive political party that adeptly used patronage.

The emergence of political parties was recognized by contemporary actors and authors who struggled to connect the past to their present, differentiate violent factions from nonviolent parties, and to otherwise assimilate the brave new political world. For example, Henry St John, 1st Viscount Bolingbroke (1678–1751), understood the *Court Party* and the *Country Party*, which were respectively for and against absolutism, as progenitors of the *Tory* and *Whig* Parties. Interestingly, as this transformation between Court and Country and Tory and Whig was made, Bolingbroke interpreted the broader agreement on religion and constitution as something of an end to parties. However, this perception has much to do with the early eighteenth century difference between *faction* and *party*. Faction had a more negative connotation, denoting a self-interested group over and above a group pressing political goals from shared ideological commitments (Skjönsberg, 2021). What Bolingbroke was observing as the proto parties of Court and Country transmogrified into Tory and Whig parties proper was the transition from violent competition among groups to controlled ideological debate within a single group.

John Perceval (1683–1748), who became the second Earl of Egmont, wrote a 1743 pamphlet entitled *Faction Detected by the Evidence of Fact*. In this pamphlet, Skjönsberg (2021) explains, Perceval respectively connects contemporaneous *Whigs* and *Tories* with *Republicans* and *Jacobites*. Whigs sympathized with republican principles and republicans gravitated to Whigs; just as Tories sympathized with Jacobites and Jacobites gravitated to Tories. Perceval, like Bolingbroke, was connecting the recognizably modern political parties of Whigs and Tories to partisans for and against Protestantism and constitutionalism on one hand, and Catholicism and absolutism on the other.

David Hume (1711–1776) had seen Court and Country parties as respectively contending for monarchical and republican interests

(Skjönsberg, 2021), the result of which was a rationalized form of monarchical republicanism that could at length foster the growth of true parties. As per Hume, parties rise in conjunction with political freedom. Hume was not unequivocally averse to parties, though he was close enough to the English Civil Wars to sympathize with, and even advocate for, absolute monarchy. He feared recrudescent anarchy. From these fears, Hume remained skeptical of political ideology, seeing it as a potential rival and replacement to religious ideology—a passion capable of fostering violent division. In this vein, Hume distinguished between party based on *interest* and that based on *principle*. *Interest* was pragmatic and therefore predictable. It might lead to corruption, but it was not volatile. By contrast, *principle* was akin to modern notions of ideology; and for Hume, political ideology was too close a cousin to religious ideology to be trusted. It bred fanaticism that stoked anarchy.[25]

7 Conclusions

In the quote above, Plumb reflects that, "By 1688 conspiracy and rebellion, treason and plot, were a part of the history and experience of at least three generations of Englishmen." In hindsight, the reader might see that generations of anarchic war and dislocating change, at length, brought a new kind of stability, one Skjönsberg labeled (2021, p. 334) *discordant harmony*, which is similar to our later reference to *dynamic stability*. Between AD1640–1688, revolution checked the power of Britain's monarch with a powerful plutocratic and oligarchic parliament to the end of insinuating faction and inefficiency into government. With time, this factional strife became ritualized and rationalized to the end of conserving liberties and forms. Less than one hundred years later, the British model "began to impress even continental observers with its capacity to wage successful wars abroad while maintaining liberty and order at home" (McNeill, 1963, p. 679). Whether instituted in the New World or the Old, governments on the British model create an institutional cage in which opposing factions can wage non-violent, ideological battle, often ending in pragmatic compromise, consistent with Ziblatt's (2017) description of *stable democratization*.

Not all contemporaries perceived this as *discordant harmony*. Some struggled to understand the significance and trajectory of change; others rejected it altogether. In early English discourse and writings, factions often are spoken of as things to be banished, eradicated, or at least suppressed. Bolingbroke held this ideal, and George III acted on it. King George's attempt to eradicate party brought reorganization and polarization. Former Jacobites, for example, within the Old Country party scattered to the winds, and a Tory party emerged that became more coherent with George III to rally around[26] (Skjönsberg, 2021). During the English Civil Wars, between-group competition predominated over within-group ideological competition.[27] Only after the Glorious Revolution, when Protestantism became sufficiently ensconced, after Ireland and Scotland's rebelliousness was contained, and after the break with Rome was sufficiently irrevocable, could British people see one another as opponents to be debated rather than enemies to be vanquished.

Notes

1. This quote is from Plumb (1967, pp. 1 & 2).
2. https://www.britannica.com/topic/Whig-Party-England
3. Relative to the latter factors, manumission is considered by medievalists to be the least influential factor in the decline of serfdom in Medieval Europe, as only a minority of serfs had enough wealth to purchase their freedom (Bailey, 2014).
4. Although famines did occur during the Early and High Middle Ages, current data indicates that crop failures associated with the Great Famine were particularly severe (Alfani & Ó Gráda, 2017).
5. These changes also favored employing female over male farm labor because herding and milking livestock require less physical strength than plowing fields and threshing grain (Alesina et al., 2011), and this purportedly led to a decline in population fertility due to the delayed marriages of female farm laborers (Voigtländer & Voth, 2013). Other historians (e.g., Edwards & Ogilvie, 2018), however, have argued that the evidence does not unequivocally support any effect of either the Black Death or pastoralism on female marriage age.

6. Parker goes further to propose that this military revolution promoted the ascendancy of Western Europe in world affairs, but this claim has been contested by others (e.g., Ágoston, 2014; Sharman, 2018).
7. Religion could "motivate altruism … by promoting behaviors that are for the good of the group and suppressing disruptive self-serving behaviors within the group" (Wilson, 2015, p. 89).
8. As recognized by E. O. Wilson who remarked that religious thinking originates from evolved brain biology which imparts "intense dislike of apostates" alongside ingroup cohesion and altruism.
9. The *Reformation* or more properly the *Protestant Reformation* severed ties with Rome, rejecting the Pope's spiritual authority and, along with it, his temporal powers of regulation and taxation.
10. Some of these sects were not internally cohesive and many were looked upon as heretical.
11. With a per capita death rate greater than both World Wars combined (Gilboy, 1992).
12. Even the Reformation's initial impetus was not purely spiritual, but rather had strong elements of Kingly sovereignty, taxation, creation of bishoprics, and control of church lands.
13. https://www.britannica.com/topic/common-law/The-feudal-land-law
14. Cromartie gives William Hakewill as an example. In his *The Libertie of the Subject Against the Pretended Power of Impositions (1641),* Hakewill denied to the King the power to tax arbitrarily even when purportedly justified as necessary to defend the realm. With the judgement of necessity lodged in the King's hands, this amounts to levying taxes at the King's pleasure.
15. Filmer's philosophy of government harkens back to Genesis. As God granted the right of dominion to Adam over the earth, Adam retained the right of rule over all the creatures of the earth, even those issuing from him through his union with Eve. This sets the template for rule, which is essentially paternalistic. The King is father and what powers he possesses are not granted to him by the people; rather he grants certain freedoms and retains certain powers as a father would over his children (Ward, 2004).
16. Hunton wrote, *A Treatise of Monarchie*, in 1643, which is thought to have provoked Filmer into writing his more famous treatise five years later. Such an exchange took place in the next century between Edmund Burke and Thomas Paine.

17. Though Hobbes is so often contrasted with Locke and Rousseau, his insistence of the right of self-preservation is one step in the direction of conditional allegiance and social contract theory; a position which brought him to the attention of the censors.
18. Absolutism's ability to arbitrarily crush dissent disallows the growth of stable political ecosystems wherein opposing sides can compete nonlethally according to known forms.
19. Ward (2004) emphasizes newfound stability granted by the Hanoverian accession of 1714. This accession brought about a period of *Whig supremacy*, exploiting the inertia of the Glorious Revolution, casting the opposition as Tory Jacobites that could not be trusted to share in political power. There was stability at this time; stability that was an outgrowth of a moderate Whig consensus, extending from the dictums of Pufendorf, about the nature of government, which avoided the extremes of radical Whigs or republicans on one hand, or divine right monarchists like Filmer on the other. The other key to the stability after 1714 and for the subsequent four decades comes from the conciliatory acceptance of constitutionalism and the ultimate rejection of divine right rule by Bolingbroke and other Tories (Ward, 2004).
20. The influence of the rising bourgeoisie played an integral leadership role, moderating the contrast between the English and French Revolutions drawn above.
21. In a short space of time, England experienced three distinct civil wars: The first war was fought between 1642 and 1646, the second in 1648, and a third between 1650 and 1651.
22. Much later, with money and without titles, Hamilton would wed America's wealthy elites to the national cause through *assumption*, a policy wherein the newly created federal government *assumed* state war debts and discharged them by issuing interest-bearing bonds.
23. Goldstone (1991) identifies demography as an important strain on the state, but even more precisely relates instability to an excess of elite sons being born into societies where they can find no positions thought commensurate with their station. Turchin (2023) recently emphasized the instability created by elite competition in his End Times: Elites, Counter-Elites, and the Path of Political Disintegration. The destabilization of elite competition is what Hamilton consciously sought to avoid by using his sinking fund to attract wealthy investors, whose fortunes would thereafter be wedded to the fate of the nation.

24. As bureaucracy proliferated, absorbing the well born, a "rage of party" resulted (Plumb, 1967; Jackson, 2005).
25. John Brown (1715–1766), in his 1757 pamphlet, *An Estimate of the Manners and Principles of the Times*, wrote of these same events without apparently recognizing the dynamism they would unleash. He may not have appreciated the distinction between *faction* and *party* with which Bolingbroke struggled to assimilate a generation before. Brown saw faction as symptomatic of weakness. He saw only the danger of the new political landscape, not its promise. Brown harkened to the strength of Ancient Sparta to counter Hume's notion that faction was an inevitable outgrowth of freedom.
26. Though as it became more cohesive, it also more obviously became a dependent faction with identity of independent country gentleman giving way to captive courtier.
27. In hindsight, it is easy to think of Great Britain as a nation, but of course it is composed of England, Scotland, and Wales; with each of these being composed of lower-level units. The Celtic fringe and other dividing lines demarcate these many groups, many of which were relevant to the dynamics of the English Civil Wars.

13

Liberalism and Conservatism in Late Modern England & Anglo-America

Steven Charles Hertler, Aurelio José Figueredo, and Mateo Peñaherrera-Aguirre

The kingdom is divided into parties, and it ever has been so divided, and it ever will be so divided
—Edmund Burke

1 Introduction

England was in the vanguard in establishing a tradition of stable opposition against absolutism—a path thereafter rapidly trod by many Western European states. Western Europe at large trended in this direction, partially as an outgrowth of a century of Enlightenment philosophy, which we emphasize in this chapter for its stage-setting function and preconditioning effects. This era witnessed the publication of several influential

S. C. Hertler (✉)
Saint Elizabeth University, Morristown, NJ, USA

A. J. Figueredo • M. Peñaherrera-Aguirre
University of Arizona, Tucson, AZ, USA
e-mail: ajf@arizona.edu; mpeaher@email.arizona.edu

S. C. Hertler et al., *The Evolution of Political Ideology*,
https://doi.org/10.1007/978-3-031-83941-2_13

Enlightenment tracts: Milton's *Areopagitica* (1644) advocating free speech; Locke's *Letter Concerning Toleration* (1689) condemning religious and political coercion; and later works of applied political philosophy, such as Trenchard and Gordon's *Cato's Letters* (serialized in the early 1720s) that casts government as a "trust committed by all…to one, or a few." This conception of governmental compact differed markedly from divine-right-rule. As reviewed in Chap. 12, the accession of William and Mary to the throne after the Glorious Revolution left England with a bill of rights that consolidated parliamentary gains made in the wake of the English Civil Wars. Nevertheless, in the early eighteenth century, opposition was still being differentiated from treason, and proto-party actors were apt to view one another as foes to be vanquished. Only as England entered into the latter half of the eighteenth century, especially under King George III, did an entrenched Whig opposition party successfully create a framework within which differences could be fought out across recognizably modern ideological poles.

As discussed in Chap. 12, some emergent English Whigs used ancient conceptions of republicanism as rhetorical and political ammunition against absolutism. Where Chap. 12 recorded its conception and gestation, Chap. 13 records liberalism's birth. This chapter considers the later emergence of classical liberalism as a philosophical and biocultural phenomenon that, at length, replaced ancient republican (extolled by Early Modern liberalism) values with Late Modern liberal values, which prioritized individual freedom from internal tyranny over and above group freedom from external domination. These developments created classical liberalism.

2 *This Frigid System*: Reacting against Republican Freedom

Historian Quentin Skinner wrote *Liberty before Liberalism*, a short work contrasting ancient republican conceptions of freedom with enlightenment liberal conceptions of freedom.[1] The two, Skinner reminds us, were very different. For example, Romans took great pride in being Roman. Be that as it may, any pride in freedoms granted by citizenship, and there was much, was intertwined with the greatness of Rome: "...the most important benefit of living in a *civitas libera* is that such communities are especially well adapted to attaining glory and greatness" (Skinner, 1998, p. 61). Skinner quotes Sallust in explaining how power is unlocked by Republican forms. As per Sallust, dynamic and powerful subjects, an object of suspicion and threat to the power of kings, are sources of pride and stability to the people of republics. Even in the Renaissance, we still find individual liberty commonly justified as a vehicle for state power. For instance, Machiavelli echoes Sallust in subordinating individual welfare to the common cause. Specifically, Machiavelli states: "...it is not the pursuit of individual good but of the common good that makes cities great, and it is beyond doubt that the common good is never considered except in republics" (Machiavelli; quoted in Skinner, 1998, p. 62). Republican conceptions of freedom overbalanced liberal conceptions of freedom, which had not yet been explicitly articulated at that time within Italian city states.

What is more, we see liberty still subordinated to power in Early Modern England. In 1656, at the beginning of the Enlightenment and before the Glorious Revolution, James Harrington[2] expounded upon perceived differences between England and Sparta.[3] The Spartans, Harrington alleged, were excessively Spartan, which actually undermined their ability to expand and conquer, and most especially to hold and capitalize on their conquests during the Peloponnesian War. The English, by pursuing agriculture, trade, and other non-martial functions, fielded larger armies, and produced surplus manpower to staff navies and populate colonies:[4]

> The Spartan, if he made a conquest, had no citizens to hold it; the Oceaner [Harrington's term for an inhabitant of his utopian vision for England] will have enow. The Spartan could have no trade; the Oceaner may have all. The agrarian in Laconia, that it might bind on knapsacks, forbidding all other arts but that of war, could not make an army of above 30,000 citizens. The agrarian in Oceana, without interruption of traffic, provides us in the fifth part of the youth an annual source or fresh spring of 100,000, besides our provincial auxiliaries, out of which to draw marching armies; and as many elders, not feeble, but men most of them in the flower of their

> age, and in arms for the defense of our territories. The agrarian in Laconia banished money, this multiplies it; that allowed a matter of twenty or thirty acres to a man, this 2000 or 3000; there is no comparison between them.

Harrington was espousing his vision of the early promise of a republicanism tempered by emergent liberalism to bring meritocracy, freedoms, legal stability, and other modern features that unlocked human and financial capital, all of which begot wealth, births, territorial expansion, and power vis-à-vis rival nations. Like Harrington, Hume extolled modern commercial republics above ancient agrarian republics, as can be seen in the following excerpt by Danford (2006, pp. 346–347):

> But moderation is more likely to be found in the kind of society to which Hobbes and Locke point the way than in Sparta or Rome or indeed Athens. Hume's rich, complex, and profound reflections on political economy were intended to persuade those with lingering nostalgia for the classical republics, or the ancient understanding called civic humanism, that modern commercial republican societies are superior.

Bertrand de Jouvenel, centuries later and with far more skepticism, nonetheless agrees with Harrington and Hume. Self-governing societies gained unprecedented powers. Whether considering the heady optimism of contemporary actors or the more staid comments of later historians, we can see correlation apt to be confused with causation. Harrington particularly seemed to have seen freedom unlocking demographic and economic powers, which, if anything, inverts the causal arrow. The argument Harrington applied to Sparta could be leveled against any other classical Greek city state. For example, Athens, though clearly more democratic, was nevertheless also unable to muster professional armies, occupy territorial expanses, or produce agricultural yields rivaling Late Modern England.[5] Thus Harrington's thesis should be understood more as nationalistic rhetoric than fact. In Chap. 9, we factually addressed, to what extent Sparta could not endure demographically as a populace as a function of pursuing their martial traditions so directly and relentlessly; by "banishing all other diversions." While it is important to understand Harrington's position, and the broader segment of seventeenth century society his modern republicanism represents, it is also important to note that, amidst a complex array of causal factors that are beyond the scope of either this section or chapter, liberal republicanism was more of a consequence than a cause of demographic and economic power. For example, Turchin discusses elite overproduction as a driver of revolutionary instability, and Ansell and Samuels (2014) discuss augmenting wealth among a broad elite as motivating revolutionary trends toward democratization. Still further, Heywood (2021, p. 10) explicitly frames liberalism as an ideology developed by a rising middle class attempting "to shape emergent industrial society." It is clear that material factors were the tail that wagged the dog of ideology. Thus, this, and the subsequent section, it should be understood, say nothing about what is either morally good or evolutionarily adaptive; neither do they make any factual claim about the nature of reality; rather they merely document the rise of liberalism, which we understand to be a more individually selected ideology; one defined as a "commitment to the individual and the desire to construct a society in which people can satisfy their interests and achieve fulfillment" (Heywood, 2021, p. 19).

3 Emerging from the Republican Chrysalis: Liberty after Liberalism

In Early Modern England, we see a gradual relaxation of group selected extremes, the justification for which, however, was superior competitiveness of the group. In other words, we do not commonly see any authors, ancient or Early Modern, arguing that a relaxation of the militant, controlling, repressive extremes of group selection is a positive good for its effects on individuals. Individual liberty, if mentioned, is merely a vehicle for state power; it is a means to an end, not an end unto itself. The individual, his happiness and freedom, only came to be justified as an end unto itself by political theorists during later modernity, with its rising tide of prosperity and its proliferation of printed philosophical pamphlets. We see the shift when comparing Harrington to Adams. Approximately a century later and three thousand miles away, John Adams echoed Harrington,[6] but reflected an evolution in thought that more fully inverted the ancient calculus, prioritizing human freedoms above all else.

Adams wrote thus of the Spartan state: "Human nature perished under this frigid system." Summarizing the judgement of Adams, in the words of J. P. Diggins (1986, p. 87) "…classical thinkers erred in making the greatness of the state, rather than the happiness of man, the object of politics." Adams could agree with Hume that:

> Ancient Sparta may have aspired to the heroic ideals of 'virtue,' but it succumbed to war and political intrigue. Such ideals distorted political understanding and stifled man's natural desires for pleasure and comfort, making a modern return to Spartan policy 'almost impossible…'. (Diggins, 1986, p. 87)

Therein, we see the chief complaint about Spartan self-suppression centering on man and his natural desires for pleasure and comfort, for freedom and individual happiness.[7] And so, the criticism was made on different grounds. No longer do we see obtusely direct and excessively severe group selection lamented for its supposed ineffectiveness, but explicitly deplored for its repressiveness. Later Enlightenment thinkers rejected extremes of self-sacrifice and subordination of self to state.[8] This is a key transition in the development of a more unabashedly individually selected society, one that no longer felt the need to justify its individualism as a driver of state power.

While we can see an inflection point, it was of course subtle. The American founding came at a time when the Republican ideal recovered in the Renaissance was waning, and the liberal philosophy of the Enlightenment was waxing. The progressive relaxation of group selected values, captured in the transition from ancient to Late Modern republics, is expertly narrated by Kalyvas and Katznelson in their *Liberal Beginnings: Making a Republic for the Moderns*. As seen in Harrington's fictional Oceana, republicanism was resurrected to challenge monarchy, but changed to the extent it succeeded:

> …liberalism as we know it was born from the spirit of republicanism, from attempts to adapt republicanism to the political, economic, and social revolutions of the eighteenth century and the first decades of the nineteenth. Between 1750 and 1830–not earlier or later-liberalism took a doctrinal and institutional form that has endured. Liberalism first became conscious of itself as a particular political and constitutional doctrine when the most

> promising and viable alternative to monarchy was republicanism. Over the course of this period, antimonarchical discourse was predominantly republican. Yet, paradoxically, just as republicans were presented with an unprecedented possibility to limit or even replace the monarchical order on both sides of the Atlantic, a dramatic form of innovation was initiated that soon transcended established republican boundaries. (Kalyvas & Katznelson, 2008, p. 4)

Kalyvas and Katznelson review the material conditions of life, such as commercial capitalism, and also changing social conditions, such as pluralistic societies, calling attention to how these functioned as catalysts to progress through the medium of Enlightenment thinkers.[9] Smith, Fergusen, Paine, Madison, and others were reaching out for ways to adapt ancient republics to modern conditions; and the solutions they happened upon became liberal tenets. Like the planks of Theseus's ship changed one by one, modern republics ceased to resemble their ancient counterparts:

> The more republicanism sought to retrofit itself for modern conditions, the more liberal it became. The more liberal republicanism became, the more its relevance was lost. From the middle of the nineteenth century, in a complex process, liberalism's entanglement with Republican thought began to bring their relationship to an end. Liberalism subsumed and transformed key elements of what previously had been a distinct doctrine of government. As a freestanding model, Republicanism disappeared. (Kalyvas & Katznelson, 2008, p. 5)

Adapted to modern exigencies and infused with Enlightenment philosophy stressing natural rights, ancient doctrines were changed until, "political liberalism burst from the shell of a republican chrysalis" (Kalyvas & Katznelson, 2008, p. 5).

It should perhaps not surprise us that the transitional period identified by Kalyvas and Katznelson (2008) of between AD1750–1830 also coincides with the time interval that witnessed the end of the Early Modern Era and the beginning of the Late Modern Era, corresponding to the end of the Little Ice Age and beginning of the contemporary trend toward global warming. Woodley and colleagues (2017) provided quantitative evidence supporting the proposition that this historical and climatic transition heralded a change from a selective regime characterized predominantly by between-group competition (generating group selection pressures) toward a regime characterized predominantly by within-group competition (generating individual selection pressures). Thus, the evolution of liberal thought and practice toward greater individualism during this period is to be expected.

4 The Consolidation of Liberalism: From Positive to Negative Liberty

Let us pause to consolidate these changes so as to better understand their political and biological significance: The polis, the empire, and the medieval monarchy gave way to post-Reformation nation states of individual citizens with recognized natural rights; authority was questioned and henceforward required to be justified; the power of the church and the Holy Roman Empire were waning; Enlightenment rationalism, capitalist forms, and population size were increasing; elites agitated for powers and freedoms that at length were broadened and shared (Klosko, 2013). Republicanism, both its revival and replacement, were caught up in this larger sequence of developments. Algernon Sidney, James Harrington, and John Milton revived republicanism, weaponizing it against absolutism, though, all the while increasingly infusing it with liberalism across multiple decades of revolution. In the subsequent century, American founders took inspiration from both ancient and English authors so that liberalism and republicanism were thoroughly merged at the founding (Dagger, 1997). The slow maturation of liberal thought under the shade of republicanism is evident when comparing the work of Machiavelli, Hobbes, Locke, and Harrington, to later works by Smith, Madison, Adams, and Jefferson. Across these authors and across more than a century of publications, individual liberty is first defended as a means, and then as an end unto itself.[10]

Republican rhetoric was apparently heartfelt and influenced the actions of these revolutionaries and their political constitutions, though at length it was largely replaced with liberal rhetoric. Yet even at the outset, and even in the writings of the most dedicated republicans, the revival of republicanism was not precisely the republicanism of the ancients. Washington's republicanism held as close as possible to ancient forms, as evidenced by his reverent reenactments of Addison's *Cato* and his extremes of material self-sacrifice and martial dedication. However, most English and American revolutionaries acknowledged differences. As we have seen, Harrington boasted about modernity's superior freedoms and Adams remonstrated against Spartan *frigidity*.[11] Both Smith and Mandeville recognized the want of self-sacrificial altruism within their social systems and consequently sought to reconcile individual interests with the good of the group. Truly from the start, from the break with Rome and the later grumblings of elites against absolutism, individualism was front and center.

The distance between ancient republican and modern liberal priorities is captured respectively in the distinction between *positive* and *negative* liberty as discussed by Kant and Berlin.[12] Positive liberty is the freeing of individuals from their own internal vices, while negative liberty is a sort of laissez-faire freedom that leaves others alone. Again, modern, negative liberty freed peoples from external interference; ancient, positive liberty freed peoples from internal vices. Positive liberty sounds almost totalitarian in its apparent unfreedom even as it is imposed as much from the bottom up by the polis as from the top down by its leadership. Plato, for example, would not trust individuals to responsibly use negative liberty. From selfish dictates and harmful appetites, negative liberty undermines group strength. Plato understood selfish and harmful motivations to *enslave* the individual; it follows that the polis, in imposing positive liberty, then frees the individual by neutralizing selfish and harmful motivations (Klosko, 2012).

This same intellectual territory was traversed by Berlin and Arendt who respectively endorsed negative and positive liberty (Hiruta, 2021). Both authors conclude that the ancients of Sparta and Athens alike had no real concept of negative liberty. Ancient positive liberty is so fully different from modern negative liberty as to amount to its antithesis. The word *liberty* means, for contemporary readers, freedom to follow one's inclination or path without interference. Liberal freedom specifically emphasized freedom from state interference. As Klosko (2012, p. 109) asserts:

> The modern conception of 'the individual,' who stands apart from the state and has a panoply of constitutionally protected rights against it, did not emerge in Western societies until the breakup of feudalism and the beginning of capitalist society, in roughly the sixteenth and seventeenth centuries.

And in addition to having a view of liberty that is positive rather than negative, and so trying to root out vice and instill virtue, ancient republics were less apt to focus on the individual as the arbiter of vice or virtue. When ancients were imposing positive liberty, they were doing this in a way that equated virtue to what is good for the group. Hard distinctions between individuals and groups, and certainly the focus on the group as a threat to the individual, were more or less foreign to the ancient mind. For as Klosko (2012, p. 109) maintains:

> ...the polis was a far smaller, more tightly integrated form of political organization than the modern nation-state...the polis was both church and state and naturally absorbed a great deal more of its inhabitants' energy and intruded more extensively in their lives than the modern state.

As Klosko more pointedly asserts, while negative liberty is usually attributed to individual agents, positive liberty is attributed to collectives, or to individuals considered primarily as members of collectivities. By contrast, in modern liberal societies, individuals are understood to exist prior to society, to have worth, and to fundamentally compose society such that society "does not exist apart from the interests of its citizens" (Klosko, 2013, p. 160). Furthermore, each form of liberty seeks to be free of different forms of control. Ancient positive liberties root out individualistic selfishness as a guarantor against external conquest. Ancient fears centered on domination by rival societies. By contrast, fear of external conquest does not figure into descriptions of liberal, negative liberties: "The individual is threatened by other people and by the state. The state protects one from other people, but poses a danger itself, and so must be held in check by constitutional devices" (Klosko, 2013, p. 111). As Lovett (2015) declares, whereas ancient liberties were ultimately guarantors of "freedom from domination," modern liberties were guarantors against tyranny from within one's own society.

5 Liberalism without Sociocultural Context

At center, liberalism is *against* rather than *for*; it *opposes* rather than *binds*; it *proscribes* rather than *prescribes*. Individual freedom and associated negative liberties are liberalism's only reliably identifiable commitments. While this is true of liberalism as an ideological abstraction, it is not true of liberalism as historical development. Herein we consider liberalism as it became unmoored from the religious ethos and republican virtues in which it was originally expressed.

Even as they were "centrally committed to the value of political liberty or freedom, where this was understood as a sort of independence from arbitrary power or domination,"[13] "Milton, Harrington, and Sidney" and "many Americans of the revolutionary and founding era such as Jefferson and Madison" reverently drew upon "Polybius, Cicero, and the Latin historians" for their emphasis of "civic virtue and political participation, the dangers of corruption, the benefits of a mixed constitution and the rule of law" (Lovett, 2015, p. 382). Thus liberalism was certainly individualistic in its advocacy for negative liberty and therefore would be unrecognizable to ancient Spartans; nevertheless, liberalism emphasized the stability-granting reliance on rationalized law, and retained something of the ideals of ancient civic virtue. Along with Christianity, tradition, and custom, ancient republican civic virtue afforded content and constraint for the emerging ideology of liberalism, without which liberalism would be more of a negative; something against domination, but not positively for anything.[14]

This has caused uncertainty about liberalism's defining features, as discussed in Steven Wall's introduction to the *Cambridge Companion to Liberalism*. Wall warns readers of the heterogeneity of liberalism, while Skorupski (2015, p. 410), a contributor to Wall's edited volume, finds liberalism to naturally incline toward cosmopolitanism, an association stemming from an emphasis on abstract values rather than parochial attachments: "agent-neutral duties of justice trump communal concerns."[15] Both the trend of not having a positive identification and not having the weight of Judeo Christian principles or ancient civic virtues is exacerbated as libertarianism grew out of liberalism, as discussed by Lovett (2015, p. 398).

> The curious thing about liberalism historically is that, for all its appeal and resilience, it has had difficulty settling on a standard overarching doctrine. Early liberals such as Locke relied on arguments from natural right that no longer find favor today. In the nineteenth century, comprehensive liberals such as Mill relied on perfectionist accounts of the human good that accommodated imperialist notions about the superiority of European civilization. In the twentieth century, liberal minimalists such as Hayek and Berlin aimed to drain liberalism of nearly any substantive commitments at all, while political liberals such as Rawls aimed to square the circle by defending neutrality substantively through a revival of contractualism. The difficulty liberalism has had in finding a secure theoretical basis betrays, republicans might argue, an underlying limitation in the doctrine itself: while certainly an important political value, perhaps toleration is not well suited to the role of foundational commitment….

John Stuart Mill and Jeremy Bentham may mark the divide between liberalism and libertarianism. The distinction is not hard or fast, we contend, because libertarianism is effectively liberalism's core stripped of its republican and religious husk. Libertarianism emerged in the middle of the nineteenth century in opposition to English socialistic policies; later in America we see libertarians unified around property rights, and thereafter in opposition to communism. Zwolinski and Tomasi (2023) authored a history of Libertarianism entitled *The Individualists: Radicals, Reactionaries, and the Struggle for the Soul of Libertarianism*. The title correctly seizes upon the core libertarian value of individualism. Libertarianism has surface content, which changed as libertarians assumed different points of principle over time. Manifest differences were evident as libertarians opposed socialism, slavery, and communism. Libertarians are only consistently against coercion. So one might see liberalism as consistently opposed to *tyranny*—authority unrestrained by law. This feature can be found in common across centuries and manifestations; though again, it is a negative, not a positive value. Contemporary libertarianism now opposes the expansion of state functions to "promote public health, education, safety and other important goals" that have radically expanded the state's "administrative structures" (Klosko, 2013, p. 405). Though associated with parties bearing the name *liberal*, the regulation and taxation of modern nations are perceived as an afront to the negative liberties of classical liberalism and especially latter day libertarianism. Where liberalism first fought against being subordinated by monarchs to state power, it now fights against being subordinated by bureaucracies to social justice. As associated religious values and republican virtues eroded, classical liberalism hadn't the ballast to resist being pulled in different directions: Some held onto liberalism's core commitment to negative freedom and took eighteenth century liberalism to the logical extreme of nineteenth century libertarianism (contemporary libertarians); but many others either embraced the transition wherein negative liberties were slowly traded for the positive values of security and equity espoused by contemporary left leaning parties (contemporary democratic liberals); still others became reactionary, recalling themes reminiscent of positive liberty (contemporary conservative republicans).

The fight for negative liberty during the Enlightenment took place in societies running on biocultural inertia. Arguably, the population at large was comparatively group selected and featured high religiosity, hierarchy, and homogeneity. Negative freedom then did not result in rampant atheism, egalitarianism, or heterodoxy. Society was self-restrained. Classical liberalism, consequently, is not associated with unrestrained social revolution and so could house Smith, Locke, and even Burke. Liberalism was articulated in a society of tradition and hierarchy and religiosity, all of which were internal constraints moderating the extremes of public corruption and private vice that negative liberties could theoretically enable; and as these waned over subsequent generations, the consequences of liberalism's negative liberties were felt more acutely. When something is a negative, it has no positive content, and is therefore liable to be co-opted. This calls to mind the ecological concept of *competitive exclusion* wherein colonization by one species keeps out a rival species. Here, religion and republicanism competitively excluded rival ideological commitments that might otherwise have been superadded to liberalism.[16]

When Faria, the aforementioned multilevel selection researcher, assails liberalism, he is assailing a later liberalism that is more like libertarianism; a late stage liberalism unmoored from initial associations with religious principles and republican virtues. Most especially, Faria sees contemporary liberalism as having structural weaknesses such as the inability for liberal societies to demographically replace their citizens. Faria (2019,

p. 7) correctly associates liberalism with the Enlightenment, claiming that liberalism sees society as a construct or contract rather than an organic aggregate. "The locus of value," Faria states, is the individual not the group. He reiterates that "moral primacy" is given to the individual. Group institutions are made to serve individual ends, allowing individuals to pursue the freely chosen and heterogeneous conceptions of *the good life* without interference from state coercion. These are criticisms reminiscent of Rousseau and Hegel, both of whom apparently saw into what liberalism might devolve.

6 Conclusions

The prior and present chapter jointly document a trend toward Liberalism discernible, but neither monotonic nor direct. In Early Modern England, a more concerted push for monarchical restraints and enumerated freedoms is made, which implicitly and sometimes explicitly employed the rhetoric of republican liberty. This established England as a *de facto* monarchical republic. This transition was accomplished by a largely self-interested elite seeking powers and freedoms, which hybridized elements of ancient republicanism and emergent liberalism. As the eighteenth century replaced the seventeenth, recognizably modern liberal thought was increasingly insinuated into political discourse, which, especially in America, inverted priorities. Rights, freedoms, and constitutional forms became guarantors of individuals for the sake of individuals, rather than as a means to unlocking state power. Natural law justified by individual rights increasingly replaced aristocratic prerogative justified by state stability. Elevated by this tide, Bayle and Spinoza, Locke and Montesquieu, and others among the intelligentsia developed Enlightenment philosophies, which, in the persons of Franklin, Voltaire, Diderot, and Beccaria, increasingly advanced practical ends and governmental reforms at odds with feudal loyalties, republican ideals, or Christian restraints.

Classical liberalism emerged after a multigenerational martial and philosophical revolution against feudal, religious, and monarchical controls. Its consolidated gains were expressed within a still conservative sociopolitical culture; yet, *negative liberty* and *individualism*, the twin products of classical liberalism's act of rebellion, had ripple effects that were difficult to contain. With liberalism increasingly differentiated from its sociocultural conditions, this loosely organized, quasi-political complement of Enlightenment thinkers both reflected and evoked more thoroughgoing social revolutionary sentiments. The modern ideological continuum was born after the victors of the English and American Revolution, now no longer bound in opposition, fell out among themselves. For some, the act of wresting away negative freedoms from monarchs was an end point; for others, it was only a beginning. This is how Edmund Burke, a Whig, became conservatism's archetype. He could support England and America's revolution, while condemning France's revolution. English Levelers, French Jacobins, and American agitators like Thomas Paine who saw the American and French Revolutions as part of an emergent social revolution that would remake the world then formed the opposite end of the ideological continuum. With the pump primed and the organs of state at stake, ever afterwards, in some shape or form, political opposition featured parties advocating for further reform, and others holding the line against them.

Notes

1. It's well worth remembering, we are not providing a historical narrative, but are, in a sense, narrating episodes in a social transition toward self-government that eventually resulted in a party system.
2. Guizot reminds us of how pervasive monarchy was becoming, which is a good contrast to considering the later rise of liberal republics.
3. It's well worth remembering, we are not providing a historical narrative, but are, in a sense, narrating episodes in a social transition toward self-government that eventually resulted in a party system.
4. It is important to recall, however, that none of these was professed Spartan objectives, so they could hardly be faulted for failing at goals to which they did not aspire.
5. The political, economic, demographic challenges forced both Athens and Sparta to independently establish long-lasting coalitions and alliances, such as the Delian and the Peloponnesian leagues.

6. Moderate Whigs, Tyrrell is here used as an example by Ward (2004), did not deny the desirability of monarchy, nor even the monarch's supremacy, though they sought to elevate the constitution above him, validating his power as its chief defendant. By contrast, radical Whigs sought to establish republican forms, as can be seen in the life of Algernon Sydney. He spent the years following the Restoration on the continent, surviving assassination attempts and attempting to stoke the Dutch and French monarchies to support an invasion campaign. Sidney erred on the side of treason and insurrection after he, at length, returned to England and ensconced himself among the opposition Whigs. Along with Monmouth, Essex, Russell, Hampden, and Howard, Sydney was a member of the *Counsel of Six,* who were all locked in the Tower following their machinations in the *Rye House Plot.* Sydney was tried and executed for treason. With only one of the required two witnesses extant, the prosecution used Sydney's unpublished discourses as an irregular alternative. Following his execution, Sydney would assume the mantle of a martyr and then his discourses would be published in 1698, which became a classic in "the radical Whig republican tradition" (Ward, 2004; p. 156). His discourses defended the right of revolution, popular sovereignty, and republicanism. Sydney advocated for a democratic republic in which powers of constitutional change and executive action were lodged in parliament. His theory is largely based on Spinoza and his studies of the Dutch Republic in the 1640s. Attracting English and American radicals of later generations, Sydney espoused natural rights, legislative superiority, the advantages of science and enlightened reason, and added to all this a species of populism and popular representation. With the Hanoverian accession of 1714, Britain experienced a period of *Whig supremacy,* which exploited the inertia of the Glorious Revolution, casting the opposition as Tory Jacobites that could not be trusted to share in political power. There was stability at this time; stability that was an outgrowth of a moderate Whig consensus, extending from the dictums of Pufendorf, about the nature of government, which avoided the extremes of radical Whigs or republicans on one hand, or divine right monarchists like Filmer on the other. The other key to the stability after 1714 and for the subsequent four decades comes from the conciliatory acceptance of constitutionalism and the ultimate rejection of divine right rule by Bolingbroke and other Tories (Ward, 2004).

7. For example, in the cold war, USSR matched US military output but not the US economy, thus it was matched in weaponry but not in martial potential. Such a ratio between economic capacity and realized military output can be seen graphed on page 42 of Ikenberry's *Liberal Leviathan* (2011) for several societies between the seventeenth and twenty first centuries.
8. James Kirchick, in *The End of Europe,* seizes on the crux of Western political development, saying something to the effect that the Western political innovation was that man was seen as an end, rather than a means to state ends.
9. Of note, while Enlightenment individualism may be natural from a Darwinian perspective, the moral and social instincts possessed by individuals, as well as the development of ultrasocial institutions, may be equally natural.
10. Of course this is not a historical review, but an illustration of how Toynbee and later historians analyzed and compared liberalism to ancient political systems.
11. Of note, the way in which these authors view Sparta is not entirely consistent with current archaeological, historiographic, and demographic knowledge. Although the Spartiates were not involved in non-martial and non-political activities, the manner in which some of these authors describe them, especially in terms of their diplomatic competencies with other polities, is very inaccurate.
12. In hindsight, Hannah Arendt, Gordon Wood, J. G. A. Pocock, Quentin Skinner, and Philip Pettit are among those eminent historians that view liberalism and republicanism as positive rivals, with the former defeating the latter.
13. Guizot reminds us of how pervasive monarchy was becoming, which is a good contrast to considering the later rise of liberal republics.
14. Lovett (2015) finds it important to separate the strands of liberalism in approaching this task, as republicanism interacts quite differently depending on the particular strand of liberalism it is compared with. It seems that republicanism is most compatible with early definitions of liberalism, but became increasingly incompatible with later instantiations of liberalism. Lovett goes on to say that liberalism all but replaced republicanism after the nineteenth century. Lovett finds self-government and the rule of law equally central tenets of ancient republicanism. He then discusses the English experiments with republicanism, saying that

after establishing a republic and having it fail, they reverted to constitutional monarchy, preserving the rule of law aspect of republicanism for the most part, while throwing over some of the self-government aspects. The English, as represented in the person and writings of Hobbes, were subdued by violence and disorder to something less utopian and experimental.

15. It's well worth remembering, we are not providing a historical narrative, but are, in a sense, narrating episodes in a social transition toward self-government that eventually resulted in a party system.
16. Of note, while Enlightenment individualism may be natural from a Darwinian perspective, the moral and social instincts possessed by individuals, as well as the development of ultrasocial institutions, may be equally natural.

14

Political Polarization in the Throes of the French Revolution

Steven Charles Hertler, Aurelio José Figueredo, and Mateo Peñaherrera-Aguirre

This unhappy Country [Revolutionary France], bewildered in the Pursuit of metaphysical whimsies, presents to our moral View a mighty Ruin…The Sovereign humbled to the Level of a Beggar's Pity, without Resources, without Authority, without a Friend. The Assembly at once a Master and a Slave. New in Power wild in Theory raw in Practice, it engrosses all Functions tho incapable of exercising any, and has taken from this fierce ferocious People every Restraint of Religion and of Respect. Sole Executors of the Law and therefore supreme Judges of its Propriety each District measures out its Obedience by its Wishes and the great Interests of the whole split up into fractional Morsels depend on momentary Impulse and ignorant Caprice. Such a State of Things cannot last.

— Gouverneur Morris[1]

S. C. Hertler (✉)
Saint Elizabeth University, Morristown, NJ, USA

A. J. Figueredo • M. Peñaherrera-Aguirre
University of Arizona, Tucson, AZ, USA
e-mail: ajf@arizona.edu; mpeaher@email.arizona.edu

S. C. Hertler et al., *The Evolution of Political Ideology*,
https://doi.org/10.1007/978-3-031-83941-2_14

1 Introduction

From party formation in the British Isles and their colonial possessions, this chapter transitions to the French Revolution with its National Assembly arrayed to the president's *right* and *left,* physically manifesting ideological division. Therein one sees a medieval kingship overturned by factional forces it was no longer capable of suppressing. In turn, without being reinforced by organically evolved cultural traditions, the decrees and constitutions hastily manufactured by the National Convention could not contain the excesses of the Third Estate. The King's dethronement and decapitation inaugurated a generation of disruptive change and sanguinary wars. While the French Revolution is addressed in its own right, this chapter is more accurately about the French Revolution's effects on political processes established in England and coalescing in its former colonies. By this time, England, more than a century away from its Civil Wars and possessed of a mature, modern government, was nevertheless plagued with dissidents, revolutionaries, and *philosophes* gathering in coffee houses, salons, and Jacobin clubs throughout the nation. When looking to Anglo-America in the 1790s, after the Revolutionary War and the adoption of the Constitution, cooperative statesmen can be seen becoming combative politicians. Though Washington had trouble recognizing as much, he and other founders became partisans of political parties as they set precedents and policies before the polarizing backdrop of Revolutionary France.

2 The French Revolution

The progression tracked across this and the previous two chapters is distilled by comparing three revolutionary slogans, which are superficially similar, but reflect an evolution in thought of profound import:

Life, liberty, property
Life, liberty, and the pursuit of happiness
Liberty, equality, fraternity

The first is from Locke and representative of the English Revolutionary impulse to constrain absolutism's interference with negative liberties. The second slogan, from the US Declaration of Independence, without repudiating property, elevates the individual, not just his negative liberties, to a place of primacy. The third strikes the ear similarly; yet only its beginning is similar. It introduces high French Enlightenment notions of *equality* and *fraternity*, both of which are understood by conservatives to violate human nature and historical reality. The conservative tradition, as here described by John C. Calhoun (1851/1995, p. 56), understands liberty and equality to be irreconcilable:

> …as individuals differ greatly from each other, in intelligence, sagacity, energy, perseverance, skill, habit of industry and economy, physical power, position and opportunity—the necessary effect of leaving all free to exert themselves…must be a corresponding inequality between those who may possess these qualities and advantages in a high degree, and those who may be deficient in them…

Combining reality and rhetoric, many English and colonial rebels believed themselves to be participating in a *conservative revolution* meant to preserve established rights (Zuckert, 1999). With some truth, they looked on the French Revolution as the most revolutionary of revolutions; as a true social revolution sapping civilization's foundations. Capturing the Revolution's scope, economic historian David Hackett Fischer (1996, p. 146) states:

> There were no fewer than four French Revolutions in 1789: a continuing aristocratic revolt against royal ministers; a bourgeois revolution against the aristocracy, a rising of urban workers against the high bourgeoisie, and a peasant insurrection against all of their oppressors.

Family was one among many primary foundations undermined by revolutionary sentiment. Revolutionary France simultaneously restricted paternity suits, and provided extended social benefits. It was now harder for mothers to pursue fathers for support, and easier to get that support from the state. At the same time, revolutionary France attempted to secure rights, including inheritance rights, to illegitimate children. Its innovative laws also undermined paternal authority and indissoluble marriage. Collectively, this might be seen as both weakening and opening the family. As Desan (2006) carped, just as revolutionaries killed the father of the nation, so they would check the father as the head of family.

The French Revolution also repudiated religion and republicanism, the twin constraints discussed in Chap. 13 as tempering Enlightenment philosophy and revolutionary inclination. Religion was attacked as an enemy of reason and as a rival source of social power. Church property was confiscated, priests were forced to marry, and the Cathedral of Notre Dame was commandeered as a temple to a deified Reason. Concomitantly, democratic preference was emphasized over and above ancient republican restrictions. When serving as United States minister plenipotentiary to France, Gouverneur Morris, as in the epigraph above, observed the French subordinating *civil liberty* (the republican constitutional forms that secure property and order) to *political liberty* (granting the body politic latitude to have its will be done). Morris saw this as mistaking the means for the ends. In other words, political liberty must actually be constrained by constitutional republican forms to protect civil liberty from arbitrary democratic impulse. Relatedly, J. W. Burgess (1915) charged French Revolutionaries with the sin of conflating *constitutional law* (law made by the nation), with *statute law* (law made by the legislature). With both distinctions confounded, France's National Assembly, inspired by a Rousseauian *General Will,* violated civil liberties and respected few constitutional limitations.

3 A Catalyzing Event: British and Anglo-American Reactions to Revolutionary France

Sometime after England's so-called Glorious Revolution, Tories became reconciled to the constitutional order and Whigs were contented by their newfound negative liberties. So, though there were now recognizable parties, there was stability. Ideological distance was moderated (Verhoeven, 2013). After this settlement, parties opposed one another, using Hume's distinction, more on rational *interest* rather than passionate *principles*

(Hofstadter, 1969). French revolutionary ideology disrupted the calm among elites then converging on a conservative constitutional order: "It is impossible to overemphasize the importance of the 1790s to the study of the dawning of political liberty and social equality in modern British society" (Verhoeven, 2013, p. 1). Verhoeven (2013, p. 1) quotes Thomas Copeland, author of *Our Eminent Friend Edmund Burke,* along with two other authors, impressing on readers that the French Revolution inaugurated, "perhaps the most crucial ideological debate ever carried on in English."

> The process of political democratization and the drastic leveling of social inequality that had been set in motion in France were now pushing the question of what legitimately constituted the nation's common culture to the top of the political agenda in Britain...the 1790s stand out as 'a decade which saw the emergence of both activist political and cultural practices (bourgeois, proletarian and feminist) in opposition to both the residual forms of courtly culture and the emergent forms of an increasingly homogenized culture industry.' ...the French Revolution debate in Britain 'marks the birth of political modernity.' ... indeed the most important debate about democratic principles, in British history.

These developments are described by Verhoeven (2013, p. 3) as a "historic transformation of its entire social and political infrastructure," sparking a "decisive transformation in the evolution of British identity," and, "a bitter confrontation between conservatives and radicals..." (Verhoeven, 2013, p. 13). The reformers, working against things as they were, shared utopian predilections of societal perfectibility, seeking to implant, "millennial ideals of liberty, equality and brotherhood firmly into the structure of British society," whereas loyalists tried to prevent reformers from, "laying the ax to the root of the constitution" (Verhoeven, 2013, p. 15). The French Revolution exerted sufficient sway over British society to expose connections between politics and their underlying moral dimensions. The phenomenon is observed by Robert Hole who understood a primarily "political, constitutional, and philosophical" debate to have shifted, by 1793, to one oriented around "control, social cohesion, morality, individual belief and restraint" (Gilmartin, 2011, p. 140).

The French Revolution swept up the masses—it engaged commoners in political debates and conscripted them into armies. The French Revolution stoked political consciousness, waking a dormant class and instilling ideological notions into a people theretofore living within a political matrix that they did not help to frame. In some measure, this process was recapitulated in England as revolutionary fervor redounded to the people at large, necessitating conservative government propaganda to descend to these same orders of men in reaction. So yes, England's conservative government was triumphant in large measure, for it preserved itself more or less intact. However, the way it won was an implicit triumph for a more moderate revolutionary trend toward democratization. The government had to court the working classes as a means of dissipating revolutionary propaganda. It therefore implicitly opened the door for their greater participation in government (Verhoeven, 2013).

Circa 1782 Franklin struck *Libertas Americana* (Olson, 2004), a commemorative coin given to Louis XVI featuring Minerva (France) shielding a cradled Hercules (America) from a pouncing leopard (England). A decade later, Revolutionary France, its king on trial, had turned leopard. It had become a polarizing ideological threat and would later, during the undeclared naval conflict known as the *Quasi War*, pose a military threat. While England was sufficiently distant from its revolutionary period to have settled into politically stable institutions revolving around ideologically narrowed debate,[2] infant America still had many precedents to set and many constitutional stipulations to live out. Quite the opposite of England, it was not distance from, but closeness to, revolution that, via the aforementioned principle of negative association (Evrigenis, 2007), provided America with relative cohesion and an ideologically constricted range; but this good will was spent as the French Revolution quickened the process of party formation that might have otherwise taken shape gradually over several decades.

We provide the next three sections, reviewing *clubs and societies* and relations among *erstwhile allies,* to further illustrate the French Revolution's effects on England and America. Here is the significance that these sections implicitly illustrate: The French Revolution disrupted England and America's stability, in part, by greatly expanding the ideological spectrum's debatable range. In effect, we have three states in different stages of development: France in the midst of revolution, America just after revolution, and England a century from revolution. In France, we see revolutionary ideas that exploded the confines of the Ancien Régime, which then brought more radical ideas into play within England's settled constitutional consensus and America's conservative revolution.

4 Clubs and Societies

A century from their Glorious Revolution, the English were living under a modernized political system with a tradition of controlled opposition. Whigs and Tories had been vying with one another, pushing the government first toward reform and then toward traditionalist royal power in an ebb and flow bounded by constitutional limits.

The French Revolution imperiled the British Empire, threatening domestic tranquility and, later, national territory through France's revolutionary wars and subsequent Napoleonic wars. The two things the conservative mind is innately oriented toward, *controlling internal chaos* and *avoiding external conquest,* were respectively imperiled by domestic Jacobin reformers and foreign French Revolutionaries. Even as the flames of Revolution did not reach across the channel in the way Paineites and Jacobins[3] hoped, there was sufficient stress to strain the system. Instead of overthrow there was reaction; and that reaction represented an end, if one partial and temporary, to controlled opposition. When a political system is threatened, and can no longer contain political extremes inside its constitutional boundaries, it either forsakes its boundaries or falls. The English system, again temporarily and partially, forsook its boundaries. By this we mean that the English ceased to tolerate opposition.

Evans (2006) documents the manifold ways in which opposition was suppressed: Some were driven to emigrate. Joseph Priestley, for example, fled to Philadelphia after having his laboratory destroyed; with him came thousands of unknown dissenters, emigrating to escape oppression[4]. Some fell under newly proliferated and strengthened sedition laws, and were fined or imprisoned. The populace was subjected to an "oppressive system of surveillance" (Evan, 2006, p. 72), which, according to radical publisher Daniel Isaac Eaton, unleashed anonymous accusations, spying, informing, suspicion, and distrust that filtered down to all levels of society. Radicals were indeed subject to surveillance, with cells being compromised by government spies. Evans (2006, p. 73) supplies names and links them to the societies that were infiltrated. For example, James Powell, a government-sponsored informant, infiltrated the *London Corresponding Society*, even rising to become its president in 1775. The breakdown of tolerated opposition was represented in newly passed laws, such as the *Traitorous Correspondence Act* of 1793, authorizing the opening of letters suspected of transmitting radical sentiments.

The *London Corresponding Society* is representative of myriad nongovernmental associations advocating for reform in the wake of the French Revolution. It claimed around three thousand active members among the strata of book sellers, shop keepers, and lawyers, with members corresponding with French men of letters to the end of bringing reform to England. These societies were not confined to London, but spread to Norwich, Newcastle, Leeds, and Manchester, numbering in the dozens (Dickinson, 2011). Conservative societies sprang up in reaction. Burning Thomas Paine in effigy and writing hundreds of proclamations to the king in support of the Constitution were among the many reactionary efforts against reform. Laws passed in response to radicalism suppressed extant freedoms. These included the *Press Act*, the *Gagging Acts*, the *Treasonable Practices Act*, and the *Seditious Meetings Act*. The authority to prosecute and imprison granted by these various acts was not in fact generally used, but their utility lay in a broad chilling effect (Dickinson, 2011). The threat of the French Revolution even led the British government to change their claims and coat of arms. George III introduced Hanover's royal cap to replace the *fleur-de-lis*, an emblem in use for centuries in association with Edward III's claims to the French throne (Black, 2006).

The Anglo-American counterpart to this was the *Alien and Sedition Acts,* described by Madison as "a monster that must forever disgrace its parents."[5] These acts were imposed by Federalists in 1798, inspiring Jefferson's famous reference to living under a *"reign of witches,"* and his authorship of the *Kentucky Resolutions*, which threatened nullification and secession (Burgess, 1915). Livingston rose to speak against the Alien and Sedition Acts, warning that the states might not cooperate with the new federal law; and then went further, arguing that the law should be disobeyed (Halperin, 2016). In reaction, Federalist Harrison Gray Otis used this as evidence to support the Alien and Sedition Acts (Morison, 1969). If a native gentleman in the government, Otis opined, can be "so easily infected, no better evidence need be required of the necessity of purifying the country from the sources of pollution" (Halperin, 2016, p. 61).[6] It became dangerous to wear the French Cockade; some spoke of emigration and others of places of refuge (Rosenfeld, 1997). During the 1790s, with Washington out of office, British impressment augmenting, and the quasi war with France raging, latent ideological divisions among elites nearly tore asunder the new-spun constitutional fabric. "Few Americans today," Warren (1942) writes, "realize the degree of alarm and semi-hysteria which inspired the Federalists in the United States during the years 1798 and 1799." "French plots, spies, and conspiracies were the constant subject of newspaper articles and Congressional debates." Lafayette, Paine, and Jefferson saw, in France's Revolution, the continuation of the American Revolution. Most especially for Paine, the American Revolution was never a conservative revolution but a part of a cosmopolitan Enlightenment liberation against Kings and the vestiges of aristocratic, religious, and feudal orders. Washington, Hamilton, and Otis, having only rebelled against ill treatment as subjects of the British Empire, saw only discontinuity and disorder, and naturally fought against, not for, France.

Freemasonry was specifically excluded from such Parliamentary proscriptions against secret societies, including the Unlawful Societies Act of 1799, which uniquely exempted the Grand Lodge of England (Kebbell, 2009). At least two of King George II's children and three of his grandchildren (by Prince of Wales Frederick Lewis), as well as six of King George III's children were Freemasons (Hawkins, 2023), perhaps rendering Freemasons somewhat difficult to persecute. In explaining the exemption, one might alternatively argue that Freemasonry was apolitical; supporting the claim by pointing to its founder, William Schaw,[7] who made the Masonic Order a haven from political and religious strife by banning all discussions of the same from the Lodge, a prohibition that has generally endured (Dickie, 2020). Still further, one could suggest that Freemasonry was buffered by its lack of enduring partisan[8] or denominational commitments or that members who participated in revolutions acted as individuals, rather than as members of a collective.[9] Some historians have claimed (e.g., Fozdar, 2011; see also Harland-Jacobs, 1999) that by the middle to late nineteenth century, as a non-denominational ideology that could potentially cut across the diverse religions of the Empire, Freemasonry had been transformed into the Imperial Cult (analogous to the Roman one) of the British Empire in spreading the humanitarian and liberal values of the "new imperialism" (Harland-Jacobs, 2003, p. 475). Famously, *Poet of the Empire* Rudyard Kipling himself was a Freemason, featuring the Craft in novels and poems.

There is some truth to all of these explanations; and yet, they are peripheral and particularistic and therefore apt to miss the instructive nature of the exception. Freemasonry was, in the main, exempted, not because it had no revolutionary import, but because it had contributed to the post-revolutionary constitutional orders of 1688 and 1776. Freemasonry, an organization devoted to individual liberty, equality, representative governance, constitutional forms, and other Enlightenment ideals, did indeed have a role in spreading revolutionary ideas throughout the world, if not in directly acting upon them in the political arena. Yet, again, these ideas, once revolutionary, were now part of what was being protected and preserved against something significantly more radical.

5 Erstwhile Allies: Charles James Fox and Edmund Burke

Recall from Chap. 5, Stenner's *authoritarian dynamic* and Inglehart's *authoritarian reflex,* which described threat-induced activation of latent authoritarian traits. Something similar was afoot in the 1790s. The French Revolution served as a catalyst, evoking a conservative *dynamic*, or *reflex*. Where Voltaire and Rousseau might be discussed broadly among intellectuals prior to the French Revolution, their names became an anathema after the French Revolution (Sack, 1993). During the French Revolution there was also a greater tendency to lump movements and peoples together as enemies to order. Deists, Catholics, Jews, dissenters and atheists, French Jacobins, German Illuminati, Italian Carbonari, and English radicals, Madame de Staël and Mary Wollstonecraft, Voltaire and

Rousseau, all found similar treatment in Tory papers. Burke's career captures the process. From a pragmatic Whig,[10] he was then an arch conservative, defending chivalry, railing against emerging women's rights movements, supporting tradition, and class distinctions. As Evans (2006, p. 16) recounts, Burke's *Reflections on the Revolution in France* reacted principally to Richard Price's claims of revolutionary light kindling in the colonies, reflected in France, and coming to illuminate all of Europe. More than repudiating any specific author, however, Burke's Reflections "grew into a generalized assault on late Enlightenment thought."

Charles James Fox and Edmund Burke were fellow Whigs, friends, and colleagues. But they took very different stances on the French Revolution. Burke famously detected danger from the outset, but Fox lauded the French Revolution from its inception, even viewing it as an event that might weaken the crown and strengthen parliament. Fox very incompletely, belatedly, and begrudgingly altered his views as French excesses mounted; but Fox originally understood France's Revolution as an imitation of England's Glorious Revolution. Burke denied any equivalence (Evans, 2006). Even before King Louis XVI was denounced and decapitated, Burke sought to correct those French Revolutionaries who celebrated England's Glorious Revolution of 1688 and who saw, in England's past, their present actions (Burke, 1790). Burke perforce admits that, in the time of the Glorious Revolution, there was a "small and a temporary deviation from the strict order of regular hereditary succession." Nevertheless, Burke says we should not draw the principle from the particular. Burke, somewhat speciously, stresses that this was an act of necessity, which could not otherwise have been avoided. Furthermore, the English Revolutionaries of the seventeenth century followed, as much as possible, tradition and the right of succession in that they neither abolished the throne nor filled it with an arbitrary occupant. Burke reminds his readers that it was Mary, the first born daughter of King James, whom parliament enthroned.[11] Fox was deaf to Burke's arguments, and praised the French Revolution as late as 1791, after which his friendship with Burke dissolved (Clemit, 2011). The storming of the Bastille was a polarizing event. It alarmed Burke (Skjönsberg, 2021), even as his friend Fox thought it, "...much the greatest Event it is that ever happened in the World!" (Hague, 2005). In April 1791,

> Fox said in the Commons that he 'admired the new constitution of France, considered altogether, as the most stupendous and glorious edifice of liberty which had been erected on the foundation of human integrity in any time or country!' By contrast, Burke argued that the French 'had completely pulled down to the ground; their monarchy; their church; their nobility; their law; their revenue; their army; their navy; their commerce; their arts; and their manufactures,' and had shown themselves 'the ablest architects of ruin that had hitherto existed in the world. (Hague, 2005, pp. 244–245)

With the execution of Louis XVI came the fracturing of the Whig Party. Burke and Fox were irreconcilable. Burke, along with forty-two other signatories, withdrew from the Whig Party and became a third-party element (Skjönsberg 2021). Charges of *apostasy* were thrown out by both wings of the Whig party, as one sought to guard against incipient Jacobinism and the other against resurgent absolutism (Skjönsberg, 2021).[12]

As Burke's falling out with Fox is emblematic of the fracturing of British Politics, Burke's falling out with Paine is emblematic of the fracturing of Western Enlightenment movement more generally. Paine and Burke were initially friends, first discussing engineering and then, ominously, politics. The political chasm was quickly identified after Paine suggested that French-British conversation would educate and improve French institutions and government. Burke thought that no such improvement in France would occur, and was convinced that the opposite effect would take place; in other words that unwanted French thought would undermine English stability. Burke cautioned against throwing over "the good, of which they had been assured possession, in favor of wild and irrational expectations" (Skjönsberg, 2021, p. 291). Later, an ebullient Paine wrote to Burke "from Paris on how well the Revolution was going, and had said that he ought to welcome it in England." To this "Burke replied grimly that the Revolution would contaminate England irrevocably, and that he did not intend to spend the remainder of his days destroying the English constitution" (Elkins & McKitrick, 1993, p. 325). "Paine was deeply hurt." Their pamphlet war ensued. Again, Burke and Paine were never precisely intimates, but were correspondents—Paine looked upon Burke as a natural ally, while Burke took an interest in Paine for his participation in the US Revolution. Paine then received Burke's reaction to the French Revolution as political apostacy and personal betrayal (Duff, 2011). As quoted by Morris (1998, p. 57), R. R. Fennessy gauged the unbridgeable distance between Paine and Burke, who offered "two appeals to English public opinion, from two entirely different and totally irreconcilable points of view." Reflecting on the Burke-Paine controversy, Duff (2011, p. 69) writes, "the effect was to raise Britain to a new level of political consciousness."

6 Erstwhile Allies: Thomas Jefferson and Alexander Hamilton

In the United States, polarization is personified in the arc of Thomas Paine's fame. Cotlar uses Paine's unwelcomed and unsung homecoming to illustrate how his fellow citizens later rejected the cosmopolitanism, French thought, radical democracy, and socialistic tendencies that Paine stood for, wrote about, and represented. Context was all. United in wresting liberty from England, Anglo-Americans celebrated Paine's *Common Sense*; but divided in controlling anarchy from France, Anglo-Americans deplored Paine's *Age of Reason*.

The relationship between Jefferson and Hamilton, even more than the fortunes of Paine, captures the breadth and depth of ideological division between quondam collaborators. Hamilton was Washington's aide-de-camp and Jefferson was governor of Virginia during the Revolution. Though fighting on the same side against a common enemy, both were temperamentally opposite. The two were fighting for different futures; Jefferson's was an agrarian federated republic, Hamilton's was a commercial, national superpower. Julius Caesar was Hamilton's hero, while Jefferson displayed paintings of Locke, Newton, and Bacon (Hertler, 2019). With such philosophical differences on greatness, it follows that the two became irreconcilable rivals. They worked together briefly under the aegis of Washington, then against one another in the cabinet, assailing each other pseudonymously, only to break in open enmity as Jefferson resigned. His formal resignation came a year after the following famous quote, taken from a 1792 letter from Jefferson to Washington:

> I will not suffer my retirement to be clouded by the slanders of a man whose history, from the moment at which history can stoop to notice him, is a tissue of machinations against the liberty of the country which has not only received and given him bread, but heaped its honours on his head.

Resigned but not really retired, Jefferson formed the nucleus of formal opposition that ended in his election, which he referred to as the Revolution of 1800. The French Revolution, and the ideological ideals it offered, was the backdrop of the story of how this relationship degenerated. It exposed different reactions to protest and rebellion, order and strength, rights and liberties. This rivalry is relevant for its representativeness. In microcosm, it captures the temperamental dialectic we have striven to expose throughout this book. Jefferson's liberal openness was such that he penned the Declaration of Independence, with its rhetoric of equality and happiness, and with all its aspirations and flourishes. Jefferson was captured by the excesses of French philosophy, and intrigued by notions of each generation forming its own government. Hamilton by contrast was a Machiavellian realist that accepted corruption and power,[13] brought out the army in force to suppress rebellions, and attempted to formalize and extend the quasi war with France. Having vanquished England, without any military threat on the continent, and having erected a stable constitutional framework, the ground was laid for formal political parties representative of rival visions; or, one might say, conflicting biologically based orientations. Federalists were more group selected, concerned with hierarchy and stability as created by strong centralized government, while Republicans were more individually selected, concerned with freedoms and rights as created by enervated, decentralized government. The French Revolution hastened and intensified this process of party formation.

7 Conclusion

After completing their elite-led, limited revolutions, Englishmen and Anglo-Americans sheathed their swords and thereafter politically wrangled within constitutional orders constrained by republicanism and religion. The French Revolution injected irreligious, progressive, and democratic ideas into these comparatively conservative constitutional orders, activating latter day levelers like Walwyn, and radicals like Paine. "The masses," Philip (2014, pp. 128–129) states, "had stepped onto the political stage at the end of the eighteenth century, and it was accepted that they could never thereafter be wholly excluded and somehow had to be accommodated and disciplined within representative institutions." This had the effect of extending the political left to more naturally accommodate the ideological spectrum's full range as it exists in the minds of men. While the French Revolution extended the left, it had the concomitant effect of consolidating the right, forcing an articulation of assumptions. Klosko (2013, p. 209) narrates the galvanization and ensuing bifurcation: "...conservative parties and parties of reform agreed on many issues, and so the stakes of political debate were lower than they would become" following the French Revolution's "categorical critiques" of the existing order.

England and its former colonial possessions were at different distances from their respective revolutions when bracing from the ideological impact of the French Revolution. In the century since the Glorious Revolution, England assimilated its constitutional edifice and increasingly wrangled within its narrowed confines over matters of interest. By contrast, the United States had just completed its revolution and drafted its constitution; the new nation still benefited from the steadying influence of Washington and the inertia of being unified in opposition to England. Though from different causes, actors within both countries rather suddenly became more acutely aware of political distances among themselves spread across an ideological spectrum broader than they had anticipated. Englishmen more fully accepted political parties as legitimate, but the conservatives among them placed farther left ideologies outside the pale. Both the acceptance of parties and the rejection of ideological extremes are seen in Burke,[14] who, in the same breath, said parties were necessary to "the full performance of our public duty," and "liable to degenerate into faction." The newness of Anglo-America's system and the strength of its republicanism blurred the line between treason and opposition, as illustrated by Jefferson's invocation of *bigots, votaries, sects, heresies,* and *apostasies* and related ecclesiastical language when writing about his ideological opponents (Hofstadter, 1969). He at least came to see party opposition as having a silver lining:

> Be this as it may, in every free and deliberating society, there must, from the nature of man, be opposite parties, and violent dissensions and discords; and one of these, for the most part, must prevail over the other for a longer or shorter, time. Perhaps this party division is necessary to induce each to watch and relate to the people the proceedings of the other.[15]

The French Revolution's failures were more salient and immediate than its successes. It was a sanguinary civil war that ended in a power vacuum filled by a dictator, [16] albeit one that ushered in an order somewhat more liberal than its monarchical precursor. It created an empirical instance of an inverse correlation between the speed and success of reform. Its abrupt break with conservative political and social moorings was an afront to life, liberty, and property, but one that gave voice to non-elite persons and egalitarian positions theretofore suppressed. Conservative elites could only meet this challenge by giving ground, absorbing its force as was later done in the revolutions of 1848 that further liberalized Western sociopolitical systems. As discussed in Chaps. 16 and 17, nationalistic, populist, and progressive third-party movements had similar effects: They were always defeated, but defeated not by negation but by absorption, having the effect of revitalizing one or another ideological pole after it had been truncated by pragmatic or corrupt elites.

Notes

1. This was a letter written by Morris to Washington, November 22, 1790
2. As discussed in Chap. 12's fourth section.
3. Here we are referencing those that adhered to the more social revolutionary convictions of Thomas Paine. Jacobin technically referred to a revolutionary club, with the term later more broadly referring to extreme elements within Revolutionary France.
4. As in Chaps. 11 and 12, we discuss emigration as a frequent strategy used by dissenters to escape persecution and prosecution.

5. This was a 1798 letter from Madison to Jefferson as quoted in Rosenfeld (1997, p. 128).
6. Of note Otis employs three references to contamination, which are known within the evolutionary and political literature to signal primal aversion to outgroups.
7. Freemasons are subject to various conspiracy theories proposed by their detractors, and have all kinds of extravagant origin stories. Some of these locate Freemasonry from as far back as the creation myths of the Biblical Book of Genesis (e.g., Anderson et al., 1734), others from the building of the Egyptian Pyramids, and yet others from the construction of the Temple of Solomon. Within contemporary Freemasonry, however, most of these origin stories are now seen as largely allegorical and symbolic, as is most of Freemasonry itself. More plausible, but still incorrect, is the widespread view that modern Freemasonry evolved from the stonemason's guilds of the middle ages; the main problem with this less extravagant view is that there were hardly any very organized guilds of stonemasons of the middle ages, mostly owing to the fact that such specialized craftsmen had to travel quite extensively for most of their work and were unlikely to form any permanent trade guilds in any fixed locations (Dickie, 2020). The opposite error is sometimes made: AD1717 is now routinely given as the inception date for modern Masonry, implicitly erasing over a century of Scottish Masonry that came before. Rather than either of these extremes, modern historians have now more accurately traced its origin to the Renaissance rather than the Medieval Era. The most accurate attribution is as follows: In AD1594, William Schaw, the *Maister o' Wark* (Master of Works) to the Court of King James VI of Scotland (later James I of England), received a commission to build a Chapel Royal at Stirling Castle that was modeled on the Biblical Temple of Solomon (ironically enough, this put at least a replica of Solomon's Temple back in the origin story) (Dickie, 2020; Stevenson, 1988). In AD1598, Schaw reassembled the master masons he had originally contracted for that job and invited a few others to join a new Masonic Lodge he founded that blended Medieval stonemason's lore with elements of Renaissance culture (such as Hermetic Philosophy as well as geometrical principles and mnemonic techniques from Classical Antiquity). This bridged the gap between what have been called operative and speculative Freemasonry, with the latter being the philosophical element that came to dominate the modern craft.

8. Consistent with the royal patronage, the Scottish organization pioneered by Schaw leaned Tory for over a century, even if it maintained an officially apolitical stance. However, as Freemasonry became increasingly popular in England, the general political orientation gradually became Whiggish. When the Grand Lodge of England was constituted in London in AD1717, it became completely saturated with the liberal ideas of the Enlightenment, which have characterized all of Freemasonry from that time onward (Dickie, 2020). Not surprisingly, these Whiggish Freemasons gave themselves the starring role in their Whig History, as seen in Butterfield's 1931 classic, *The Whig Interpretation of History* (this has been characterized as a form of historiography that depicts history as a progressive march from a past of despotism, ignorance, and superstition toward a glorious present of enlightenment and freedom).
9. Much has been written about the prominent role of Freemasons in both the US Revolutionary War and the French Revolution, but it appears that these influential figures generally acted as individuals and not as Freemasons, *per se*, with only 68 out of 241 Founders being Masons, for example (York, 1993). It has also been proposed that the much-touted role of Freemasonry in South American Wars of Independence has been overestimated (Zeldis, 1998).
10. Again, Burke was a Whig, not a Tory. Thus, Burke was not reflexively against reform, though he countenanced a species of moderate, evolutionary reform. Nevertheless, Burke still had to work through the tensions between his status as a Whig Parliamentarian and his authorship of the Reflections, as described by Gilmartin (2011, p. 132): "…the way the *Reflections* worries over how the past should condition the present suggests that Burke was wrestling with his earlier Whig and reformist commitments, rather than providing the bearing for High Church and Tory positions that came to dominate reactionary Britain."
11. A supporter of the Anglo-American but not the French revolution, Burke was charged with inconsistency by Fox and other former Whig friends and allies.
12. See p. 304 in this reference, from which this sentence is closely paraphrased.
13. Hamilton certainly accepted corruption as a realist. He could also be said to have, not simply tolerated corruption as an inescapable aspect of human nature, but to have harnessed it, as in *assumption,* his economic system which funded the federal government via elite investment.

14. Burke, as quoted in Hofstadter (1969, p. 32).
15. Jefferson, as quoted in Hofstadter (1969, p. 115).
16. Napoleon filled the power vacuum and effectively ended the most revolutionary phase of the Revolution. While he indisputably undermined self-government, he embraced many revolutionary principles and was considered liberal by monarchs and defenders of the pre-revolutionary order.

Integrative Metacommentary: Part V. The Formation of Faction

Aurelio José Figueredo, Steven Charles Hertler, and Mateo Peñaherrera-Aguirre

The Formation of Faction

Part V presented three chapters on the evolution of political factionalism. Chapter 12 started with the cascade of environmental shocks, including global cooling, famines, and plagues, that progressively destabilized the Medieval feudal system. These set the stage for the beginning of the Early Modern Era, which saw the gradual decline in slavery and then serfdom in Western Europe, followed by the rise of the urban middle classes, as the Late Medieval towns and cities were enriched by the extensive trade networks generated by the Levantine Crusades. The revolution in weapons and tactics that also ensued further disrupted the feudal system that altered the nature of warfare in ways that no longer advantaged the old military aristocracy with respect to the peasants and the bourgeoisie.

Chapter 12 went on to describe how these socioecological preconditions set the stage for political reform and revolution, which limited

A. J. Figueredo • M. Peñaherrera-Aguirre
University of Arizona, Tucson, AZ, USA
e-mail: ajf@arizona.edu; mpeaher@email.arizona.edu

S. C. Hertler (✉)
Saint Elizabeth University, Morristown, NJ, USA

traditional royal prerogatives and expanded the bounds of parliamentary authority. These developments facilitated a gradual transformation in the way England and other European states dealt with dissent; transitioning from a policy of brutal suppression to one of increased toleration of political opposition. These changes were not achieved without certain costs, including several violent political revolutions and the military suppression of various separatist revolts.

Chapter 13 explained how this tolerated factionalism evolved into a system of political parties. The Conservative parties (*Tories* in Britain), of course, stayed conservative and more or less defended the continued validity of the sociopolitical institutions of the old regime. The Liberal parties (*Whigs* in Britain) experienced more of a transformation during the transition from the Early to Late Modern Eras, from an ideology of more group-selected republicanism, based on the models of classical antiquity, to one of more individually selected libertarianism. The virtues of the former were extolled as ultimately serving the interests of the *republic*, even while permitting greater freedom, whereas the advantages of the latter were touted as instead serving the interests of the *individual*, eventually framing this as a matter of natural right and the construction of an orderly society as something contractual and contingent rather than primal and essential. By the time of the US Declaration of Independence, individuals were free by Divine Right in the same way that monarchs were originally sovereign.

Chapter 14 considered how these evolving accommodations in the British Isles and their former colonies were at once energized and threatened by violent upheavals produced by the French Revolution(s). Certain panic-inspired, intolerant reactions ensued, such as laws passed against any nongovernmental associations or secret societies that pushed past conservative Enlightenment constitutionalism designed to enfranchise and protect the property rights of a newly expanded elite. Although there was some sympathy for the democratic ideals of the newly formed French Republic, most contemporaries were horrified by the murderous excesses of the Jacobin Reign of Terror that followed. The French Revolution had the effect of fully modernizing political systems by articulating far left ideas suppressed in all prior revolutions, and by forcing conservatives to politically organize in order to defend the *status quo*.

Part VI

The Maturation and Institutionalization of Faction

15

Party Machines in the United States of America

Steven Charles Hertler, Aurelio José Figueredo, and Mateo Peñaherrera-Aguirre

Legislate how you will, establish universal suffrage, if you think proper, as a law which can never be broken. You are still as far as ever from equality. Political power has changed its shape but not its nature. The result of cutting it up into little bits is simply that the man who can sweep the greatest number of them into one heap will govern the rest. The strongest man in some form or other will always rule. If the government is a military one, the qualities which make a man a great soldier will make him a ruler. If the government is a monarchy, the qualities which kings value in counsellors, in generals, in administrators, will give power. In a pure democracy the ruling men will be the wirepullers…Changes in the form of a government alter the conditions of superiority much more than its nature.

—Sir James Stephen[1]

S. C. Hertler (✉)
Saint Elizabeth University, Morristown, NJ, USA

A. J. Figueredo • M. Peñaherrera-Aguirre
University of Arizona, Tucson, AZ, USA
e-mail: ajf@arizona.edu; mpeaher@email.arizona.edu

S. C. Hertler et al., *The Evolution of Political Ideology*,
https://doi.org/10.1007/978-3-031-83941-2_15

1 Introduction

At this juncture, at the end of one section and the beginning of another, it is helpful to survey the field and locate our place. Four chapters within Part II described genetic, neurobiological, and temperamental differences disposing persons to liberal or conservative values. The next three chapters, comprising Part III, detailed the evolution of temperamental antecedents to political ideology within primate troops and pre-state societies. Chapter 8, the final chapter in that third section, described an evolutionary push and pull between individual and group selection responsible for maintaining temperamental, and therefore ideological, diversity within human minds. Part IV's three chapters detailed group-selection responses from ancient societies beset by external competition and medieval societies threatened by internal heterodoxy. Part V's three chapters then detailed how the ideological range within human minds came to correspond with the ideological range within social systems: Chap. 12 witnessed elites securing liberal freedoms and republican forms; Chap. 13 witnessed diffusion of these values into the lower orders and prioritization of individuals over groups; Chap. 14 witnessed the further development of the left and a coalescing of the right. In turn, this and the next two chapters, grouped together in Part VI, illustrate the fate of ideology in political systems: Chap. 15 describes the amplification or suppression of ideology by elite-led political machines; Chap. 16 describes the reassertion of ideology within political systems that coalesce around a narrowed band of disagreement, leaving a discontented far right or left outside the political pale; Chap. 17 describes the populist, nationalist, and progressive movements capable of realigning political platforms and political discourse. Having climbed the hill and scaled the tree to locate our place, let us pursue our immediate course.

Considering *political machines*, which are effectively coalitions or parties organizing, collaborating, and competing for political power, this fifteenth chapter looks past the ages of revolution and constitution formation wherein political ideologies were first represented in political processes; it considers how political ideology figures into established political landscapes as parties matured and competed for power. More than any other, this chapter distinguishes between ideology as a feature of mind, and political parties as vehicles for gaining and maintaining power. Hearkening back to Hume's distinction between *principles*, genuine ideological commitments extending from individual temperamental differences, and *interests*, practical desire for power and resources extending from material desire, we can see that political machines are creatures of interest. What political machines do with principled ideological commitments depends on how they advance or detract from practical interest in obtaining power and resources. As discussed, political machines alternately stoke and suppress ideological principles as they grant or thwart political interests.[2]

2 The Politics of the Ordinary and Extraordinary

It might first be helpful to cite Andreas Kalyvas's distinction between the politics of the *ordinary* and *extraordinary*. As can be seen in the quote below, while extraordinary politics activate ideological principles and the general will to constitute states, ordinary politics emphasizes elite interests and power to control states:

> …in extraordinary times the sovereign people wake up to reaffirm their supreme power of self-determination and self-government. During these periods, there is an intensification of popular mobilization, an extensive consensus around some fundamental constitutional changes, and the formation of spontaneous, informal movements that directly challenge the status quo. All of these developments become crystallized in a new general will, which seeks to speak directly, against the ossification and inadequacies of the instituted political system. In these singular moments, the people decide to introduce radical constitutional changes. Rather than expressing private interests, they press for pivotal modifications in the name of the public good.

During these times, as celebrated political scientist Bruce Ackerman states, "…a mobilized majority hands down new principles to guide the polity" (Ackerman as quoted by Kalyvas, 2008, p. 164). The most extraordinary of all extraordinary politics were the English, American, and

French Revolutions that respectively ushered in Liberal, Republican, and Democratic governmental elements (Auslander, 2009), creating, formalizing, and expanding modern political processes and forms. To a lesser degree, the third-party movements considered in Chap. 16, and the populist, nationalistic, and progressive movements considered in Chap. 17, can sometimes approach or even become extraordinary politics. However, political machines, and the interests they are organized around, which we are considering in this fifteenth chapter, are expressions of ordinary politics. In ordinary political times, as theorists as different as Carl Schmitt and Bruce Ackerman report, there can be "depoliticization and civic apathy," the "invisibility of the sovereign people," a "retreat of the masses from actual politics." In contrast to great nationalist movements during war, or expressions of the general will during revolt, ordinary politics sees the masses fragmented, "devoid of any collective project that could unify the popular sovereign around some concrete fundamental issues." Instead, politics is dominated by "narrow, political interests." This is the time for elite domination, and the time when political ideological differences simmer and principles take a back seat to interests; ordinary politics are ripe for political machines (Kalyvas, 2008, p. 164).[3]

3 Political Parties and Political Machines

Political parties antedate political machines. Machines are products of professionalized parties. Hard and fast distinctions are difficult to make, but the term *machine* implies extremes of organization and exaggerations of elite power centralization. In illustration of this distinction, consider definitions of *political party* and *political machine*. Noel (2013, p. 13) borrows the following definition of political party from E. E. Schattschneider: "an organized effort to gain political power." Noel (2013, p. 14) then expands on this definition, stating that:

> …parties are instrumental in achieving policy goals, and that is why parties form. Policy demanders look around the political landscape and find allies with whom they can win. Once the parties are formed, new policy demanders will choose which alliance to join on the basis of who will have them.

Parties, in this view, have the job of managing all of these constituent groups and their interests (Karol, 2009). The platform that the party advances under this framework is the result of the compromises and bargains among the groups that united to form the party.

Contrast this with definitions of political machines as:

> …a party organization, headed by a single boss or small autocratic group, that commands enough votes to maintain political and administrative control of a city, county, or state…The term refers to their ability to elect candidates or enact measures with mechanical efficiency and predictability.[4]

These definitions are different more in degree than in kind. There is at least a quantitative increase in centralization from broad party coalitions comprising elected officials to a smaller, less visible cadre of orchestrators operating behind the scenes. Again, increasing in degree more than kind, machine politics pursue power ever more unabashedly for its own sake. A citizen can more defensibly pledge allegiance to both a party and a state, seeing the ideological commitments of the party to advance the ultimate ends of the state. This is less plausible in the case of a political machine which more nakedly advances interests of a narrow elite by pandering to subgroups in ways that are recognizably at odds with the aggregate, long-term interests of the state.

4 Coalition Formation

Political machines call to mind glad handing, graft, spoils systems, backroom deals, pork barrel politics, ballot harvesting, and related machinations. Even as they include varieties of corruption apt to be morally condemned, these are mechanisms of coalition formation that, as Sir James Stephen's quote above implies, are ineluctable features of self-governing systems. There is a perhaps inevitable progression from statesmen, to members of political parties, to cogs in a political machine. This is reminiscent of inevitable evolutionary trajectories like the dulotic progression discussed in Chap. 9, or as seen in examples of what are called *Red Queen Effects*, which produce extremes within species locked in coevolutionary arms races, such as that between the cheetah and gazelle whose predator and prey relationship pushes both to excesses of speed. In other words, as one political party builds a political machine, its rival is compelled to do so to remain electorally competitive.

As is well known among both farmers and evolutionists, the egg output of chicken coops is dependent, not only on individual chickens, but on the interactions among them.[5] Not knowing anything about either fowl or multilevel selection, one might gather the most prolific egg layers from many coops and put them together in one coop. The consequence, rather than being a super-generative coop, would be rampant individual competition, and in consequence, reduced egg output. This example of egg laying among chickens parallels party nominations among politicians. The election of Benjamin Harrison illustrates the point. Harrison was part of an Indiana faction that was turned to, in the end, after the Ohio faction failed to secure the nomination. William McKinley sacrificed his own nomination, deferring to Sherman, his senior and chosen son of the Ohio faction. When Sherman's nomination momentum stalled, and even when it became clear that he would not get the requisite votes, he favored *defeat rather than retreat*. In other words, as the McKinley biographer H. Wayne Morgan explains, Sherman did not defer to McKinley. McKinley, exemplifying loyalty, drafted a statement denying his candidacy in unequivocal terms. Had Sherman acceded to McKinley, that faction might have secured the nomination. They had one of two prerequisites of a good coop, a man willing to subordinate himself. As political parties birthed political machines such party discipline and organization would become more commonplace. In this way we can then increasingly recognize altruism of a sort in political machines, which may appear paradoxical given the negative reputation of machines. However, the negative reputation of machines relates to their relationship to the state at large. Altruistic collaboration among actors within the political machine shows allegiance to a lower level group (the machine) to the potential detriment of the higher level group (the state). The following quote, taken from Bertrand de Jouvenel's *On Power* (1993, p. 303), captures the transfer of power and allegiance from state to machine:

> The further the organization of parties is pushed and the greater the part that is played in winning elections by … 'the machine,' so much the more complete is the subjection of the member of congress or parliament to that machine, which is the real holder of his seat. Parliament is then no longer a sovereign assembly in which an elite of independent citizens compare freely formed opinions and so arrive at reasonable decisions. It is now only a clearing-house in which the various parties measure their respective parcels of votes against each other's.

What Hamilton and Burr did in throwing around influence during the election of 1796 was taken to a new level a generation later by men like Martin Van Buren and Thurlow Weed. They professionalized party politics and influence to such a degree as to render obsolete the independent statesman above the fray. In speaking of Van Buren and the nationalization of the New York machine, Hofstadter (1969, p. 246) states that, "party unity was the democrat's answer to the aristocrat's wealth, prestige, and connections." Van Buren is maligned as a *magician*, a protean man of many forms, but Niven (1983) doubts the accuracy of such representations. The greatest defense against such charges is that Van Buren was a Jeffersonian Republican in spirit and that this ideological commitment more or less actuates his conduct. Ideological principle was at least in the background. There was ambition and party and spoils, but not indiscriminate pursuit of aggrandizement at the public expense. There is a fundamental difference, however, in the action of Van Buren and someone like John Quincy Adams. Adams's allegiance was directly attached to the nation; Van Buren's allegiance was directly attached to the party, and only through it, to the nation. Where Adams stressed unity of the nation, Van Buren stressed the unity of the party. Many of the latter's maneuvers stemmed from the desire to preserve party unity—he reserved judgment, organized, denounced, rallied, inspired, and spoke to this end (Niven, 1983). In these ways, Van Buren's politicking behind the scenes during the Jackson campaign and administration is therefore discussed as the emergence of a more recognizable political machine, earning Van Buren the moniker, *little magician*.

5 The Fate of Ideology

Henry Sumner Maine (1976, p. 53) speaks of the *wire pullers*, persons that harvest power and votes from the masses. In modern democracies, the devolution of power to *the many* has fractionalized power into infinitesimally small units, units which are too small to be widely valued: "The morsels of power are so small that men, if left to themselves, would not care to employ them." In steps the political *chiffonnier*, pulling the wires to collect the votes (Sumner Maine, 1976). This has more to do with the reins of power than ideology *per se*, though Sumner Maine describes ideological manipulation as a means of aggregating votes. The main method is to keep the party spirit at *white heat*. "The wire puller is not intelligible," Sumner Maine (1976, p. 54) notes, "unless we take into account one of the strongest forces acting on human nature—party feeling." With excellent intuition, the author remarks that, "party feeling is probably far more a survival of the primitive combativeness of mankind than a consequence of conscious intellectual differences between man and man" (Sumner Maine, 1976, p. 54). He likens it to that same sentiment that drives civil, tribal, and national conflicts, designating it a universal human impulse. Maine's characterization of ideology in the hands of machine politics is apt, but only one of several possibilities. Machines can amplify party spirit, but will take the opposite course if that better secures power.

Political machines exist in the outside world. They can embody ideology, but at the same time encompass temporary concerns, practical issues, propaganda, special interests of persons, regions, foreign interference, and the like. Again, the modern machine adopts a utilitarian stance toward ideology, suppressing or amplifying it to the extent that it advances or detracts from obtaining power. The reality of ideology as a feature of mind then ensures that it will not be written out of the picture entirely even in the most corrupt systems, as discussed by Sowell (2002, pp. 7–8):

> The political battles of the day are a potpourri of special interests, mass emotions, personality clashes, corruption, and numerous other factors. Yet the enduring historic trends have a certain consistency that reflects certain visions. Often special interests prevail to the extent that they can mobilize support from the general public's responsiveness to visions which can be invoked for or against a given policy. From the standpoint of personal

> motivation, ideas may be simply the chips with which special interests, demagogues, and opportunists of various sorts play the political game. But from a broader perspective of history, these individuals and organizations can be viewed as simply carriers of ideas, much as bees inadvertently carry pollen-playing a vital role in the grand scheme of nature while pursuing much narrower individual purposes.

Still, genuine debate, at least within parties, wanes as machine discipline waxes. In essence, this is the function of a machine. It is to quell internal debate and orient outward. It is the political equivalent of a warring nation that silences dissent through the suspension of habeas corpus so as to present a unified challenge to the enemy.

Now debate and ideology are different. Just because a machine curbs debate, does not mean that the machine and the party it dominates will no longer faithfully represent an ideological pole. Much of the discipline a machine imposes is nonideological, such as agreements about who will wait to run, and who will step down. Yet, of course, some is ideological, and that ideology promulgated by the machine may well be of an identifiable left or right variant. What is curbed is internal dialectic and debate. Internal heterodox, extremists, mavericks, and the like are not tolerated. Given these facts, the party line the machine pursues can drift away from the starting point, into a territory orthogonal to right or left. Thus, a party machine is as apt to suppress as to channel ideology. What it does about ideology it does as it helps or hinders the acquisition of power. As further explained in the subsequent section, machines are dispassionate pursuers of power and they will advance ideological ends only as long as those ideological ends are pathways to power.[6] Political parties originally gel around ideological differences, though the party machines they create sometimes actively suppress ideology, as was observed by Hofstadter.

6 Elite Sorting

As stated in this chapter's opening, *what political machines do with principled ideological commitments depends on how they advance or detract from practical interest in obtaining power and resources.* Matthew Levendusky, political science professor and author of *The Partisan Sort*, elaborates on this point, contrasting two strategies, one which moderates partisanship by courting centrists and one which exacerbates partisanship by courting ideologues.

To begin with, there is the *Downsian Model*, emphasizing the importance of the center and the swing voters that populate it. The moderate center gains attention to the neglect of the party base, whose votes are taken for granted. At least in the United States, the two-party system, with its primary and general elections, induces elites to first fight among themselves, pandering to the base for the party nomination; but, under the Downsian Model, elites thereafter move to the center to capture moderates. This is a very different model than that of Europe with its coalitions that can form strong bases around positions further right and left.[7] By contrast, two-party systems have classically avoided pandering to ideological extremes in their need to create broad coalitions encompassing half the country. The United States, however, has seen a shift away from the Downsian Model to *Huntington's Strategy*. Articulated in 1957 by political scientist Samuel P. Huntington, his eponymous strategy advised base mobilization, which, rather than endeavoring to capture swing voters, is meant to turn out the base, getting them to vote in higher numbers (Levendusky, 2009). The tradeoff is then always between a strong base activated and engaged around a shared ideology versus a broader coalition, but one that is less ideologically pure and politically engaged. Attempting to balance within- and between-group differences by creating the broadest possible coalition that will remain coherent, party machines use ideology judiciously, tacking toward the center right or left when situational factors favor a Downsian Model, and do the opposite when situational factors favor Huntington's Model. In either case, whether pursuing one or the other strategy, political machines perforce alienate some voters who find their ideological position unrepresented. Centrists will be empowered and accepting of a Downsian political landscape, but feel caught between extremes in a Huntingtonian political landscape. The opposite will be true of those that are highly ideological—whether leaning to the right or left they will feel well-housed within a Huntingtonian political landscape, but frustrated by a Downsian political landscape.

Ideology and party have fluctuated over time, overlapping, drifting apart, and even inverting. In modern US politics since Reagan's election, liberal left ideology has resided more reliably in the Democrat Party while conservative right ideology has resided more reliably in the Republican Party. By the 1990s, elites had separated along more definite party lines. Voters absorbed elite messaging, with ideological shifts resulting, and voter behaviour following from the ideological shifts. Circa 2004, with ideology and party continuing to align, there was a transition from a Downsian to a Huntington strategy. This shift from a Downsian to a Huntington model might have further exacerbated polarization. In seeking to motivate their base, as per the dictates of Huntington's Strategy, elite messaging became correspondingly extreme and ideological.[8] Increased partisanship is evidenced by reduced *split-ticket voting* and increased *party voting*. This means that issues and candidates are being decided, less on their respective merits, and more on their party affiliation. With this process now in play, Levendusky describes the hollowing out of the ideological centre, with a bimodal distribution taking the place of the normal distribution. Thereafter, to continue the statistical analogy, activity would be at the extremes of the distribution rather than at its center. From a classic bell curve resembling the single hump of the dromedary camel, the electorate would fall out along a bimodal distribution resembling the twin humps of the Bactrian camel with a gulf of middle ground centrism separating the ideological right and left.

Before leaving off, we should differentiate *sorting* from *polarization*.[9] Whether pursuing a Downsian or Huntingtonian strategy, political machines are engaged in sorting, which is a sort of coalition building. Some believe elites to have fundamentally reshaped electorate ideology, though Levendusky finds a *sorting effect*, which can augment polarization, but which is distinct from it. This is why Levendusky encourages his readers not to equate polarization and sorting, because the former relates to ideological change and the latter relates to a shuffling of the electorate along ideological lines; these are different processes, both important, but distinct. Again, both processes are happening, but sorting does not fully create polarization. Elite-driven sorting, in short, creates polarization partly by causing some level of ideological change and partly by changing outward voting behaviour. A sorted voter votes along party lines whereas a polarized voter, while also voting on party lines, more genuinely embraces the extreme views of the party. The most ideologically volatile landscape would feature a polarized electorate that is also sorted by elites.

7 Conclusions

Van Buren described political parties as promoting the public good; as instruments to energize civic spirits and check corruption. And even when corrupt, political parties can convert the potential energy of ideology into the kinetic energy of political action. But these favorable views of political parties became less defensible as political parties nurtured political machines. Men of conscience like John Quincy Adams, when they voiced independent thoughts, were counseled to keep their peace. Any departure from the party platform, rather than being seen as principled, was seen as disloyal. Relentlessly emphasizing party discipline and loyalty, machines ensured that power became an end unto itself, rather than a means to an end. In effect, latter day political parties with full-fledged political machines became groups unto themselves that demanded loyalty over and above any loyalty bestowed on the nation. Consequently, political machines routinely engage in electoral sorting by emphasizing shared ideology and interest within their parties and exaggerate differences with rival political machines. In this way, political parties with modern political machines oppose one another politically, as rival nations oppose one another militarily. Both alike strive to suppress within-group competition as a means of promoting effective between-group competition. As has been seen most explicitly in Chap. 10 which featured the ebb and flow of losses and gains in the Christian effort to control Jerusalem, competition induces aggregation and increased political complexity. Similarly, Chap. 11 related how sociopolitical groups often seek to suppress internal dissension by forcing ideological conformity, sometimes violently and sometimes not. Writ large, warfare, as well as economic rivalries, has *made the state* so to speak; meaning that martial competition drives sophistication and organization among rivals. We see this same dynamic here in this chapter on political machines, only the competition is nonviolent rather than violent, and takes place within rather than between groups. The result is a runaway Red Queen effect with rivals driving increasing sophistication among one another, and political machines are the products of this political sophistication.

As the size, power, and organization of party machines grew over time, machines became powerful groups within the national whole; and with their growth, party machines gained the power to redirect or suppress ideological divisions within the body politic. Party machines, such as those created by "Boss" Tweed, are power-oriented, not ideologically oriented, with the consequence that biologically informed ideological differences are emphasized or suppressed as they help or hinder in the pursuit of political power. As we will see in the subsequent chapter, this can leave more extreme ideological party members apathetic and effectively disenfranchised, creating the conditions for the realignments, third-party movements, and the populist reactions discussed in Chap. 16.

Notes

1. As quoted in his Liberty, Equality, Fraternity by Sumner Maine (1976, pp. 52–53).
2. Of course Hume's distinction between principle and interest, though valid and useful, is not always hard and fast. Principles can extend from interests, as when a wealthy entrepreneur advocates for libertarian, laissez-faire policies.
3. In Chap. 4, we discussed the nature of ideology as *visions* and *moral emotions*, personality-driven biases and predilections. Ideology is at once a "principled set of ideas" and "a construct developed to justify either interests or personal psychological dispositions" (Noel, 2013, p. 65). This description of ideology is consistent with Converse's (1963) notion of ideology as a system of constrained beliefs (Gerring, 1997; Knight, 2005). Constraint is the key element—what one believes on one issue predilects what one believes on others because specific beliefs are expressions of lower-level worldviews.
4. https://www.britannica.com/topic/political-machine
5. The example of egg laying among chickens has been used to illustrate multilevel selection by Haidt and D. S. Wilson among other authors.

6. Lewis (2019) discusses the bidirectional interaction between ideology and party machines. Lewis assumes politicians, of all ideological stripes, want to expand and exercise their power. Second, he assumes that ideology can drive action, as is customarily thought, but, and to Lewis's point, action can drive ideology. The relationship is therefore bidirectional. This bidirectional relationship clarifies inconsistencies between out-of-office ideological stances and in-office political action. The consequences of these two premises are that inconsistencies become rather regular and expected. At the end of his book, Lewis (2019, p. 170) reiterates his two main points, stating that (1) "ideologies change," and (2) that "the experience of holding power changes them." The political party out of power will gripe about its exercise, and then tend to exercise power when it accedes to office. For example, Jefferson complained bitterly to the point of disunion in the Kentucky Resolutions about the Alien and Sedition Acts, which he later used rather than abolished when he was president.
7. Centrism can also be created by corruption, as extremes are given over for remunerative moderation.
8. Thus, the parties of Anglo-America might come to resemble the coalition of Europe, only there would be more polarization because there would be no coalitions in the center. Huntington's Strategy would make an American two-party system like removing all the centrist coalitions from the European landscape.
9. These distinct processes are treated in different chapters. Sorting is more properly the subject of this chapter, whereas polarization is discussed at length in Chap. 19.

16

Political Realignments & Third-Party Movements in the United States of America

Steven Charles Hertler, Aurelio José Figueredo, and Mateo Peñaherrera-Aguirre

By definition, third parties represent dissatisfaction with the two-party system. A diversity of reasons—intense concern with particular issues, factionalism within the major parties, and unorthodox ideological views, among others—have produced dissent from established parties and from the general consensus in which they have operated. … Massive defections to new parties that raise or exploit certain issues, therefore, indicate the failure of the system to integrate the emergent political demands of significant portions of the electorate

—M. F. Holt[1]

S. C. Hertler (✉)
Saint Elizabeth University, Morristown, NJ, USA

A. J. Figueredo • M. Peñaherrera-Aguirre
University of Arizona, Tucson, AZ, USA
e-mail: ajf@arizona.edu; mpeaher@email.arizona.edu

S. C. Hertler et al., *The Evolution of Political Ideology*,
https://doi.org/10.1007/978-3-031-83941-2_16

1 Introduction

Historically, dissenting factions have been suppressed. Banishment, execution, and incarceration controlled dissenting factions in monarchies, empires, and dictatorships. Ancient self-governing societies sought to avoid factions by remaining small, an axiom countenanced by Montesquieu's mid-eighteenth century political writings. Dismissing Montesquieu's warnings, James Madison believed that *ambition could be made to counteract ambition*, and, on that predicate, a large and variegated republic was formed, lacking formal provisions capable of controlling factions or directing party spirit. Madisonian theory was influenced by David Hume, the Scottish Enlightenment philosopher who indeed had some sanguinary notions about controlling faction with faction, but only insofar as party lines were demarcated by *interests* rather than *principles*. Patronage, spoils, appointments, foreign policy, and monetary policy are examples of interests that can be fought over without severely testing the constitutional cage and political process. Power sharing, office seeking, and practical politics push parties toward the political center. On average, political machines and elite corruption, too, can temper extremes of conscience and political moralism. Yet, a fever pitch of quasi-religious fervor ensues when party divisions are drawn across *principles* rather than *interests*. In the parlance of modern political theory, party divisions based on principles transect *cultural cleavages*.[2] Political principles transecting cultural cleavages are volatile because they trace natural fault lines dividing individually selected liberal from group selected conservative ideologies.

It must also be kept in mind that large political parties are coalitions, not unitary blocks. Constituent factions within major parties can defect or reorganize, switch allegiance between larger aggregates or, alternatively, split off on their own to form third parties. A good example is the progressive faction within the current Democratic Party, which is often in a state of tension with the mainstream party and thus frequently threatens to defect. As discussed in Chaps. 10 and 11, the existence of subfactions within factions is not a unique feature of modern political parties, and the risk of defection by subfactions is a perpetual threat to any multilevel sociopolitical organization.

Both interests and principles are at play in all political systems, but the ratio between the two varies. Third-party movements working outside extant party systems, like political realignments working from within-party systems, are revolts against the pragmatic centrism of interest-based politics. When pragmatism, corruption, machine politics, and power seeking temper political climates dominated by elite interests, realignments, and third parties reinvigorate ideological commitments that are otherwise suppressed and scattered, and which can end up orthogonal to the political divide or outside the Overton Window. At least in the American two-party system, these movements and realignments regularly come from the far Left where they are associated with labor and resource distribution, or from the far right where they are associated with decline and existential threats.

2 Reactions to New Political Forces

Recall from Chap. 2, comparative to any other politically relevant metric, *party affiliation* has the lowest aggregate genetic loading, and is furthermore subject to change when measured at different times (Závecz & Littvay, 2022). Genetically informed political ideologies and political party platforms can drift far apart or align as the moon and sun align on a solstice. As Chap. 15 reviewed, Huntington's Strategy can fan the flames of ideology by alienating the center in the attempt to activate the party base. The opposite stance has an opposite effect; tacking toward the center, as in the Downsian Model, alienates one's base, but captures moderates and independents from the center. In these ways, political parties and their machines stoke or suppress ideology as it helps or hinders electoral success: When stoked, ideological differences are organized, concentrated, and channeled, resulting in heightened political tensions; when suppressed, ideological differences are scattered, diffused, and diluted, resulting in reduced political tensions. In this latter scenario, when Downsian approaches are used to court the center, opposing political parties might differ modestly on monetary policy, wealth transfers, geopolitical positions, and the like, but without being substantially different with respect to positions most relevant to evolved ideological features of mind, such as immigration, hierarchy, tradition, and law and order. When these ideological features of mind are absent from debate, temperatures[3] sometimes simmer, and stability can ensue. However, when those on the right feel threatened by immigration, decline, and disorder, or when those on the left feel threatened by intolerance, injustice, and tyranny, and neither party speaks to that political threat, parties are apt to be challenged from without by third parties, or changed from within by realignments. Both political realignments and third-party movements can therefore refocus political debate to ideological faultlines. Capturing undecided centrists is a winning strategy only when committed ideologues vote the party line in lieu of better alternatives. Extant parties may *co-opt, stigmatize,* or *dismiss* electoral threats of emergent parties.[4] Such pragmatic, interest-driven, moderate stances adopted by party leaders often leave more extreme ideologues feeling unrepresented and disaffected. These political energies sometimes sputter and fume to no end, or are ineffectually signaled in protests; though at other times they are organized and channeled to the end of realigning existing parties or creating third parties.

Several authors, frameworks, and concepts inform this chapter's discussion of political change. Holt, Elkins, McKitrick, and Hesseltine are representative of important authors documenting the tumult of political history, including third-party movements, party transformations, factionalism, and sectionalism. Grande (2008, p. 330) outlines three ways in which countries and their political parties react to emergent political forces: (1) new parties agitating *transformation from outside*; (2) extant parties managing *transformation from within*; and then there is (3) *absorptive change*, a process wherein political pressures are absorbed within existing party structures in such a way that ideological commitments are not substantively changed. To simplify, Grande is really saying that political force can effect change or be absorbed; and secondly, that political force can come from within or without. Shifting and narrowing of the *Overton Window*, a sort of political strike zone delimiting an acceptable and active zone of political debate, is also relevant to discussions about political realignments and third-party movements. The Overton Window can narrow, stifling those with strong ideological commitments, either to the left or right, which can elicit a third-party movement. Or the Overton Window, instead of narrowing, can shift, making new political stances debatable and old political stances undebatable, which can be both a cause and an effect of political realignments.[5] Parties and their machines can attempt to size and shift the Overton Window, but its ultimate parameters and position extend from broader sociocultural forces. The work of Fowler and Smirnov (2007) on political mandates is also relevant. Post-election, parties adapt to electoral outcomes. Adaptation is greater for the losing side, which attempts to shift its platform toward the center, as demarcated most clearly by post-election results. The winning side may also shift, not toward the center, but toward their ideological ideal. Of course, modest shifts generally occur, but when there are clear mandates, when a party wins by overwhelming majority, political realignments can occur (Fowler & Smirnov, 2007). This straightforward relationship is complicated by several factors detailed by Fowler and Smirnov, one of which is *electoral volatility*, an index of noise around a median. Electoral outcomes, much more than polling, provide an honest signal to both parties on where the center lies, but even this is not perfect; and its accuracy fluctuates. Fluctuation is indexed, not exclusively, but importantly, by voter turnout. As voter turnout increases, so does confidence in electoral results as honest signals of the median.

3 Political Realignment

Objective change can combine with attitudinal change and broad cultural forces to effect political realignments. This stochastic complexity has proved difficult to model and predict by any overarching quantitative theory (Fenton, 1982; Fogel, 1992), though individual instances of realignment are well documented. For example, Russell J. Dalton, the author of *Political Realignment: Economics, Culture, and Electoral Change,* cited in this book's introduction, documents the way in which cultural issues have risen to prominence in affluent, Western societies, causing party realignments that traverse party lines previously established by economic policy. Lubell (1970) and Phillips (2014) were among those detailing the Nixon era shift that brought forth a recognizably conservative

Republican Party (Ladd et al., 1971). Fogel (1995) discussed a politically relevant religious resurgence in 1990s, which is part of a larger set of cyclical religious revivals bringing significant political change. Huston (2011) analyzed immigration, slavery, and temperance, three interacting issues that augured political realignment in the 1840s and 1850s. Barber and Pope (2024) discuss educational attainment replacing economic earnings as causing a political realignment. Gilbert (2024) reviews Kennedy era civil rights inspired realignment. Jenner (2010) discusses war and isolationism during Franklin D. Roosevelt's tenure in office as an instance of realignment around foreign policy. Weed (1989) reviewed this same time period, focusing on political realignment among Republicans imposed by overwhelming electoral defeat. These are just a sampling of the literature on US political realignments, which of course is part of a broader, international whole. Rather than creating any compendium of political realignments, we seek only to feel for the pulse of political ideology in all of this.[6]

As reviewed by Schofield et al. (2003), Schattschneider (1960) and related authors emphasized elite-led political realignments; though more recently authors such as Mayhew (2000) have seen political alignments coming from the bottom up with elites simply bending with the winds of political change.[7] As Schofield and coauthors explain, however, there is truth in both positions, and there is reciprocal influence between the electorate and elites. Schofield et al. (2003), in their analysis of the 1860 presidential election, consider ideology in connection with economics and regional coalitions to explain the resulting political realignment that brought the new Republican Party to power. More broadly, these authors provide figures representing the 1860, 1896, 1932, and 1964 political realignments that feature ideological cleavages cutting across a political field populated also by economic and regional issues. Notice that the election years identified as realignments were years of turmoil, respectively triggered by the Civil War, the Panic of 1893, the Great Depression, and the Civil Rights Era. Realignment often follows crisis, whether its initiation or conclusion.

Political realignment can follow the waxing and waning of ideological contention. Consider first the waning of ideology. Once the original Jeffersonian Republican party vanquished Federalism, it grew to encompass most active politicians during the *Era of Good Feelings* (Dangerfield, 1986). With the election of James Monroe in 1817 coming on the heels of sixteen years of Republican dominance across Jefferson and Madison's presidencies, it seemed that one party replaced the other for all time. So, rather than a temporary transition in party dominance during an election cycle, this seemed, to some during this early time in US politics before the two-party system was clearly entrenched, to amount to a stable equilibrium.[8] Of course this was a fantastical hope. Even during the Era of Good Feelings, political differences were latent.[9] However, the political convergence experienced during the Era of Good Feelings was indicative of waning ideological contention, allowing the Republican Party to encompass a broader swath of sentiment after softening its founding values of radical dissent which had shaped it during twelve years of opposition to Federalist dominance. Representative of this change, John Quincy Adams, son of John Adams, became secretary of state to President James Monroe, successor of Jefferson and Madison. A generation before in the contentious 1790s, John Adams functioned as the *de facto* head of the Federalist Party opposed to Thomas Jefferson who headed the Republican opposition. So, it is significant that the son and successor of Federalist and Republican rivals found themselves working together in a newly moderated and expanded Republican party. The Era of Good Feelings serves as a valuable counterexample to the more common crisis-driven engine of realignment. This political realignment came on a tide of prosperity and stability. Thus, this is an example of good times bringing a waning of ideological contention, quelling partisanship, and ushering in a political realignment.

As good times can quell ideological divisions, bad times can augment ideological divisions. This is the tenor of Watts (2021), an author who characterizes the decline and fall of Rome as *a dangerous idea* because it, and decline generally, can drive fear, which augers cessions of power to demagogic leaders. He documents some early Roman leaders artificially stoking declinist anxieties to grab power. Nevertheless, Watts just as often documents real cases of decline, showing leaders that minimized realities or simply forged ahead with affairs of state. Whether it was the specter of decline stoked by demagogues or the reality of decline handled by well-meaning rulers, the result was often some form of *innovation*[10] that undermined forms, norms, and institutions. In this author's warning, we see the opposite of the Era of Good Feelings. Crisis generally, whether brought on by decline or whether indicating the possibility of decline, can arouse fears among those on the ideological right, akin to that discussed by Stenner as the *authoritarian dynamic* and by Inglehart as the *authoritarian reflex*. By contrast, ecological abundance, economic growth, and wealth, the absence of existential threat and like factors quell partisanship and limit between-group difference and ideological difference within the body politic. In either case, both the quelling and stoking of ideological divisions change the political landscape and can usher in a political realignment.

4 Third-Party Movements

The Greenback, Free Soil Party, People's Party, Progressive Party, Socialist Party, Reform Party, Libertarian Party, and the Anti-Masonic Party are examples of the United States of America's[11] many third-party movements (Critchlow, 2015; Holt, 1992). Moderate positions are not often shared by leaders of third-party movements, who tend, instead, to be driven by rigid principles. Third parties therefore are apt to be spirited, but small. Adherents are activated by an ideological principle or slate of principles to which they adhere zealously; yet that zealotry disallows moderation and compromise, both of which are prerequisites to building a broader coalition that might actually be competitive to one of the two major parties.[12] This tradeoff is common to all groups; cohesion and size are pitted against one another. Just as a political party can be strongly ideological but consequently small, so a state can be genetically homogenous but consequently restricted in terms of population and power, and a religion can be devout and exacting but consequently limited in terms of devotees.

In charting the history of third-party movements, Hesseltine (1957) mentions the Workingmen's Party, the Equal Rights Party, the Anti-Monopolists, the Honest Men's Party of Massachusetts, and the Readjusters of Virginia as examples of those many short-lived, localized third-party movements that never gained national influence. More successful third-party movements "threatened the existing parties and forced lasting change of policy" (Hesseltine, 1957, p. 10). Powerful, national third parties cohere around a neglected ideological issue that is thrust into national consciousness to the end of threatening one or both prevailing parties. Third-party ideological commitments are too narrow and dogmatically argued to succeed directly, but the third party functions like a siege engine, propelling a neglected ideological commitment into play, thereby shifting or expanding the Overton Window around which the two dominant parties debate. Again, this is so often the function of a third party; rather than actually displacing one of the two prevailing parties, they highlight issues which only get addressed by the major political parties because of the threat of the third party. To the extent that the issue raised by the third party is addressed, it weakens the third party. Third parties, in effecting change within the two-party system, dissipate their force, sometimes to the extent that they disappear. For example, communist parties were divided and weakened when many apparatchiks took jobs within Roosevelt's New Deal bureaucracy. Before that, the same fate befell progressives who attempted to mount a third party run under the leadership of Wisconsin Governor, Robert LaFollette; a run that captured millions of votes and brought progressive concerns about monopolies, inequality, abuses of power, and corruption into the political fore. Though the progressive party failed to gain power, progressive party sentiments were an expression of left leaning ideology that was to some extent adopted both by the Republican and Democrat parties (Hesseltine, 1957). A rising tide of wealth caused many to turn away from LaFollette and the progressive platform. With *a chicken in every pot and two cars in every garage,* and with the promise of upward mobility and pre-depression stock gains, progressive reform was less urgent and interesting, and oligarchic control seemed less corrupt and problematic. Progressivism followed the fate of many third-party movements. It was a sufficiently strong force to invoke change, but spent its force on pushing a major party in its preferred direction, and died out as a third party thereafter, with a temporized version of its agenda absorbed within Roosevelt's New Deal Coalition (James, 2000).

Third-party stalwarts, including the socialist historian and political aspirant Hesseltine whom we have been citing, take umbrage at what they perceive to be defection. When one of the two major parties change in reaction to a third-party threat, and thereby absorb some of the third party's voters and leaders, this is often seen as a Machiavellian move to preserve dominance by those that remain loyal to the third party. Of course, this interpretation can be true, or the ideological shift could be genuine; but either way we see the group dynamics in the same way. Standing parties can only meet the threat of being overtaken by third parties by co-opting some measure of their party platform, and thereby, some number of their voters (MacKay, 1972, p. 14). In reflecting on the dissolution of the progressive party after its 1924 bid, and thereafter the movement's probable sociological inertia in ushering the Roosevelt's New Deal in a nation reeling from depression, MacKay (1972, page 262) wrote thus:

> As a rule, only a minor party can afford the luxury of expecting defeat at the polls; it can afford to be recklessly courageous. Our weakest parties are our most unequivocal parties. In contributing new issues, the third parties become the trial horses of the American party system. Hence the value of a third party must be accepted in terms of its effects upon the other parties rather than by an exclusive history of its own efforts and achievements.

Extreme ideological commitments are no basis for broad coalitions, precluding the actual success of third parties; but the power of committed ideologues, especially when organized into third-party movements that seriously challenge prevailing parties, can cause one or both prevailing parties to realign, with that realignment generally causing parties to more completely overlap with innate political ideological divisions. Third-party movements challenge entrenched party power. In doing so, they often die off, but not before injecting "much needed vitality into old parties" in a sort of "splendid antidote for political indolence and stagnation." As F. E. Haynes noted, "…the chief function of third parties has been to bring new issues before the people; they force new policies upon the older parties, and after accomplishing their work, they pass away" (MacKay, 1972, p. 262).

5 From Anti-Masons and Free Soilers to the American Party and its Antithesis

The life and death of a bricklayer from Western New York inaugurated the Anti-Masonic Party (McCarthy, 1901), arguably the United States of America's first and most powerful third-party movement. William Morgan, the bricklayer in question, signaled a forthcoming publication exposing Masonic practices and secrets. Morgan was arrested twice for an unpaid debt and ostensible theft. He was then released and disappeared. In a vacuum of verifiable fact, conspiracies arose. Many of these had Masons in some form involved in Morgan's original arrest, later disappearance, and in the cover up of both by highly placed Masonic members (Adams, 1847). The Anti-Masonic movement is representative of the plebeian populism of many third-party movements, which act in opposition to perceived elite corruption (Holt, 1992).

The Free Soil Party was a patchwork of practical and ideological motives. Though overtly against slavery, free soil roots extend back through the 1840s and encompassed northerners working against black freeholders alongside those hoping to contain slavery. David Wilmot, the Pennsylvania congressman who introduced the *Wilmot Proviso* to exclude slavery from territory acquired after the Mexican American War, in a single line written in 1852 to Franklin Pierce, captured the motive power and cohesive force of the Free Soil movement: "I am jealous of the power of the South." Wilmot describes the longstanding south-shifted balance of power, extending back to the early days of the Republic. That was a time when there was the so-called *Virginia Dynasty*, broken up only by one New Englander, John Adams. Otherwise, four of the first five presidents were Virginians. So Wilmot speaks of humbling the slave power, though he is speaking more of the South generally than of slavery particularly. In short, the Free Soil Party was a sectional party. The motives to join the Free Soil Party, Holt states, were shared by those thereafter flocking to the standard of the subsequently formed Republican Party. The Kansas-Nebraska Act polarized Democrats, driving third-party moves, attracting Van Buren, Preston King, David Wilmot, Francis Blair, Gideon Welles, and Hannibal Hamlin, none of whom could see a way toward "checking the slave power within the party" (Holt, 1992, p. 75). Fault lines emerged across class and section, with landholding Southern aristocrats controlling the Democratic party to the increasingly intolerable frustration of middling Northern bourgeois.

In an effort to further illustrate the dynamics of realignment and third-party movements, we focus on the *American Party*, a nativist party bonded by deep worries about Irish and German Catholic immigration that *shattered* prior political alignments in antebellum America (Critchlow, 2015, p. 54; Baum, 1978). The American Party, also known as the *Know Nothings* because they began as a secret society before becoming a political party, garnered ideological energy from those affronted by rapid social change, those who feared immigrant slums, those who deplored easily organized immigrant voting blocks, those displaced wage workers, and those whose small hamlets were swelled with newcomers. The backdrop was middle nineteenth century, a time that had imparted a dizzying and disordering degree of immigration, with waves of European Catholics from Ireland and Germany stoking local resentment, especially among native born Protestants. Formed around an outgroup threat not addressed either by Whigs or Democrats, the American Party was driven by fears of displacement and decline. It pitted natives against immigrants, and had an innately conservative ideological bent. Know Nothings evinced, "ferocious hostility toward both old major parties and the selfish and unresponsive politicians who led them" (Holt, 1992, page 9). Holt (1992, p. 9) stresses the absence of recourse offered by either the Whig or Democrat parties:

> ...hostility to both the Whig and Democratic parties derived in large part from the prior disappearance of issue-differences between them that made both organizations seem hollow, meaningless, and spoils-oriented machines that now seemed to threaten, rather than protect, the cherished American ideal of republican self-government.

In explaining the rise of the Know Nothing movement, Holt cites Rutherford B. Hayes, who attributed the movement to "a general disgust with the powers that be." Similarly, James A. Garfield asserted, "I am exceedingly disgusted with the wire-pulling of politicians and the total disregard for truth in all their operations." The American Party, though it did not elect a candidate, *gutted* the Whig party, leading to its collapse and eventual replacement by the then newly formed Republican Party. Many Know Nothing Party members were disillusioned when their party, after having gathered power and garnered votes, absorbed more traditional societal elites like Simon Cameron and Millard Fillmore. Through a combination of patronage and nativist rhetoric, the Republican Party capitalized on these developments, wooing disillusioned Know Nothings by taking on some of the American Party's trappings and speaking to persistently neglected immigration policy.

6 Conclusion[13]

Twenty years after the Anti-Masons helped shape the Whig party, the Know Nothings helped destroy it. The Republican Party successfully cobbled together most constituents from these two major third-party movements, neither of which was sufficiently powerful to capture control of the presidency and other top federal posts. As Holt notes, the Republicans did this by playing to the antipathies that spawned each movement. Former Anti-Masons were sung a song of "moralistic antagonism against slavery," while former Know Nothings were sung a song of "virulent anti-Catholicism." To each, there was, in this new Republican Party, the opportunity to viably oppose the Democrats and, in the process, this had sectional implications. The realignments successfully captured political power, but reorganized the electorate in a sufficiently sectional way so as to precipitate civil war.

Within this broader story of nineteenth century US history, there are some general lessons specific to understanding political realignments and third-party movements. First, large political parties are coalitions, not unitary blocks. Defections and reorganizations can be viewed as based on the behavior of the constituent factions within major parties. Such smaller factions can either switch allegiance between larger aggregates or, alternatively, split off on their own to form third parties. Political realignments and third-party movements, "by definition," signal discontent with the prevailing parties. Third parties, gelling around neglected ideological issues, "indicate the failure of the system to integrate the emergent political demands of significant portions of the electorate" (Holt, 1992, p. 88). When sufficiently powerful, third parties force prevailing parties to absorb their pressures by incorporating, in part or full, the third party's ideological grievance. Third parties are of interest in that the issues around which they coalesce are so often some mainstay of the liberal or conservative temperament, which has become neglected or unrepresented within the two-party system. Third-party movements arise when both prevailing political parties neglect a mainstay of each respective side's temperamentally informed agenda. Elites often co-opt parties and use them for self-serving ends, causing political parties to become misaligned with political temperaments. Third-party movements are then seen as a corrective, bringing the party and temperament in line. Yet, as the examples within this chapter implicitly suggest, significant realignment and national third-party movements are powered as much by group competition as by ideological difference. Anti-Masons were organized around class cleavages, fueled by opposition to elite corruption; Free Soilers were organized around sectional lines, fueled by opposition to southern dominance; Know Nothings were organized around nativist security, fueled by opposition to immigrant power. In this way, movements strong enough to threaten extant parties with realignments or third-party alternatives have combined the *between-group competition among interest groups* and the *within-group competition among ideological positions* discussed in Chap. 8 as the two sides of the political.

Notes

1. Holt (1992, p. 88).
2. Comparative to economic cleavages, cultural cleavages have explosive potential, threatening the structural integrity of political confines because they cut across the ideological divide.
3. Levendusky (2023) uses the term *affective polarization.*
4. An analogue of this third factor, absorbing the political thrust of a third party, had operated outside the party system in earlier days of revolution such as when, during the English Civil Wars, Leveler radicalism was partly dissipated by acceding to demands, or when, during the revolu-

tions of 1848, monarchies and other existing governments reactively or proactively adopted constitutions, granted rights, or in other ways gave ground as a way to check revolutionary force. Very often if not inevitably, this fractures the movement, whether it is a third party or a revolution. Public opinion can be appeased and enough within the radical or third-party movement can be appeased to negate the threat. While this may leave a core of disgruntled faithful, their ranks are weakened, and their power broken.

5. Shifts in Overton Windows can take place over decades. This point is made by Brick and Phelps in their concluding review of *Radicals in America.* What was once radical can become debatable. "Radicalism," Brick and Phelps (2015, p. 311) state, "becomes invisible, paradoxically, in its victories." By way of illustration, the authors point to George Herbert Walker Bush's signing of the *Americans with Disabilities Act* in 1992. At that time, few were thinking of the radical agitators that pushed the Overton Window in the direction of making this possible, including Hellen Keller, Randolph Bourne, Edward Roberts, and so many other individuals and groups stretching back to 1911. Radicals rarely find themselves in power, but work as agitators. As they push to tear down and rebuild from the ground up, these forceful visions can spend their power down, agitating against the mainstream, but retaining sufficient inertia to create a practical effect on a prevailing party or person.
6. Motyl (2012) provides a textual analysis of party platforms, documenting their changing usage of moral emotionally laden terms.
7. There is great debate about political realignments, especially whether they are sudden or slow, and come from the top down or the bottom up.
8. Realignments come after the end of *stable equilibria* (Schofield, 2006, 2007, 2009).
9. As Holt explains, internal party tensions extended from the expansion of the Republican Party.
10. *Innovation* is used here in the older sense of the term as a pejorative. It denotes the perversion of an ideal or practice.
11. Third-party movements assume a two-party system, and so we are here focusing on the United States of America.
12. We have generally focused on the Anglo-American political systems that divide along a single fault line, creating a two-party system. Continental European politics often feature a multiparty system that allows multiple interests to be represented, which changes the dynamics of party formation, party machines, and power sharing. The system of constructing

large-scale political parties in the USA from smaller constituent subfactions is actually not too different from the system used by many European parliamentary systems, the main difference being that, in the latter, the political parties participate in the election separately and only afterwards form coalitions to create a functional government. In the USA, the subfactions amalgamate into the two major political parties *first*, before participating as such in the election, such that the coalitions are already prefabricated.

13. This was neither a compendium of political realignments nor third-party movements. We have not discussed them all, or any in depth. We can write a chapter in place of a book on political realignments and third-party movements because our aim is merely to further distinguish between interior features of mind and exterior features of party.

17

Populism, Progressivism, and Nationalism

Steven Charles Hertler, Aurelio José Figueredo, and Mateo Peñaherrera-Aguirre

Of those two developments of which we have spoken, and which constitute the fact of civilization, the development of society on the one hand and of humanity on the other, which is the end, which is the means? Is it to perfect his social condition, to ameliorate his existence on earth, that man develops himself, his faculties, sentiments, ideas, his whole being?—or rather, is not the amelioration of the social condition, the progress of society, society itself, the theatre, the occasion, the *mobile*, of the development of the individual, in a word, is society made to serve the individual, or the individual to serve society? On the answer to this question inevitably depends that whether the destiny of man is purely social; whether society drains up and exhausts the whole man; or whether he bears within him something intrinsic—something superior to his existence on earth.

—Francios Guizot, The History of Civilization

S. C. Hertler (✉)
Saint Elizabeth University, Morristown, NJ, USA

A. J. Figueredo • M. Peñaherrera-Aguirre
University of Arizona, Tucson, AZ, USA
e-mail: ajf@arizona.edu; mpeaher@email.arizona.edu

S. C. Hertler et al., *The Evolution of Political Ideology*,
https://doi.org/10.1007/978-3-031-83941-2_17

1 Introduction

An extension of the sixteenth chapter, this seventeenth chapter explores political realignments and third-party movements as they have been specifically harnessed by progressive populism associated with the political left, and nationalistic populism associated with the political Right. Populism, though frequently associated with the right, is not ideologically valanced. As discussed by Ostiguy (2017), populism is, however, invariably a movement of the masses, even when it is exploited by charismatic demagogues. With impatience for institutional norms and impersonal authority, populism pits plebeians or proletariats against urbane, learned, and cosmopolitan elites. Given that populism is merely an index of mass revolt against elite power and processes, we consequently explore two populist movements, one originating on the political left and one on the political right. First, progressive populism was both a social and political movement broadly calling for reformation, rectification, and restitution. From inception to end, we can see the banner of American progressivism begin as a social movement in the hands of Jane Addams and Jacob Riis, continue as a political movement influencing Republicans under Theodore Roosevelt and Democrats under Woodrow Wilson, and then briefly become strong enough to support Robert La Follette's presidential bid as a third-party progressive candidate before the movement's force was finally absorbed by FDR's New Deal, which moderated America's ethic of self-reliance (Howe, 2009) with a slate of socialistic programs. Whether as a social movement advocating reform, a political force operating on extant parties to secure voting rights, or as an independent party deploring child labor, multilevel selection allows us to see the unity in the progressive movement's advancement of individually selected rights. Second, nationalistic populism began as sentiment among the masses and variously manifested as a movement and as a party. As reviewed in the prior chapter, the nativist *Know-Nothing* American Party challenged the Whigs from without as a third-party movement, bringing about a realignment to homogeneity, nationalism, traditionalism, and other mainstays of the conservative temperament. Also considered are war, immigration, geopolitical strength, and related conditions (Wimmer, 2013) that elicit nationalistic sentiments and elect nationalistic leaders.

Both the progressive and nationalist movements have suffered the same fate as other third-party movements: Their ideological force was absorbed by an existing party, necessitating some movement toward one or the other ideological pole; either to individually or group selected ideologies latent in the body politic.

2 Populism

Populism offers a discontented and disenfranchised people (1) an exaggerated distinction between themselves and elites; (2) a clear grievance; and (3) a rapid and powerful, if sometimes extralegal, means to rectify wrongs to those that have little political power. Populism can of course be channeled or stoked by sincere or demagogic leaders, but arises from a groundswell of sentiment. It differentiates patricians, elites, oligarchs, or some similarly small and powerful segment of society from some essentialist core that is taken to be the people, the citizenry, the masses, or some similarly large and disenfranchised segment of society. Populism is not pure protest; it is also an effort on the people's part to take back the power delegated to a ruling class perceived as wayward, treasonous, or corrupt (Mudde & Kaltwasser, 2017). Once again, it is a bottom up movement placing "the people in a morally charged battle against the elites" (Gidengil & Stolle, 2022, p. 443). This is evident in Gidengil and Stolle's (2022, p. 443) further description of populism:

> All populists, no matter their background, argue that certain groups among the elites are corrupt, uninterested in the well-being of the people, and seek power for their own ends. Thus, we find a strong anti-elite sentiment in populist discourse that includes anti-establishment sentiments directed at prior or current power-holders and anti-expert opinion. Populists use moral language to condemn mainstream political power.

Populist movements are often associated with young males of lower educational attainment, though these correlates are attenuated or eliminated when focusing broadly on populism, rather than only on right-leaning populist movements (Gidengil & Stolle, 2022). For Runciman, populism is composed of four ingredients: *absence of war, financial hardship, technological advances*, and *augmenting inequality* (Brown, 2022). In Bugaric's (2019) words, populism is *Janus-faced* in that it can be right or left leaning. Populism is a movement, a sort of pre-revolutionary temper (Oswald, 2022). As Gidengil and Stolle (2022, p. 443) state, populism is a "thin ideology that can be molded to various political goals of both the left and the right." Their use of the word *thin* does not go far enough. Again, it is difficult to conceive of populism, if having proper historical and cross-national perspective, as an ideology at all.

To assimilate these observations into a more explicitly multilevel selectionist framework, one could conclude thus: Populism is a class-based social cleavage pitting the people against elites.[1] This cleavage inevitably emerges on some ideological grounds; if elites are perceived to err on the side of hierarchy and stratification, left-wing populism may arise, whereas if elites are perceived to err on the side of liberalism and pluralism, right-wing populism may arise. It is possible, as has been seen in recent American politics, that elites drift and perceptions of corruption pull for twin populists movements, one from the right and one from the left. Hereafter, we focus on what could be considered two forms of populism: Progressivism which is left-leaning; and nationalism which is right-leaning.[2]

3 Progressivism

Chapter 16 referred to progressivism in the form of a political movement that sometimes influenced one or both prevailing parties, or otherwise set up on its own when it could not bear the "compelling force of compromise and concession" (MacKay, 1972, p. 14). In 1924, Senator LaFollette's run garnered five million votes, providing an alternative to embedded operators within dominant parties. This progressive ticket drew and combined "populists, single-taxers, socialists, writers from the East, farmers from the West, railroad men, Bull Moosers, Greenbackers, politicians, college students, dreamers and opportunists" (MacKay, 1972, p. 12). Yet, America's progressive movement was not merely a time-bound third-party run, but a particular manifestation of individually selected psychology and its concomitant liberal ideology, as can be seen implicitly in

Merle Curti's *The Growth of American Thought*, wherein progressivism is connected to "faith in man's dignity, in his natural rights to liberty, to equality of opportunity, and to happiness...."[3]

Along with Walter Lippmann and Walter Weyl, Herbert Croly was a scion of progressivism (Stears, 2006). Croly placed his finger on the pulse of social reality and past history, drawing attention to the founding and the frontier. In effect, Croly argued that Americans were once content with a passive government that allowed talent to chase wealth. Nevertheless, where this governmental form once brought independence and opportunity, it thereafter failed to correct the growing inequality of an older America in which natural, free resources are claimed, and class disparities were augmenting (MacKay, 1972). The American frontier filled up (Turner, 1921). It no longer functioned as a safety valve. Masses of citizens and immigrants swelled cities. From thence came contention between capital and labor; strikes and scabs, muckraking investigations of insalubrious working environments and usurious working arrangements, Sinclair's *Jungle* and Steinbeck's *Dubious Battle*. Variously, Social Darwinists (Sachs, 2011), capitalists (Hurst, 2010), patriotic leagues, and vigilante corps colluded with government (Lovestone, 1923) and law enforcement to counter destabilizing and potentially revolutionary combinations of laborers and labor unions (Millan & Saluppo, 2021). There were *adjusters* and *stabilizers*; there were hundreds of strikebreaking organizations employing thousands of informants and other operatives to an estimated expense of eighty million dollars (Levinson, 1937). While their opponents were looking to government to compel labor and crush strikes, progressives alternatively exhorted government to shed its small, delimited, passive role that had classically folded its arms and allowed markets to reward meritocratic effort. The only way to bring back the "loyal adhesion of the economically less independent class," progressives argued, was to acknowledge and fulfill demands that "the state...help him to conquer a higher standard of living..." (MacKay, 1972, p. 19).

The progressive rejection of laissez-faire individualism in favor of paternalistic policy is well captured by Woodrow Wilson:[4] "To let the individual alone is to leave him helpless as against the obstacles with which he has to contend" (Purdy, 2009, p. 183). Freedom and individualism, progressives argued, no longer realistically determined outcomes within increasingly complex and impersonal late modern societies (Smith, 2017). Pestritto and Atto (2008), authors of an edited volume on progressivism, contrast the classical liberalism of the founders and the constrained constitution they created as an impediment to progressive policy implementation:

> Progressives had in mind a variety of legislative programs aimed at redistributing private property in the name of social justice. The Constitution, if interpreted and applied faithfully, stood in the way of this agenda. (Pestritto & Atto, 2008, p. 3)

Progressivism pointedly illustrates how parties and ideology can become misaligned, in that progressivism was a left oriented ideology that was largely unrepresented by either Republicans or Democrats; and furthermore, when progressivism did rear up, it affected and altered both the Republican and Democrat parties. Progressivism existed at that time as a movement appealing to both parties, but in Theodore Roosevelt (a Republican who later ran explicitly as a progressive) and Woodrow Wilson (a Democrat) we can see progressivism embraced on different grounds that reflect their different political ideologies. Wilson embraced progressive foreign policy as a moral doctrine, seeking to *serve not subdue* (Smith, 2017), while Roosevelt embraced progressive foreign policy as an imperial justification. America would have to gain power and impose its will and vision through direct and indirect means so that, at some future point, nations like Cuba and the Philippines would be ready for democracy. This reasoning made Wilson, for practical purposes, similar to Roosevelt.

As Wilson shared these hopes of proselytizing the world with Theodore Roosevelt (Smith, 2017), the same point is illustrated by Herbert Hoover and Franklin Roosevelt, another pair of rivals from different political parties that were both nevertheless influenced by progressivism. Hoover fit some definitions of a progressive in emphasizing faith in technical administration of society by expert fiat. As he was rising to prominence, Hoover lived the life of a self-made man; his was not precisely of a Horatio Alger story for he did not start so low, but a life marked by the ethos of American Individualism. However, once situated with power and wealth, he became paternalistic and believed that those so placed had obligations to society.

Especially after his activities in World War I's Belgium relief efforts, he showed a zeal for planning that emphasized top down expert bureaucratic control that always seemed pure of motive, but which sometimes erred toward socialism, and which always smacked of paternalism (Burner, 1979). Franklin Roosevelt was possibly less temperamentally enamored with progressivism, but more readily incorporated progressive platforms into his new Democratic coalition. Progressive leaders pressed progressive agendas on FDR, which were accepted in exchange for their support in his reelection bid. The La Follette contingent accepted the alliance, even as many among them intended to build on electoral wins and position progressivism as a stable third party on the national scene (Scott, 2000). Of course, the arrangement with FDR undermined the latter goal; the two were antithetical. Progressives were simultaneously being courted by the Republican Party, and so FDR had to make more sincere shifts left-wards to retain them within the orbit of the Democratic Party. This involved FDR dissociating himself more earnestly from corporate interests that had theretofore been politically expedient. This policy shift was advocated by Edward J. Flynn of New York, who counseled Roosevelt to attract unattached voters from inside and outside the progressive movement, many of whom were members of racial and religious minorities. The bait was "radical programs of social and economic reform" (Flynn as quoted by Scott, 2000, p. 204). With this strategic shift, FDR was able to add to the *solid South*, "Progressive Republicans, labor, [and] ethnic big-city machines" (Scott, 2000, p. 204). This proved a reliable reelection coalition.[5]

Progressivism has been critiqued in J. Passmore's religiously conservative, *The Perfectibility of Man*. Therein, progressivism is cast as the most modern instantiation of human hubris in a long line "beginning with the Renaissance," which "began to maintain that in their relationships to their fellow-men, rather than in their relationships to God, lay their hope of perfection" (Passmore, 2000, p. 260). The danger of a secular society pursuing progressive causes is that those progressive causes become sacred cows; the people worship at the fount of progressivism, and so are worshipping themselves (Passmore, 2000). Unlike Passmore's reactionary critique, other detractors singled out progressivism from earlier forms of progress. Presaging criticisms leveled by modern democrats, libertarians, and conservatives, Grover Cleveland effectively equated progressivism with paternalism, charging that it took once *self-reliant* citizens and sent them scrambling for *sordid gains*. Paternalistic policies stupefy "…every ennobling trait of American citizenship" (Purdy, 2009, p. 62). Cleveland's strict constructionism contrasted with progressives like Teddy Roosevelt and Woodrow Wilson who had less stringent constitutional scruples. As one example among others, Roosevelt had the war department survey and subsequently maintain the Statue of Liberty, which Cleveland neglected for want of constitutional authority to do otherwise. As progressivism, in the name of *paternalism*, rode roughshod over forms thought vital to containing government, it also made, in the name of *interventionism*, individuals instruments of freedom as an abstract ideal, rather than its beneficiaries (Fischer, 2005).[6] As Pestritto and Atto, authors of *American Progressivism*, state, progressives were impatient of divided power, which Wilson saw as an impediment to the general will, as well as constitutional limitations, which Croly thought, "tend[ed] to undermine the foundations in human nature and human will upon which the whole superstructure of a progressive democrat society must admittedly be built." Conservative critiques of progressivism complain of the progressive rejection of historical values, constitutional forms, and founding myths. For example, progressives were, as Terrence Ball (2001, p. 110) discusses, embarrassed by America's Puritan founding, which was deemed *hypocritical, peculiar, embarrassing, illiberal, conformist, orthodox*, and in other ways inimical to "the exuberant optimism and individualism of American's liberal past and progressive present." Similarly, Ball (2001) also shows the way in which progressives were critics, debunkers, and myth exposers, citing the famous example of Beard's economic history of the constitution which framed the constitution as tool of the elite rich to capture the reins of government after the Revolution.[7]

4 Nationalism

Renaissance humanists and Enlightenment philosophers looked beyond national borders when corresponding and philosophizing. Turgot and Condorcet, for example, saw all humanity as the *object* and *agent* of authority. Kant thought likewise, and even "viewed a world community as the outcome of a gradual progress to the point of making states and nations redundant as the vanguards of cultural diversity." Reason and justice would progress, allowing "spontaneous cooperation, unrestrained by any boundaries" that "served the higher ends of a true human community on a global scale" (Bartelson, 2009, p. 166). Globalism seeks to dissolve borders and nations to a "universal community of all mankind" (Bartelson, 2009, p. 172). All the above defines nationalism by contradistinction. These global communities extending from individually selected, left oriented ideologies are not in any way like nationalism writ large. They are not "ideologies of empire" (Bartelson, 2009, p. 179); they have nothing to do with the drive to extend political dominion or spiritual dominion. Rather than being an Enlightenment value, nationalism is often associated with Romanticism. "The primary conviction of Romantic nationalism," Guibernau (1999, p. 23) writes, "is that culture, a particular way of life and the more important social institutions are essentially formed and shaped by the nation." As discussed in Chap. 1, here we see the nation reified; Romantic Era nationalists conceived of the nation as an "expression of a unitary force," speaking of the "soul, mind or spirit of a people" as an entity unto itself.

Opposite Enlightenment cosmopolitanism, nationalists celebrated the uniqueness of their national history, language, and achievements. While this seems benign enough, critics saw a short step between celebration of difference and claims of superiority. Nationalism is routinely associated with might and chauvinistic conquest; its "vulgarity and atavism" is cast as incompatible with "tolerance and reason" and as "essentially incompatible with the liberal attitudes and civility required of a truly civil society" (Goldman, 2003, p. 112). Eric Hobsbawm, John Breuilly, Benedict Anderson, and Ernest Gellner are emissaries of an increasingly dominant *social constructivist* approach to nationalism studies, which is effectively a critique of Romantic Era nationalism or what is often referred to as *primordial* nationalism. As exemplified in Anderson's notion of *imagined communities*, social constructivists depict both nationhood and nationalism as comparatively recent, top-down fabrications. Distinct from social constructivist perspectives, primordial nationalism encompasses "kinship and descent, language, religion, and customs, as well as historical territory"; all features which have ancient roots and have a place in time before history, and a place in mind outside of rationality. Primordial nationalism, evident in Herder, Fichte, Lord Shaftesbury, Vico, Montesquieu, and the latter works of Rousseau, captures the ineffable sense of community that inspires "mass fervor and self-sacrifice" (Smith, 2000, p. 25). Primordial nationalism runs counter to social constructionist narratives, especially those with postmodern elements that are "unable to grasp and credit the emotional depth of loyalties to historical nations and nationalism" (Smith, 2000, p. 62). Certainly there is, in primordial accounts, some degree of reification and modification, but not wholesale fabrication, as social constructionists charge.[8] Smith concludes that nationalism rests on six points, which include common names, myths, memories, shared cultural practices, territory, and sentiments of solidarity (Guibernau, 1999). He provides a compromise position between social constructivists and primordial nationalists.

Smith's compromise position seems to have satisfied neither the social constructivists nor the primordial nationalists. Maxwell (2020), for instance, charged Smith with perpetuating "mischaracterizations, fallacies, and contradiction" in his article entitled *Primordialism for Scholars Who Ought to Know Better: Anthony D. Smith's Critique of Modernization Theory*. Adrian Hastings (1997), who most explicitly fashions his conceptualization of nationalism in counterpoise to that of the social constructivist narratives now in vogue, argues that Smith only goes part of the way: "We see this in Anthony Smith, a modernist whose consideration of ethnicity has forced him to abandon the full orthodoxy of late eighteenth-century origins but who refuses quite to grasp the nettle of medieval nations" (Hastings, 1997, p. 54). To those social constructivists who see nationalism as nothing but a nineteenth century fabrication, Hastings answers with the "anthropological constant." He finds nationalism in the deeper past. Nationalism does not need a territorial state replete with governmental bureaucracy; it requires only the idea of a nation, with a people and their myths and traditions. The *anthropological constant* is a bridge of sorts, sanctioning the antiquity of nationalism by seeing it as extrapolated from tribalism. To Hasting's point, nations can exist prior to states, as in Alfred's England and Renaissance Italy (Hirschi, 2011, p. 34). From the credibility of nationalism's origins, we now consider the durability of its form.

Immediately, within the first few sentences of the introduction, Montserrat Guibernau, in her *Nations without States*, seizes on the levels of selection problem as it applied to political organization. The modern nation state is asked, Guibernau notes, to increasingly surrender sovereignties to the global community, even as nationalism surges in Scotland, Catalonia, Quebec, Flanders, and other subnational regions. "Contemporary politics," Guibernau asserts, "are grounded upon a constant tension between pressure exerted by sub-state movements which put forward nationalist claims, thus questioning the state's legitimacy, and the growth of supranational organizations which require the state to surrender relevant attributes of its traditional sovereignty" (Guibernau, 1999, p. 1). In this passage, the author explicitly identifies nation states as mid-level organizations under attack from above and below.[9] Guibernau (1999, pp. 14–15) further explains her position by contrasting the terms *state* and *nation*. A state is defined by the monopolization of power in a territory, whereas a nation is a "human group conscious of forming a community, sharing a common culture, attached to a clearly demarcated territory, having a common past and a common project for the future and claiming the right to rule itself." When used in conjunction, as in the term *nation-state*, both properties are shared; that is, there is a legitimate use of force monopolized by governmental power, which is expressed within a discrete region where citizens or subjects are both *united* and *homogeneous*. By contrast, nations without states persist when homogenized unity of a people is mismatched with territorial boundaries. Guibernau pointedly pursues the conflict between nation-states and nations without states in the context of globalism. This implicitly sets up a dynamic across levels of selection from region to state, to supranational federation. In the context of modern globalization, nationalistic regions frequently surge toward independence from nations, in reaction to their nation's cession of autonomy to global organizations. Thus, Guibernau (1999, p. 158) writes

> The demise of the traditional nation-state involves the partial loss of control over its territory, some matters being dealt with at a supranational level, while others are devolved to regional institutions which often put forward specific claims based upon cultural and historical claims over their territory.

In the concluding section of her book while referencing J. Rex, a contributor to an edited volume she previously published, Guibernau highlights the distinction between public and private domains made by post-traditional states. Post-traditional states attempt to cement societies exclusively within the public domain of law, politics, economics, and education, while tolerating, acknowledging, or actively encouraging heterogeneity in the private domain, which subsumes "moral education, primary socialization and the inculcation of religious beliefs." These features of the private domain, however, were central to the formation and maintenance of premodern nation states; and these are among the features activating group selected psychology, allowing allegiances evolved in relation to the kin group and tribe to be transplanted to the nation or state. Obviously, pluralistic nations have survived the dissociation of public and private domains, though vulnerabilities arise whenever cohesion and inclusion are pitted against one another. The main vulnerability may prove to be one of self-defense. Vulnerabilities in war might arise from asymmetries between polyglot territorial states and traditional nation states that can simultaneously draw psychological allegiance from a homogenized citizenry feeling both a cerebral loyalty from public motives and an emotional loyalty from private motives.

This same theme is taken up by Ikenberry (2011) who describes a process wherein civic institutions and abstractions assumed the unifying functions once exclusively shouldered by shared religion and heritage. This is a double edged sword. Whatever one thinks of it in terms of value judgements, from an evolutionary perspective on power vis-à-vis other groups, it can have twin effects. First it can strengthen the group by expanding to embrace disparate peoples (this was precisely the power of proselytizing religions), but this comes with the probable effect of weakening the group internally. This generates a group that is larger, but less unified. It includes members that might have been defined as the outgroup, but has brought in these members by eschewing notions of ethnic and religious identity. Nationalism is sometimes a resurgence of these earlier, ethnically or religiously informed foundations of unity; and when it operates in pluralistic societies, it often entails identification with an original core that are the "true citizens" and an explicit or implicit disidentification with the others perceived as interlopers or outsiders. Again, this is not an indispensable feature of nationalism, but does characterize some nationalist movements within pluralistic states.

5 Conclusions

We have flown low over the literature on populism, progressivism, and nationalism. At this chapter's end, we try to gain a more properly aerial perspective.

This chapter singled out populism as an ideologically unvalenced phenomenon. It is a class-based contention between elites and non-elites. The contention is not inherently ideological, but in practice takes on left or right tendencies to counterbalance elite drift to the center. We then provided the examples of progressivism and nationalism, both of which are enduring ideological positions which have effloresced respectively into left and right leaning populist movements. In doing so, progressivism and nationalism, different as they were in content and ideological position, signaled discontent to a ruling class and with a political system that failed to represent the range of latent ideological sentiment. Actuated by individually selected, left-leaning ideology, progressivism signaled discontent with governmental elites that had too often colluded with capitalist elites, to the abuse of laborers and consumers. Actuated by group selected, right-leaning ideology, upwells of nationalism signaled discontent with threats to the nation; for example from being *hollowed out* from above by global organizations or from below by transnational corporations (Bevir & Rhodes, 2010).[10] Especially in times of prosperity and peace, where political elites and their parties and machines pursue power, progressive or nationalistic sentiment can be neglected and remain quiescent. Truncated political choice becomes misaligned with the fuller range of latent ideological positions. By turn, elite corruption and rising inequality can evoke progressive sentiment, while immigration and financial hardship can evoke nationalism.

Progressivism's concerns with economic inequality, greed, and corruption extend from our species' history of meat sharing and altruistic cooperation; they call to mind aspects of human nature highlighted by behavioral economics studies wherein participants spite themselves to punish unfair division of resources (Baumard, 2010), or anthropological observations of egalitarianism and *demand sharing* in small-scale societies (Peterson, 1993). Nationalism's concerns with group integrity, unity, uniformity, and boundedness extend from our species history of tribal organization and between-group competition; they call to mind aspects of human nature highlighted by primatological studies wherein troops vie for territory and food resources (Koenig, 2002), or historical observations of raiding and warfare in pre-state societies (Cohen, 1984). Progressive or nationalistic populism represents corrections to such misalignments that ignore these salient aspects of our evolved psychology. They are sub-revolutionary sentiments that can be channeled into political movements that force realignment. While both progressivism and nationalism can function as healthy correctives, in excess, either can be destabilizing: progressivism can undermine republican forms, federalist principles, meritocratic ideals, and constitutional limitations in its pursuit of social justice, while nationalism can undermine pluralistic tolerance, individual rights, liberal norms, and democratic freedoms in its pursuit of group strength.

Notes

1. Though it is beyond the scope of this chapter, there are important socio-cultural and geopolitical factors that evoke populism and nationalism as discussed variously by Wimmer (2013), Hirschi (2011), and Ikenberry (2011).
2. We mean to associate progressivism with the left and nationalism with the right only as a generality. Also, by right and left, we are indicating right and left leaning temperaments; this is a distinction about political ideology, not political party.
3. This quote was taken from MacKay (1972, pp. 45–46).
4. Wilson saw the fourth of July as a time to reassess the path to *happiness* and *safety* rather than a celebration of founding principles (Pestritto & Atto, 2008).
5. A point about progressivism to make relates to stability. All else being equal, the status quo is inherently superior because it is a known entity. Change, both in concept and reality, limits predictive power and creates

uncertainty. However, there can be change that eliminates instability. For example, a law may prescribe tighter parameters for taxation or prosecution, and so make it more obvious what should be paid to the government and who the government should arraign. In this way, the change purged instability. So putting aside whether a law is just or unjust, its introduction is destabilizing; but if it stabilizes a system, it can reverse this effect. The extreme example is Roosevelt's New Deal. One biographer of Roosevelt, trying to find coherence amidst the many headed hydra that was the New Deal, emphasized stability; that is he considered that all these policies, at least in their intention, were meant to stabilize an economic system that had become unstable. For example, FDIC insurance, allowed deposits to be made with the certainty that they would be honored.

6. For Fischer (2005) the Wilson administration seems inconsistent, promulgating liberal principles and authoritarian coercion at one and the same time. The conscription and intervention of WWI again is a prime example. Fischer (2005) states that nine of Wilson's *Fourteen Points* were oriented more toward the destiny of nations than to individual freedoms.
7. Beard was perhaps most radical. He cast the founding as a ruse of sorts that had no democratic features to build on or revere. Beard is simply one among a school of American Progressive historians, which also include James Harvey Robinson, J. Allen Smith, and Herbert Baxter Adams (Pestritto & Atto, 2008).
8. Beyond the scope of this chapter, readers can pursue the last chapter of Smith's *The Nation in History*, for his use of *ethno-symbolic nationalism,* which reconciles the realities of nationalism with the charges of social constructivism
9. Rather than disappearing into Kantian globalism as Marx and Durkheim predicted, a resurgent nationalism persists, which is sometimes ethnic and exclusionary, but at other times cultural and political.
10. As these authors assert, this comes in the backdrop of increasing government expansion, which disallows vertical integration and effective action, leaving states ever weaker.

Integrative Metacommentary: Part VI. The Maturation and Institutionalization of Faction

Aurelio José Figueredo, Mateo Peñaherrera-Aguirre, and Steven Charles Hertler

The Maturation and Institutionalization of Faction

Part 6 presented three chapters on the maturation and institutionalization of political factionalism. This maturation process constituted the evolution of political heterodoxy from merely tolerated dissent to an accepted social institution.

Chapter 15 started by showing how political parties evolved mechanisms to maintain internal cohesion. These mechanisms are required because political parties at the level of the nation-state are necessarily large-scale coalitions of smaller subfactions that are often in a state of dissension and low-level competition. Thus, just as nation-states and international alliances required mechanisms to maintain social cohesion to effectively compete with other such polities, as illustrated in Chaps. 10 and 11, political parties require some way of maintaining unity to

A. J. Figueredo • M. Peñaherrera-Aguirre
University of Arizona, Tucson, AZ, USA
e-mail: ajf@arizona.edu; mpeaher@email.arizona.edu

S. C. Hertler (✉)
Saint Elizabeth University, Morristown, NJ, USA

effectively compete with other political parties. Because these mechanisms need to be nonviolent within an orderly state, this function became served by what is known as a party machine. This latter is an ultrasocial institution that relies on peaceful methods of coercion (the metaphorical "carrots" and "sticks") to enforce coordination among subfactions (the metaphorical "arm twisting" within "smoke-filled rooms"). Party machines are owned and operated by party elites.

Chapter 16 followed through on this logic by examining the history of third-party movements within the dominant two-party system of the United States of America. The dynamics described for the emergence of third parties is that they are typically breakaway subfactions of the larger coalitions represented by the two main parties. The typical strategy for reincorporating them within one or the other of the two main parties is to co-opt or somehow incorporate into the overall party platform whatever might have been the specific issues or grievances that led to their secession from the coalition. As party elites gravitate toward the extreme right or left, the need for reabsorbing and reintegrating such breakaway subfactions serves as a centripetal force that pulls each of the two main parties toward the center while remaining somewhat to the left or right of it. A major coup can be appropriating a disaffected subfaction from an opposing party.

Chapter 17 takes this point even further to examine the dynamics of numerous populist, progressive, and nationalist movements that have swept through the United States of America at various points in its history. Generally, these movements signal upwellings of disaffection from the lower strata of society with the elites' management of any given issue or set of perceived injustices. Such issues might be political, social, cultural, or economic in nature. Furthermore, these movements might be associated with either the political left or right. Paradoxically, although they purportedly advocate for the interests of the common man, the movements are often led and spearheaded by individuals that are instead drawn from the elite strata of society. For example, Theodore Roosevelt was considered a Conservative whereas his fifth cousin Franklin Roosevelt was considered a Liberal, but both were scions of the same old, prominent, and wealthy East Coast family tracing their descent to the seventeenth-century Dutch colony of New Amsterdam.

Part VII

The End of Tolerated Opposition

18

From Ritualized Political Conflict to Revolutionary Violence in Early America

Steven Charles Hertler, Aurelio José Figueredo, and Mateo Peñaherrera-Aguirre

Political revolutions, being violent crises in the careers of institutions, engage closely the attention of historians. The sudden blaze of lurking passions, the explosion and the incendiary propagation of principles which had been working underground, the rocket-like ascent into importance of new men, the play of characters in brutal and violent action, the monstrous outbreaks of the mob in which the serious faces of men about their business are soon no longer seen but only the terrifying visage of hate and animal cruelty—here indeed is matter to inflame the writer and to give the peaceable reader at his fireside the shudders!

—Bertrand de Jouvenel[1] (1993)

S. C. Hertler (✉)
Saint Elizabeth University, Morristown, NJ, USA

A. J. Figueredo • M. Peñaherrera-Aguirre
University of Arizona, Tucson, AZ, USA
e-mail: ajf@arizona.edu; mpeaher@email.arizona.edu

S. C. Hertler et al., *The Evolution of Political Ideology*,
https://doi.org/10.1007/978-3-031-83941-2_18

1 Introduction

This eighteenth chapter is the first in this book's final section. Recall, the second half of this book is divided into four sections, documenting the fate of ideology through a sociopolitical arc starting with *a time before tolerated opposition*, onto the *formation* and thereafter the *maturation* of faction, and finally ending in this seventh section entitled *the end of tolerated opposition*. More specifically, as seen in Section IV's chapters, bottom up and top down controls suppressed ideological dissent. As seen in Section V's chapters, as suppression relaxed, ideological differences were subsequently ensconced within a nonviolent political process. Most recently, in Section VI's chapters on party machines, political realignment, third-party movements, populism, progressivism, and nationalism, we saw ideology sometimes straining against, but always acting within, the political process.[2] By contrast, Section VII's chapters come full circle, following the fate of ideology as it defies political processes and is expressed in imperiled or decadent constitutional orders.

In this eighteenth chapter we continue to consider ideologically motivated dissent, here specifically focusing on crises that more acutely threaten to sunder constitutional constraints and destabilize the state. We take pains to consider the function of political ideology in determining these crises, which, of course, also often interact with nonideological group interests. Multilevel selection theory is particularly helpful in this process because it is fundamentally a theory about group dynamics, explaining when and why groups form and fight. Thus, ideological determinants and group determinants can be accommodated within this multilevel selectionist framework. Even strong third-party movements and populist realignments generally garnered at least some energy from nonviolent conflicts between the major political parties; and so it should be no surprise that truly revolutionary events only come from the combination of ideology and faction. Thus, again, we see the interaction of what we have called the two sides of the political, only one of which is purely ideological. In illustration, at this chapter's end, we briefly review representative historical events wherein ideological and sectional fault lines converge.

2 Taking Constitutional Cages for Granted

Runciman's *The Confidence Trap* (2018) addresses the issue of overconfidence of self-governing societies. Runciman ably attempts to reconcile pessimists that see democracies as an experiment ready to collapse, and optimists that see democracy as an inevitability that is replacing its alternatives. Declinists indulge their pessimism when discussing rioting, depressions, recessions, wars, or any other event in the present that can be framed as proof of impending collapse; where optimists point to the resolution of these same events, and continued existence of self-governing societies, as a means to dismiss their counterparts. Runciman's point is that optimists have so often been right, they cannot see when they are wrong. Self-governing societies have an adaptability that allows them to meet these crises which support the optimists, and counter the pessimists. At the same time, this breeds further optimism, creating overconfidence, which makes democracies less responsive, prepared, and proactive for the next crisis, which could in fact be the last crisis.

Overconfidence can come from being blind to the kinds of external competition Skocpol discusses in her *States and Social Revolutions: A Comparative Analysis of France, Russia, & China*. "In dire periods before the Revolutions," Skocpol states (1979, p. 51):

> ...each of these regimes [Bourbon France, Romanov Russia, and Manchu China] ...found itself in a situation of intensifying military competition with nation-states abroad that possessed relatively much greater and more flexible power based upon economic breakthroughs to capitalist industrialization or agriculture and commerce.

Modern, democratic regimes are not immune from these external pressures. It is especially easy to dismiss Hobbesian fears within the stability provided by the *Pax Americana*,[3] the relative world peace that came in the wake of World War II (WWII) and which was solidified by the fall of the Soviet Union. We know from the industrial organizational psychology literature that humans don't positively appreciate ambient environmental conditions such as heating and cooling, but of course bemoan their absence. The same seems true of peace, order, and safety. So, when conservatives clap the tocsin, the sound of the alarm can fall on deaf ears. Purdy (2009, pp. 89–90) exemplifies such a conservative trying to arouse fellow citizens to threats that seem incongruent with prevailing peace:

> In this snug, over-safe corner of the world we need it [war], that we may realize that our comfortable routine is no eternal necessity of things, but merely a little space of calm in the midst of the tempestuous untamed streaming of the world, and in order that we may be ready for danger.

On one side, terror attacks, war, and other existential threats can cause states to tack right, while surfeits of resources and power vis-à-vis other states cause states to tack left. The Pax Americana's unusual record of stability converted, what was an ebb and flow between peace and war, into a sustained peace, leaving generations with no experience of major wars, and consequently unable to heed Passmore and his ilk.[4] This, in turn, enabled a long march left.

As modern self-governing societies are not immune to external competition, neither are they immune to insidious augmentation of internal dissent. There is, as it should also be noted, an inverse relationship between external threat and internal dissent. Internal dissent is a danger emphasized by Betrand de Jouvenel. Revolutions "rend the air with denunciations of tyrants." They seem to be the ultimate check to power. These David and Goliath dynamics, however, can deceive. Regimes that are overthrown, even if they would be tyrannical, and often toothless; they are senescent and inefficient, which is why they can be thrown over. Conversely, revolutionary movements often have roots beyond their flowering. Switching metaphors, Jouvenel then likens revolutionary events to rapids, while reminding us that we should carefully observe the stream of power before it entered into those rapids, and after it has left them. In doing this, one will find power always increasing downstream of revolutionary events.

In response to such pressures, regimes can unravel rapidly, succumbing to conquest or revolution. This sudden collapse to lower levels of complexity coming after slow decline is illustrated by concepts like *critical mass* and the *tipping point* described by Thomas Schelling to show the influence of *micromotives* on *macrobehavior*. Choices of individuals affect behaviors of groups; but they do so nonlinearly. Axelrod's writings on the evolution of cooperation and game theory generally allow us to see the bottom-up effects of individual action on defection or cooperation. This is similar to a rout in war, wherein some panicked soldiers have no effect on morale until a tipping point is reached, causing the troop at large to retreat in disordered haste. So, there is a point in which groups can act as one, but are undermined by individual dissent, which has little impact until reaching critical mass, after which the group, formerly a cohesive body, becomes merely a mass of self-interested individuals. In this vein, think of the political as ritualized violence. The taming of contention, and its constraint to nonviolent expressions within constitutional orders, is an achievement of modernity. We saw the arduous development of this reality over the course of this book. Once in place, however, any given generation, especially one overconfident and blind to alternative histories, can take their constitutional order for granted.

3 Extended Phenotypes and Niche Construction

We now want to make an important point that we will briefly recollect at the end of Chap. 19, and explore thoroughly in Chap. 20: Societies are, at their founding, more apt to be conservative. We mean this in all three senses discussed in Chap. 8: *genetic, developmental,* and *facultative.* Founding populations had a preponderance of genetically informed group-selected temperaments,[5] which were socialized within more group-selected societies, and who responded in group-selected ways to harsh conditions. Group-selected persons vault the space under which individually selected persons can flourish. More precisely, it is largely due to the martial efforts of group-selected societies, acting in group-selected ways, led by group-selected leaders that stable national borders are erected, and order maintained within them. Securing social stability against the threat of external conquest or internal anarchy represents a foundational need akin, on the personal level, to those basic needs for food and shelter within Maslow's pyramid.[6] It is only because the group-selected have established these preconditions that the individually selected can flourish, multiply, and concern themselves with various aspects of social justice that, on a personal level, equate to those humanistic needs for generativity and self-actualization within Maslow's pyramid. A society never progressing beyond external security and internal order is then like a person never progressing toward generativity and self-actualization; alternatively, a society that achieves generativity and self-actualization at the expense of external security and internal order is then like a person reaching the peak of the pyramid at the expense of the base. In short, mature states created the precondition of secure *cages* within which ideological dissent could be vented, verbally and politically.

In this vein, we recall the idea of the *extended phenotype*. As described by Dawkins (1981), the extended phenotype is a behavioral extension of the organism that serves much the same function as an observable anatomical feature. Compare a turtle's shell to a beaver's dam, colorful plumage to an ornamented bower, thick fur to a warm nest, adipose tissue to a scatter hoard. In each of the latter instances, a behavioral disposition induces the organism to procure or make that with which he is not endowed. In some way, coats, homes, and refrigerators are all part of the complex extended phenotype of humans, which in most cases broadly overlaps with technology and culture. However, the extended phenotype of humans is not confined to securing one's person or home but extends to the larger environment. People light and pave streets, educate and instruct, for instance, and this can be understood as forms of collective action and enculturation, which in turn may be thought of as part of the larger extended phenotype. But there are constraining as well as constructive expressions of the extended phenotype. Certainly, courts, laws, and prisons are agents of constraint that are part of any mature state, and which are necessary to avoid anarchy. This provides the citizens against broken bones and picked pockets. However, much of the constraining, not to say coercive, expressions of the extended phenotype are laws, policies, customs, and mores supportive of conservative values. These are captured in Haidt's *authority, ingroup,* and *purity,* the *binding foundations* of the conservative temperament, meant to enforce conformity and cohesion. Other expressions of extended phenotypes in stable societies include ultrasocial institutions that enforce cooperation and punish cheaters via *strong reciprocity*, a concept in the evolution and economics literature wherein punishment is meted out that is costly to both punished and punisher (Hechter & Horne, 2009; Fehr et al., 2002; Gintis, 2000).[7]

All of this is relevant to the next chapter, which takes up the unraveling of senescent states and weakened constitutional orders. However, putting aside the question of state vitality, it is also germane to this more general discussion about challenges to the constitutional order from insurrection, revolt, and civil war. Extralegal violence has classically come from a revolutionary left chafing at an overly severe, tyrannical, and arbitrary regime. Extralegal violence can come from a populist right chafing at an overtly decadent, corrupt, and cosmopolitan regime. These ideologically driven tensions, if anything, play second fiddle to tensions within societies. As this next section will show, it is generally only when intrasocietal tensions are added to ideological tension that a constitutional order is threatened with disseverance, revolt, or revolution.

4 American Illustrations

Henry Sumner Maine (1976/1885, p. 49) speaks of *irreconcilables*, which are secular factions within the state at odds with the fundamental ethos of the state.

> There can be no more formidable symptom of our time, and none more menacing to popular government, than the growth of Irreconcilable bodies within the mass of the population. Church and state are alike convulsed by them; but, in civil life, Irreconcilables are associations of men who hold political options as men once held religious opinions. They cling to their creed with the same intensity of belief, the same immunity from doubt, the same confident expectation of blessedness to come quickly, which characterizes the disciples of an infant faith. They are doubtless a product of democratic sentiment; they have borrowed from it its promise of a new and good time at hand, but they insist on the immediate redemption of the pledges, and they utterly refuse to wait until a popular majority gives effect to their opinions.

This passage is rich with insight. There is wariness about democratic excesses degenerating to *ochlocracy*[8] that were shared by Gouverneur

Morris, Alexander Hamilton, John Adams, and many other American founders. There are concerns about hubris and zeal; exactly the principled ideological commitments that made David Hume a qualified and cautious supporter of self-government. Paired with the concerns about ideological zeal within a democratic populace is an emphasis on the dangers of faction. We here review some episodes in American history that combine Henry Sumner Maine's tripartite fears. The Alien and Sedition Acts, the Hartford Convention, the Nullification Crisis, the Civil War—these are times when the political process broke down, or threatened to do so.

The *Alien and Sedition Acts* (Hofstadter, 1969; Smith, 1966) were passed in 1798 specifically to advance the group-selected, conservative ends of subordination, hierarchy, and group homogeneity by restricting speech and citizenship. Hamilton, though in many ways the archest of the arch-Federalists, did not actually endorse the Alien and Sedition Acts. In fact, he took pains to moderate them and cautioned other ultras, probably Theodore Sedgwick and Fisher Ames among others, not to "establish a tyranny" (Ferling, 2003, p. 426). While some Arch-Federalists thought that they could crush the opposition with the Alien and Sedition Acts, Hamilton intuited that such extreme measures would produce extreme counter measures. More specifically, he knew that the Alien And Sedition Acts had the potential to "give to faction body & solidarity" (Ferling, 2003, page 426). Just as the usurpations of the crown united north and south, conservative and rebel, patrician and plebian into one unified body which came to oppose it, so the Alien and Sedition Acts would aggregate the loose body of Republicans into something closer to a modern party with all the output of an organized mechanistic body. Hamilton's predictions were prescient. The Virginia and Kentucky Resolutions, written in protest of the Alien and Sedition Acts, were milestones in the formation of a Republican opposition to Federalist dominance. We see two enlightenment scions of liberalism, Thomas Jefferson and James Madison, defending free expression from the Federalists. Though Southern Republicans most vocally protested these Acts imposed by a Northern Federalist administration, ideology dominated section, as can be seen in the dissent of Matthew Lyon, Edward Livingston and Abraham Baldwin, respectively from Vermont, New York, and Connecticut (Warren, 1942; Rosenfeld, 1997).

Madison drafted the Virginia Resolutions when part of an emerging Republican opposition. He acceded to the presidency in 1809. Commercial warfare using tariffs and embargoes, and the War of 1812, made Madison unpopular among Northern Federalists who now found themselves in opposition, and resorted to the same threats of disunion by late 1814. This took the form of secret meetings rather than published manifestos. In what became known as *The Hartford Convention*· northern Federalists from Vermont, Connecticut, Massachusetts, Rhode Island, and New Hampshire discussed states' rights and secession.[9] It was then hard to see the forthcoming *Treaty of Ghent* diplomatically ending the war, and the decisive victory of Andrew Jackson at the *Battle of New Orleans*; and it was even more difficult to envision the Republican party, not only retaining power, but ushering in the Era of Good Feelings discussed in Chap. 16. It was then Andrew Jackson who presided over the *Nullification Crisis*. This time, we see the South, most specifically South Carolina, straining under tariffs, bombastically dubbed the *Tariff of Abominations*, meant to protect northern industry. This caused a constitutional crisis and threatened civil war as South Carolina representatives declared federal law null and void, and Jackson responded by federalizing the South Carolina State Militia.

Then, of course, there was the Civil War. Here we see, not a state, but a section, attempting to leave the union. The American Civil War combined the issues present in the aforementioned crises. Both Madison's and Jackson's administration imposed federal tariffs that disproportionately burdened states and sections; similarly, Southern aristocrats felt economically threatened by Northern elites. Like the response to the Alien and Sedition Acts and the Nullification Crisis, the American Civil War pitted state against federal power. And as with all prior crises, the American Civil War, both in its precursors and execution, exposed fault lines inherent in America from its founding. Slavery was the cause of the Civil War in a proximate sense.

Following the work of David Hackett Fischer's *Albion's Seed*, we can see that there is more than ideology at play, and that in fact these tensions can be traced back to the distinct peopling of Massachusetts (Midlands and New Englanders) and Virgina (Tidewater South and Southern Highlands). Fischer's demonstration of the nonrandom patterns of migration that resulted in the thirteen colonies being nonrandomly settled by different European nations had lasting significance as seen in Colin Woodard's *American Nations.* Thus, we can see effectively two nations in one federal union, interacting with more purely ideologically based tensions that transcended sections and states. In these select cases, albeit to different degrees, both determinants, ideology and intranational competition, were operative. Again, this illustrates something of a maxim; that states are rarely seriously threatened by rebellion and revolution except where intranational conflict aligns with ideological differences.

5 Conclusions

In Chaps. 16 and 17 we saw political realignments, third-party movements, and populists surges gain sufficient reach and power only when they combined both sides of the political; that is, when competition among interest groups was added to competition among groups with different ideological positions. Relative to realignments, third-party movements, and populist tumults, rebellion and revolution are more extreme forms of protest, and so we should not be surprised that they, too, require the combined energies of competition among territorial regions of the same polity and competition among political factions having different ideologies.[10]

Recall from prior chapters, most especially Chaps. 5 and 8, that political ideology extends from genetically informed and neurobiologically grounded personality variation distributed within all groups. Some groups can have higher or lower means; they can have a greater or lesser proportion of persons exhibiting group or individually selected traits. Nevertheless, all groups contain a diversity of personality-based variation, which is productive of variation along the ideological continuum. Mechanistically, this was discussed in Chap. 5 as the product of negative frequency-dependent balancing selection. Thereafter, Chap. 8 took pains to describe selective pressures responsible for creating and maintaining temperamental and ideological variation within populations. This is the reason why ideology, in and of itself, is rarely productive of rebellions, revolutions, and civil wars. Unlike cultural mores, genetic ancestry, and

economic modes of living that can be concentrated in states and sections, political ideologies are naturally distributed even within lower level groups such as towns and families. At present, we have significant ideological differences, but these are no longer amplified by sectional strife. Heeding Runciman's warning about overconfidence and combining them with the lessons of this chapter, however, there is some evidence in recent electoral maps of a newfound coincidence between a rural-urban divide and a conservative-liberal divide. This is artificially increased by surges in politically motivated migration and the ability to substitute geographic for media-based concentrations of ideologically like-minded persons. We can see to some extent in the liberals of progressive cities marching for an expanded set of rights, a strong contrast with the conservatives of rural militias preparing to defend their vision of an older order.

The next chapter takes up the same topic, but as it is expressed in the context of declining states. In other words, like Chap. 18, Chap. 19 will also consider revolutionary political outbursts, but do so specifically as they are expressed in transaction with the declining state with its weakening political processes and institutional integrity.

Notes

1. This is quoted from Jouvenel's *On Power: The Natural History of Its Growth*; specifically page 237 of the 1993 reprint published by the *Liberty Fund.*
2. Rather than becoming a persistent part of the political landscape, the third parties discussed in Chap. 16 effect change when their representatives, constituents, and agenda are absorbed by established political parties. The populist movements discussed in Chap. 17 are more explosive manifestations of the same discontent; but even as they more severely test the political process, they are generally absorbed by it.
3. The Pax Americana can be seen in the following description: Post-WWII, the United States "engaged in the most ambitious and far-reaching liberal order building the world had yet seen" resulting in a "liberal hegemonic order" that assumed "the duties of building and running an international order, organizing it around multilateral institutions, alliances, special relationships, and client states" (Ikenberry, 2011, p. xi).

Post-WWII power vacuum, American dominance could be measured in gold reserves, shipping, invested capital, manufacturing, and gross domestic product (Ikenberry, 2011; Leffler, 2010). Hegemony is ultimately power-based, though the American system palliates and temporizes hard power with capital, order, and rhetoric. This long peace is also discussed by S. Pinker in *The Better Angels of our Nature* (2011).

4. As discussed by Haidt, conservatives are probably more sensitive to these external threats than liberals.
5. This is consistent with what have been referred to as tribal social instincts (Richerson & Henrich, (2012).
6. Abraham Maslow constructed a hierarchy of needs in a pyramid, with basic needs for food and shelter forming the base and less physiological needs, such as belonging, populating the upper portion of the pyramid. Self-actualization was at the top, wherein people thrive and grow toward ideals. Kenrick et al. (2010) recently updated Maslow's pyramid based on evolutionary principles.
7. Jeremy Brecher once said, "revolutionary movements rarely begin with a revolutionary intention; this only develops in the course of the struggle itself" (Skocpol, 1979, page 16).
8. From the theory of Anacyclosis, ochlocracy relates to mob rule, and is the degenerate form of democracy.
9. https://www.britannica.com/event/Hartford-Convention
10. For example, the archetypical French Revolution added between-group competition among social classes to within group competition among ideological positions; specifically, it pitted the third estate against the first and second, and radical left ideology against entrenched conservative societal values.

19

Polarization and Revolution as a Function of the Lifecycle of the State

Steven Charles Hertler, Aurelio José Figueredo, and Mateo Peñaherrera-Aguirre

There is the moral of all human tales;
Tis but the same rehearsal of the past,
First freedom, and the glory-when that fails,
Wealth, vice, corruption-barbarism at last.
—Lord Byron[1]

1 Introduction

In explaining the process of decline and internal degeneration, Vico spoke of shirking civic responsibilities in favor of individualistic goals; Toynbee (1951) described being morally routed or experiencing a loss of élan or moral courage; Spengler (1991) referred to inexorable societal

S. C. Hertler (✉)
Saint Elizabeth University, Morristown, NJ, USA

A. J. Figueredo • M. Peñaherrera-Aguirre
University of Arizona, Tucson, AZ, USA
e-mail: ajf@arizona.edu; mpeaher@email.arizona.edu

S. C. Hertler et al., *The Evolution of Political Ideology*,
https://doi.org/10.1007/978-3-031-83941-2_19

degeneration (Farrenkopf, 2001). Each author intuited that the proportion of individually selected societal members increases over time as states become decadent. Specifically, in the twilight of a state, individually selected ideology augments, both from evolutionary and social processes. This allows individually selected programs and peoples to increasingly exercise power by electing leaders and controlling resources. All this time, the decadent state is weakening vis-â-vis rival states. The group selected become marginalized, both in terms of their proportions and projected power. At the same time, the fear of internal disorder and external conquest that the group selected are primed to perceive becomes increasingly activated by the real prospect of decline. And so, as a function of decline, the individually selected become more numerous and vocal, whereas the group selected become more fearful and desperate, all of which creates a climate of political polarization. This contrasts with the relative harmony of mature, but not yet decadent states, which retain a high proportion of group selected members that have carved out a stable and secure territory from the larger world, which buffers societal members from conquest. As discussed in this nineteenth chapter, it is precisely this stability which increases the proportion of individually selected persons and allows them the latitude to voice dissenting opinions and redresses basic injustices. Thus it follows that group selected persons and values, so integral in state formation, give rise to a stable society wherein fairness and freedom can flourish as the power and proportions of individually selected societal members rise. Nevertheless, the process, coming full circle, ends in decline; not necessarily decline generally, but decline by the standards of power and influence against rival groups and nations, as when declining Greek societies fell under Roman rule.

It must be kept in mind, however, that when many authors write of the *decline* and *degeneration* of societies, they are referring to decreases in the integrity of the *nation* or the *state,* and not necessarily to the fates of the *individuals* within those polities. This kind of language is obviously value-laden, and betokens an allegiance to group-selected values on the part of those who choose to use this kind of vocabulary. In reviewing this literature, we paraphrase the authors in a manner faithful to their original descriptions without necessarily endorsing (or rejecting) such value judgments. Nevertheless, it is important to consider that other historians (reviewed in Cline, 2024) have speculated that sometimes a decline in sociopolitical complexity, *via* the disintegration of higher levels of the social hierarchy, may serve the self-interest of individuals by liberating them of the burden of supporting that sociopolitical superstructure.

2 Political Polarization

Polarization and decline are distinct phenomena. Political polarization has its unique literature and causal phenomena as does the decline of civilizations. We hold no grandiose goal of surveying the literature associated with either phenomena, but instead, briefly review political polarization in abstract; then, in the remaining sections, consider political polarization as a symptom of decline.

Reagan's revocation of the *Fairness Doctrine,* a Federal Communications Commission policy imposing centrist balance on news outlets since 1949 (Kruse & Zelizer, 2019), can be counted among political polarization's many causes. Also creating a fractured media landscape was the rise of cable news (Birdsong et al., 2014), talk radio (Berry & Sobieraj, 2011), algorithm-driven news feeds (Feezell et al., 2021), and social media (Yarchi et al., 2021). Beyond media, there are compelling demographic explanations of polarization, such as assortative mating[2] (Klofstad et al., 2013) and non-random migration to liberal cities and conservative towns (Strickler, 2016). Cultural factors can intensify polarization. For example, declining social capital reduces the frequency of local level bipartisan interaction. Also, polarization derives from the presence of *groups within groups* wherein religious (Claassen et al., 2021; Perry, 2022), ethnic, or other subgroups form that singly or in combination challenge the state. Governmental forms can sometimes increase polarization, such as when there is increased centralization, when there is no separation of church from state, when there is more money in politics, when the moral gets merged into the political, and when principles come into play as much as interests. Economics must be added to the list. As mentioned in Chap. 5, economic growth and controlled inequality (Stewart et al., 2021) can serve as a social lubricant, and it follows that the opposite conditions have the opposite effect, leading to polarization as a symptom of rising between-group conflict. And of course, there is the aforementioned literature on the polarizing effects of geopolitics discussed in Chap. 14, and of *partisan sorting* discussed in Chap. 15 (Hout et al., 2022; Brown & Enos, 2021; Ciuk, 2023; Lang & Pearson-Merkowitz, 2015; Kevins & Soroka, 2018; Bonikowski et al., 2021; Fioroni et al., 2022).

Similarly, indexes of political polarization are instructive. Whether caused by income inequality (Wheeler, 2017; Chaker, 2017; Bowyer & Kahne, 2017), media fragmentation (Duca & Saving, 2017), or misinformation (Ribeiro et al., 2017), increasing political polarization is well documented (Shi et al., 2017; Willis, 2017; Gooch, 2018; Chen et al., 2018). We see this graphically[3] when looking to decadal representations of political polarization, which is evident among all, and extreme among the *politically engaged.* Contemporary polarization is illustrated by Peter Turchin's *Ages of Discord* (2016), the cover of which depicts negative correlations between *political stress* and *well-being* indexes, that, today, mirror those observed at the end of antebellum America. Tempers were high just before the Civil War, and so they were just after. The sectional struggles of the post-bellum era with its vacillation between a South subordinated to the North during reconstruction or set at defiance during the Jim Crow era (Woodward, 1981, 2008) are times wherein polarization seems greater than at present. Still, extreme polarization is not confined to the present and the Civil War era. We do well to recall the 1790s (Elkins & McKitrick, 1993) with its battle between Federalists and Republicans (Rosenfeld, 1997) waged in the wake of the French Revolution (Dunn, 1999; Cotlar, 2011). More recently, Lyndon Johnson's Great Society (Bornet, 1983), the civil rights era (Harvey, 2012), and a string of government shutdowns[4] exposed comparable political polarization. What has been will be again, what has been done will be done again; there is nothing new under the sun. At the same time, we should observe that civilizational collapse is part of this process.

3 At the Intersection of Polarization and Decline: *Inefficiency*

In exploring how political polarization can come in consequence of civilizational decline, we begin by relating America's post New Deal regime to Tainter's model of decline through inefficiency. America is arguably on its third regime, the federated republic falling to a consolidated state after the Civil War, with that consolidated nation state transitioning to a paternal state following progressive policies and FDR's New Deal (Primus, 1999; Ackerman, 2000). The story of America's three regimes is complicated, but its effects at least can be summed up in two words: *centralization* and *growth*. "The centralization story," Bevir and Rhodes (2010, p. 151) write, "tells us there is often a gulf between intervention and control. Central agencies intervened, but rarely to great effect." Government growth begot bureaucratic inefficiency. Even massive growth and spending were insufficient as demonstrated by extra-governmental bodies, from think tanks and lobbyists and consulting firms, offering policy advice to governmental staffers. This comes at a cost, as these extra-governmental bodies are implicitly paid for their services by acquiring influence and indirectly affecting policy (Bevir & Rhodes, 2010). Government growth is correlated with centralization, but distinct from it; one can say that America's age of centralization followed the Civil War when there was a transition between semi-autonomous states federated into a union that had circumscribed powers, to a centralized state that had relegated states to districts. Alternatively, the age of growth came after the progressive era which promoted an interventionist foreign policy and a paternalistic domestic policy.

J. A. Tainter's *The Collapse of Complex Societies* (Tainter, 1988; Hertler et al., 2020) defines collapse as a loss or substantive diminution of sociopolitical complexity, hierarchical organization, occupational specialization, integration and centralization, behavioral control and regimentation, architectural and artistic efflorescence, central to peripheral connectivity, trade activity, and general coordination. By this definition, which is really more consistent with commonplace uses of *decline*,[5] America's transition from the first to the third regime resembles growth more than collapse. At the same time, we can see that, as America grew to solve collective action problems which no individual could undertake, and problems of scale which no smaller agency could efficiently solve, it became the kind of complex system that is subject to collapse. Centralization and growth bring complexity, and all complex things are subject to entropic decay. Tainter draws on J. F. Powers, Polybius, Kenneth E. F. Watt, Jack Goldstone, C. F. Volney, Charles Gray, Alfred Kroeber, and Pitirim

Sorokin in producing a multifactorial model of decline that accounts for resource depletion, catastrophes, rivals, intruders, chance, and social dysfunction, among other factors; though his treatment of complex societies as energetic systems is both most novel and relevant. Tainter speaks of the point of diminishing returns, wherein higher upon higher levels of complexity become very costly to achieve and maintain. Expanding territory and increasing populations demand energy inputs, resources, bureaucracy, and infrastructure, resulting in an unraveling described thus:

> The population, meanwhile, must contribute ever more of a shrinking productive base to support whatever projects the hierarchy is still able to accomplish. Many of the social units that comprise a complex society perceive increased advantage to a strategy of independence of the hierarchy. Behavioral interdependence gives way to behavioral independence, requiring the hierarchy to allocate still more of a shrinking resource base to legitimization and/or control. (Tainter, 1988, p. 121)

This inefficiency creates an overburdened, underperforming regime that becomes increasingly tyrannical in its efforts to gain sufficient resources to overcome its augmenting inefficiencies. The inefficiency itself, and the alienation of its populace from attempts to overcome inefficiency, jointly renders declining states prone to collapse from exogenous system shocks, such as war with rivals or financial strain.

Political polarization can be partly understood as a function of the inefficiencies Tainter describes in his rational approach to decline, inspired by physics and economics. A complex society that has become sclerotic and inefficient offers less and demands more, undermining allegiance and creating tensions between elites and nonelites, between primary payers of tax revenue and consumers of tax revenue, and between those ideologically wed to small government policies and those favouring their opposite. America's first regime was particularly resistant to Tainter's decline through inefficiency, mainly because it precluded the creation of the kind of complexity that was subject to inefficiency. The founding generation was wary of tyranny and jealous of its autonomy. It first created the Articles of Confederation, which were too inefficient and ineffectual to solve the collective action problems that states must meet. The

Constitution replaced the Articles of Confederation, but retained the federal nature of the new union and carefully limited and balanced new powers. The Federal government then remained relatively weak, poor, and divided. These checks on tyranny proved to be bulwarks against complexity, which thereby protected the country from becoming the kind of complex system subject to collapse under its own weight. Once it completed its evolution from the first to the third regime, the American government had repudiated these founding principles and thereafter became both inefficient and complex. Contemporary political polarization can be partially understood as a result. From a small, decentralized, federated government that efficiently delivers its circumscribed share of services, America arguably transitioned to a large, centralized, national government that inefficiently delivers on a wide range of promises. In this transition one can see the polarization between small government conservatives and libertarians who want to claw back the complexity of the state, and progressive liberals who want to make it deliver on an expanded set of obligations.

4 At the Intersection of Polarization and Decline: *Demagogues and Dependents*

Tainter's efficiency problem is reminiscent of the trophic loss of the sunlight's energy as it is photosynthesized by plants, as plants are eaten by herbivores, and as herbivores are eaten by carnivores. Yet even assuming absolute efficiency, centralization and growth increase the "lootability of a resource" (Le Billon, 2004). Thus, we have the self-evident term, *petro-violence* (Allen, 2013), and self-evident phrase, *conflict diamonds*. Animals, like humans, display aggression in competition for concentrated resources (Craig & Douglas, 1986; Estevez et al., 2002; Donohue et al., 2013). Large, centralized governments garner high, concentrated revenues, which then mimic these geological and ecological concentrations of resources. America's third regime, by virtue of its growth and centralization, concentrated resources, thereby potentiating polarization.

There are two consequences of centralizing power and money within the federal government, one relating to representatives and one relating to citizens. First, the enormity of the federal budget and the vast sweep of its powers set office seekers to grabbing at the levers of power and wealth. This incentivizes influence pedaling, rent seeking, spoils system dynamics, lobbying, and other elite activities that can nonrandomly foster self-interest over and above service. These conditions foretell corruption among leaders, which in turn fosters alienation or populist revolt among citizens. These spoils, positions, and related attractants that inherently entice self-interested elites, directly create tensions between the ruling class and the remainder of the country. They are also apt to create bad policies, the consequences of which also foster polarization and decline. The effects of centralized bloat are not limited to elites, but affect the electorate. Like bird seed gathered in a feeder rather than scattered across a field, power and tax revenues concentrated in Washington DC place hitherto private and local matters into national debate, which inherently and unavoidably strains the political process and polarizes the populace. Educational and zoning law debates, for example, are polarizing when nationalized. Finally, centralized big government imparts an interaction effect between elites who are incentivized to act the part of demagogues and citizens who are incentivized to act the part of dependents.

5 At the Intersection of Polarization and Decline: *Disintegrating Bonds*

As per social anthropologist Ernest Gellner, there is a further consequence to expanded and centralized modern governments that would exist even were they both efficient and uncorrupted. Gellner sees government's ability to absolve citizens from governance and buffer them from threat as a sort of enervation that breaks down cohesion. He develops this theme by contrasting primitive herding society with "atomized and un-cohesive, trustless populations" living in cities. These herding societies are operating at a lower level of social complexity, and more closely approximate hunter-gatherer, tribal organization, knitted together by custom, kinship, and ever-present external threat. The delegation of authority to professionalized governments, as per Gellner, precipitates decline by absolving groups from local collaboration (Gellner, 1987; Hechter & Horne, 2009, p. 304). This is reminiscent of Robert Putnam's narrative of declining social capital, but applied to the national level. In Felix Morley's *Freedom and Federalism* (1981), we see a radical federal policy in early America, in its first regime, wherein functions devolved upon the lowest level of government, being delegated upwards from individual to municipality, from municipality to state, and from state to federal government only as necessary. Only a highly circumscribed set of responsibilities, such as interstate commerce, foreign relations, and national defense, were arrogated to federal authorities. This then caused these lower level conglomerations to work amongst themselves to govern, which brought the affiliation, societies, and groups that Tocqueville found so interesting.

In the course of Gellner's writings, one can discern a second explanation. Societal bonds are not only loosed by delegating authority to a ruling class, but from the protection provided by that ruling class. Existential risk strengthens societal bonds. This insight can be seen in Toynbee's concept of the *stimulus of blows.*[6] Most recently, Evrigenis (2007) developed this concept into a theory of *negative association* in which divergent peoples and interests are nevertheless welded into groups by the "fear of external threats." To show the antiquity of negative association, Evrigenis (2007, p. xi) cites Publius Scipio Nasica who, contra, Marcus Porcius Cato, understood Carthage, in its threat and rivalry, to bind Romans to one another. Validating these fears, J. G. A. Pocock (1975, p. 208) observed that "...after the Romans had conquered all their enemies and had ceased to fear them, they began electing to office, not those best fitted for it, but those most skilled at gaining the favor of others."

6 At the Intersection of Polarization and Decline: *Liberalism*

The great empirical historian Arnold Joseph Toynbee thought Gibbon's *Decline and Fall of the Roman Empire* something of a contradiction in terms in that the transition from republic to empire was itself a symptom of decline. German-American philosopher Leo Strauss felt similarly about Western liberalism. Hannah Ardent and Eric Voegelin were with Strauss in countering the idea of the *liberal achievement*. These emigrees, Ball (2001) explains, had just witnessed fascism infiltrate and eradicate the internally and morally weak Weimar Republic. The charge was that, having no values makes a culture open to invasion. Strauss claimed that "…liberalism is a uniquely self-subverting political philosophy" that "…can tolerate almost anything, including its own demise" (Ball, 2001, p. 114). Liberalism's celebration of individual freedom, pluralism, diversity, and tolerance are negative values, which, Strauss argued, cannot resist strong ideological pushes from the farther right or left. This invokes a sort of social analogue to the biological concept of competitive exclusion referenced in Chap. 13. For example, commensal flora competitively exclude pathological bacteria from colonizing mammalian intestines. An empty gut, like a value-neutral state, is prone to invasion.

In *A World After Liberalism* Matthew Rose reviews right-leaning political ideologies that grew in the shade of liberalism, and which supply those who we would describe as more group selected with the positive content that liberalism's negative values cannot. Rose reviews the work of five *rehabilitated* philosophers of the far right: Oswald Spengler, Julius Evola, Francis Parker Yockey, Alain de Benoist, and Samuel Francis. For Rose, this is a vehicle for describing a shift in the Overton Window that increasingly accommodates "nationalists, populists, identitarians, futurists, and religious traditionalists" (Rose, 2021, p. 3). Juxtaposing Ball's emigrees with Rose's philosophers stimulates one to think about liberalism at different stages of society. The fears of these emigrees and the discontents of these philosophers would not apply at liberalism's inception. Though this grossly generalizes, we can see liberalism, as articulated in England and America, to have empirically demonstrable correlations with growth and wealth. Liberalism was then expressed in a society with much positive content derived from Christianity, tradition, and classical respect for civic virtues. The loss of positive content discussed by Ball and Rose was then only absent from government, but remained present in religion, tradition, and local institutions. Positive values were not stripped from society, only divested from government. Church and state might be separated, but this did not secularize society, it merely tolerated denominational differences in still strongly Christian societies. This point is explored by John Patick Diggins. Diggins (1986) discusses the weaknesses of Lockean liberalism when stripped of the positive liberties of "classical political philosophy" and when secularized and thereby deprived of religion's unifying function.[7] After commenting on declining civic virtue and remarking that "we no longer have Calvinism to save liberalism..." Diggins (1986, p. 16) frets about "A liberal society dedicated to nothing more than the pursuit of self-interests and the pleasures of consumption."

Thus, in reflecting on all these authors from Strauss to Diggins, we can see liberalism as an expression of center left, individually selected values, which were originally expressed within an ascendant West, still strongly bound by Christianity, and strongly impressed by the recovery of its classical heritage of positive values. The resulting dynamism only increased growth and wealth, tending to bind society, narrow the Overton Window, and reduce ideological polarization and between-group competition. Advance time's arrow two centuries to an older, arguably decadent, West and we have liberalism operating alongside less robust civic and religious handmaidens. Demographic and economic growth stalls and lower level group factions command more allegiance than the state. This breeds polarization on the right and left, while eroding the center. This reality can be seen in the switch to Huntington's Strategy, discussed in Chaps. 15 and 16, which more or less abandons the center to court the extremes. The group selected psyche is activated by decadence and decline[8] to abandon liberalism for the classical world's positive values or religion's homogenizing effects. At the same time, the individually selected psyche, lacking the temperamentally driven angst about erosion in the West's heritage of positive liberty or religious moralism, continues pushing further left, crossing the boundary between libertarianism to a new form of left-leaning positive liberty that leverages state authority to guarantee equality of outcome.

7 Conclusions

Media, demography, culture, subgroups, economics, geopolitics, and partisan sorting can drive or reflect polarization. Rather than reviewing polarization's vast literature, this chapter considered polarization as a symptom, not a driver. It further narrowed the scope of review, considering polarization as symptomatic of decline. That narrowed scope was then explored through four points of discussion: Inefficiency, demagogues and dependents, disintegrating bonds, and Western decadence. None of these factors quickly or singly cause collapse to lower levels of sociopolitical complexity. Instead they signify a hollowing out and weakening. They are like a chronic disease that contributes to death from acute infection, or like a rot in a tree that topples from a storm. All four factors are apt to expand the Overton Window to incorporate far right and left ideologies into political discourse. Also, all four factors expose groups within groups and set them competing with one another for resources and influence. A society in decline, with weakened cultural, religious, and government institutions, fails to subordinate lower level allegiances. Its stalling growth and economic austerity deprive it of a valuable social lubricant, one that, for example, frustrated Marxist appeals to the prosperous working class in early nineteenth century America.

Declinist narratives typically have the one-two punch of internal dissolution and external threat that can wend and work on group selected minds like a canker. Just as group selected minds are prone to declinist narratives, individually selected minds are prone to melioristic narratives. Those on the left tend not to share those concerns emphasized by Strauss or Diggins. Each side of the ideological spectrum has different thoughts and feelings about the West's fading extra-governmental heritage. Judging from Strauss, Ball, and Rose, those more group selected contemporary conservatives see liberalism as a cut flower, with its beauty fading from being severed from the plant, whereas contemporary liberals see liberalism as a child full grown, no longer needing the parental womb or teat. Strauss's charge that there is something uniquely enervating about liberalism and the broader theme pursued in this chapter's last section will be taken up in the next chapter.

We end this chapter with a caution about understanding these processes of polarization in declining states through a purely sociocultural lens. Gellner, like Toynbee and students of decline and fall, drew on fourteenth century Arab philosopher Ibn Khaldun's concept of *Assabiyah,* a corruption of societal vitality extending from sedentary living and wealth. In our 2020 monograph *Multilevel Selection: Theoretical Foundations, Historical Examples, and Empirical Evidence,* we describe the biocultural dynamics behind small cohesive group selected societies that established polities and wrought stable conditions out of the state of nature. These might be controlled from the top down or the bottom up, and may be in many ways unjust or tyrannical, but they tend to solve that most basic Hobbesian problem of anarchy that can come from internal dissent or external conquest. Gellner among others recognized that overcoming anarchy can undermine cohesion, but what is not often recognized by Gellner, or any other sociocultural theorist, is that this taming of anarchy changes the selective pressures to which a society is subject. The stable society not only slowly allows individually selected ideology a voice, but it also fosters selective pressures that can increase the proportions of the individually selected within a society.

Notes

1. This is quoted in J. P. Diggins' The Lost Soul of American Politics: Virtue, Self-Interest, and the Foundations of Liberalism (1986, p. 48).
2. If people assortatively mate based on ideology this can create households with less ideologically diverse discussion and offspring.
3. Attributing it to the Pew Foundation, this image taken from *Cassandra's Legacy,* a site run by Ugo Bardi, a professor of physical chemistry at the University of Florence: https://cassandralegacy.blogspot.com/2016/05/trump-unavoidable.html
4. The Ford Administration witnessed what seems to be the first official government shutdown, which was followed by two consecutive, and highly salient, shutdowns under the Clinton Administration. After coming close to a shutdown in 2014 and 2015, another shutdown commenced in 2016 under the Obama Administration. In all, there have been approximately twenty government shutdowns, all of which occurred on or before the bicentennial of American Independence.

5. To further specify, Tainter's use of the word *collapse* might invoke genocide, eradication of peoples, the loss of territorial integrity, or some such similar loss of a society or its peoples. However, Tainter uses the term to denote a collapse from a higher to a lower level of complexity. Recall, this was discussed during the course of delineating groups in reference to a figure placed in section four of Chap. 1. Also of note, as Tainter discusses in his book, polarization does not lead to collapse.
6. The stimulus of blows was a term used by Toynbee to show how outgroup pressure, such as that imposed by war, could produce ingroup solidarity. Thus, the "blows" are the pressure and their "stimulus" induces unity.
7. Positive liberty was discussed in Chap. 13.
8. This is not dissimilar from the Authoritarian Dynamic discussed by Stenner.

20

Liberty after Liberalism: A Precarious Balance Derived from Dynamic Tension

Steven Charles Hertler, Aurelio José Figueredo, and Mateo Peñaherrera-Aguirre

It has been the search of the ages to find a political system, the travail of the ages to construct one, in which Government and Liberty shall be reconciled, in which each of these all-comprehending means of civilization shall strengthen the other and in which finally each shall be the fulfillment of the other. Down to the present moment millennial equilibrium has not been fully attained and mankind always has been, and still is, in danger of diverging from the true path which leads to it, towards despotism on the one side or anarchy on the other.

—J. W. Burgess's (1915) introduction to the *Reconciliation of Government with Liberty*

S. C. Hertler (✉)
Saint Elizabeth University, Morristown, NJ, USA

A. J. Figueredo • M. Peñaherrera-Aguirre
University of Arizona, Tucson, AZ, USA
e-mail: ajf@arizona.edu; mpeaher@email.arizona.edu

S. C. Hertler et al., *The Evolution of Political Ideology*,
https://doi.org/10.1007/978-3-031-83941-2_20

1 Introduction

In preceding chapters we reviewed the ways in which persons are evolutionarily pulled in two directions by multilevel selection's sometimes opposing forces—individual and group selection.[1] Rather than a dichotomy evolving, under the dictates of balancing selection, persons fall out along a continuum of temperaments from which their political ideologies derive. On one end of the continuum are more group-selected, conservative minds marked by pessimism and closed-minded traditionalism, and on the other end of the continuum are more individually selected, liberal minds marked by optimism and open-minded progressivism.[2] Thus, individual and group selection operate simultaneously to create and maintain a broad range of variation between these extremes. Describing this sociobiological continuum was the function of Parts I to III's eight chapters. Thereafter, Parts IV to VII's eleven chapters explored how this continuum played out at the societal level in different historical contexts. Just as individuals can vary along this continuum, so can societies, which can fall between the extremes of *collectivistic despoticism* at one end, and *individualistic libertarianism* at the other.

This twentieth chapter explores intersections between ideological minds and sociopolitical systems—between the individual psychology reviewed in Parts I to III and its socio-political manifestation reviewed in Parts IV o VII. Avenues of exploration are effectively endless; however, we develop a specific thesis focusing on liberal republics and conservative ideology. For clarity's sake, we state and restate our thesis, first emphasizing its core claim, and then adding context. Most pointedly, our thesis runs thus: Classical Liberal republics,[3] meaning Modern Republics incorporating liberal enlightenment ideals into classical republican models, are products of biocultural evolution that can amount to *evolutionarily stable strategies*—strategies that stably coexist in competition with rival strategies. Considered as evolutionarily stable strategies, it then follows that liberal republics will remain stable only insofar as they, and the selective regimes in which they evolved, remain stable. Stability relies on preserving geographic and material preconditions undergirding liberalism, remaining martially competitive with illiberal rivals, and moderating the rate and extent of change. In connection with this last point, the rate and extent of change, we emphasize the importance of conservative ideology: Conservatism served as a brake on liberalism, ensuring that it did not go too fast so as to outpace what was possible given geopolitical and biocultural realities; and now that it does not go too far, so as not to undermine its preconditions and extend beyond bounds of possibility given perennial risks of internal dissolution and external conquest. In what follows, we enumerate and elaborate this thesis:

1. Self-governing liberal republics are products of biocultural evolution.
2. Liberal republics can, under the sociopolitical conditions of the modern era, amount to *evolutionarily stable strategies*.
3. The emergence of liberal republics depended on, or benefited from, geographical and material preconditions.
4. The persistence of liberal republics depends on political pragmatism created by dynamic tension between liberal and conservative individuals.
5. The basic logic of this dynamic tension evolved in the risk/reward calculus of pre-state societies.
6. The counterweight of conservatism becomes increasingly important during times of societal decline and war, but also in the modern culmination of liberal societies which reached an inflection point beyond which the pursuit of fairness and freedom no longer augments power and stability.

In developing this thesis, we ask a series of questions, all variously related to liberal republicanism's stability and trajectory. More specifically, we ask, in Sect. 2, *can liberal republicanism be considered an evolutionarily stable strategy;* in Sect. 3, *is liberal republicanism reliant on preconditions*; in Sect. 4, *is liberal republicanism vulnerable to threats*; in Sect. 5, *is liberal republicanism subject to limits*? In the confines of this chapter, we can only ask these questions rhetorically, rather than answer them empirically.

2 Can Liberal Republicanism Amount to an Evolutionarily Stable Strategy?

Among evolutionists, there is a range of opinion concerning liberal republicanism. For example, as per Faria (2019), liberalism is *strikingly unsustainable,* notwithstanding its dominance over Western political thought for several hundred years past. The choices people freely make, *contra* Smith's *Invisible Hand*, don't reliably redound to national strength as per Faria. As Gibbon attributed the decline of Rome to Christianity, so Faria attributes demographic decline and national weakness to liberalism. Notwithstanding Faria's relevance, the emergence and success of liberal societies suggest Mandeville's Bees[4] could in fact carry the load given over by Spartan self-abnegation and Roman civitas.[5] In some way, their existence is tentative proof of concept.[6] At least thus far, there is no evidence that liberal societies follow the course of decline and fall more quickly (or more slowly) than, for example, their monarchical or despotic counterparts.

By contrast, Fukuyama famously cast history as an evolutionary process tending toward liberal democracy. This thesis is an article of faith explicitly premised on claims to democracy's purported ethical, political, and economic superiority.[7] We find Fukuyama's optimism groundless. We reject any notion of teleology because evolution has no ultimate aim or end, and also reject Fukuyama's more limited claim that liberalism will spread to fixation, displacing other forms of government. Were we to take Fukuyama at his word by seriously applying evolutionary reasoning to liberal democracy, we would have to acknowledge that all adaptations are adaptive in relation to some specified set of environmental conditions. Evolved adaptations only persist in the face of stability,[8] and only spread in the wake of similarity. To illustrate the point, consider complexity. In evolution, there is a trend toward complexity supported by the fossil record. All else being equal, in many phyla and classes, evolutionary arms races and other mechanisms selected for increasingly complex organisms replete with cells, tissues, and organs. Nevertheless, complexity has not affected all organisms in all places, and the trend is sometimes halted, and at other times reversed. As with the evolution of complexity, we do see a trend of increasing liberal republicanism, but this trend is not inevitable, irreversible, or independent of spatial and temporal change. In a word, the evolution of anything, whether an organismic characteristic or a societal structure, will arise and persist only under certain favorable conditions.

Fukuyama and Faria see liberalism tending in different directions: For Fukuyama, liberalism will spread to fixation; for Faria, liberalism will end in extinction. Yet, both authors see inevitable directionality, where we see none. Indeed in evolution, some traits have such clear advantages that they are driven to fixation, while deleterious mutations can have such clear disadvantages that they are driven to extinction. Nevertheless, as discussed in Chap. 5, many other traits, personalities (Penke & Jokela, 2016), and blood types (Meyer et al., 2018) among them,[9] are subject to *balancing selection*, a concept that is an apt model for understanding liberal republics in geopolitical context. Recall from prior chapters, under balancing selection, two or more adaptive solutions evolve, rather than a single optimum. Sex determination serves as a ready example. Reproductive success favors a female in a mostly male population, and a male in a mostly female population; and so, more or less balanced sex ratios evolve. Morphic variation serves as another common example. Morphs, of which salmon[10] and side-blotched lizards are exemplars, are rival, alternative strategies—behaviorally and physically different types that are nonetheless contained within the same sex, species, and locale. When one adaptive solution, evolved via balancing selection, maintains its existence in competition with rival adaptive solutions, it is then considered an *evolutionarily stable strategy*. Barring substantial ecological change, or in this case geopolitical change, evolutionarily stable strategies (by definition) cannot be displaced by rival strategies once they are established within a certain ecology.

It is in this context that we see liberal republics. They are products of biocultural evolution emerged in modernity and existing for several hundred years in competition with monarchies, dictatorships, and other forms of sociopolitical organization. To the extent that liberal republics represent an evolutionarily stable strategy, like any other evolutionarily stable strategy, they should *not necessarily* be expected to progress to fixation, as Fukuyama believes they will. To see liberal democracies exist for some time in some places does not mean that they will endure for all time and spread to all places. At the same time, we should not, like Faria, expect that they will systematically extinguish themselves; at least as long as they remain essentially the same, and their external geopolitical environment remains similar to those past geopolitical environments in which liberal republics rose and throve.

Now, having explained that multiple adaptive solutions can be maintained in competition with one another via balancing selection, and having likened liberal republics to evolutionarily stable strategies balanced in competition with rival sociopolitical strategies, it follows that we must consider the conditions for continued balance. In other words, what are the conditions upon which liberal societies depend, without which they would cease to be evolutionarily stable? In the subsequent three sections, we review facilitative ecological conditions and historical events alongside the role of conservative minds, which slow and curb liberalization. As this chapter develops its three remaining sections, it increasingly focuses on conservatism's role in maintaining liberal republics as evolutionarily stable strategies, such that, by the conclusion, one can consider *how continued dialectic across the ideological divide might be necessary to walk the narrow corridor; to strike that precarious balance that maximizes liberal republicanism without eroding its internal foundations of order and wealth or its external bulwarks against conquest and dissolution.* In other words, within what parameters can liberal societies continue to be evolutionarily stable?

3 Is Liberal Republicanism Reliant on Preconditions?

We first ask, *are there geographical, agricultural, industrial, or economic material conditions necessary for the emergence or maintenance of liberal ultrasocial*[11] *institutions?* Some candidate preconditions invite interesting conjecture. For instance, the United States is possessed of a unique history and geography, which ostensibly enabled its experiment in self-government to work. In this vein, one can consider the importance of geographical space,[12] as did Frederick Jackson Turner (1921) in *The*

Significance of the Frontier in American History. "So long as free land exists," Turner (1921, p. 32) declared, "the opportunity for a competency exists, and economic power secures political power." Free land facilitated freedom. Certainly its distance from the mother country, affording 150 years of colonial self-government, provided a foundation for enduring independence. Were England adjacent to the United States, it might have more directly governed its colonies from the start, and thereby defeated or suppressed any independence movements. Treatises and books containing the cultural software of self-government crossed the Atlantic more easily than the munitions and men necessary to stifle self-government.

The Atlantic buffered America from Europe, as the Channel had buffered England from the continent (Macfarlane, 2003). Between 112 and 21 miles wide,[13] the English Channel arguably allowed English constitutionalism itself to wend its way through history, accreting from bottom-up without being extinguished by, for example, the Spanish Armada (Hutchinson, 2014), the Battle of Trafalgar (Bennett, 2004), or Operation Sea Lion (McKinstry, 2014). Notwithstanding these intriguing correlations which have some irrefutable causal reality, geography seems neither a necessary nor sufficient precondition. Demonstrating the point, though an island nation like England, Japan did not follow the path of liberal republicanism at this time, while several continental European states did. Similarly, as Adrian Hastings (1997) observed, England's precociously established liberalism plausibly stems from geography, being both small and delimited; though Hastings quickly notes that Ireland, sharing both geographical features, did not follow the same course. Moreover, philosophical liberalism derived from, or is at least associated with, Enlightenment thought, which was cosmopolitan in extent.

The British Agricultural and Industrial Revolutions have greater claims to the status of preconditions. The former created people and the latter created energy. Together, the British Agricultural and Industrial Revolutions liberated first England, and then other societies, from Malthusian[14] constraints (Clark, 2007). From mere subsistence, societies secured the bodies and energy inputs necessary to beat back existential threats stalking populations from within and from without. Populations with full bellies and growing families, living in states with secure borders and full coffers, can more realistically think along liberal lines that value individuals intrinsically and orient toward progressivism implicitly. While there were examples of ancient democracies and republics, classical liberal republicanism itself, and liberal democracies, seem not to have existed prior to the British Agricultural Revolution beginning circa 1750 in Britain. Unlike a war or a political coup, this event cannot be dated with precision, and indeed some have argued that truly revolutionary agricultural output came in the nineteenth century; however, Overton (1996) documents agricultural output increasing from 1500 to 1850 in a long view of change that was revolutionary only in its effect. While the Agricultural Revolution's early date plausibly casts it as a precondition, the Industrial Revolution's later date, often said to have begun circa 1750 (Toynbee, 2011), suggests that it could have been an accelerant only. Some early inventions and the use of fossil fuels, however, were in place by 1700 (Clark & Jacks, 2007; Spear, 2014). Coal, for the heating of houses and for the smelting of ore, coincided with demographic increase, and more broadly ushered in a *mineral economy*[15] less reliant on the muscle of beasts and men (Landers, 2003). While neither food nor coal is necessary for liberal thought, it may be necessary for functional liberal republics that incorporate substantial portions of their population into the body politic.

These factors, *abundance, energy,* and *security*, collectively reorient significant portions of the population away from base *deficiency needs,* allowing pursuit of those *growth needs* at the top of Maslow's *Hierarchy of Needs.* Such conditions allow demographic growth, capital investment, vocational specialization, and related societal change, which makes room within lives for civic participation. These material factors are also the midwife of mass literacy and education, which may well be more direct preconditions. Prior to the coalescing of such preconditions, individuals existed as links in a *great chain of being,*[16] but thereafter were increasingly understood as intrinsically valuable, autonomous agents having a right to the labor of mind and body. Liberal freedoms, coinciding with the British Agricultural Revolution, and thereafter the Industrial Revolution, stoked individual ambition and group dynamism to create unparalleled wealth, which, in turn, allowed demographic booms, further expansion of trade, colonization, infrastructural improvements, capital investments, and taxable resources with which to wage war. Contrary to historically naïve or unrealistically optimistic hopes, liberal republicanism may well have requirements, vulnerabilities, and constraints. In these ways, a historically informed evolutionary perspective suggests that liberal progress has depended on, and cannot extend beyond, certain biocultural and material conditions.

4 Is Liberal Republicanism Vulnerable to Threats?

Next, we ask: *can liberal societies cope with non-liberal rivals?* Though this is ultimately a matter for future empirical investigation, we tentatively return an affirmative answer, but register a caveat. Because they are more heterogeneous and adaptable, it has been argued that liberal societies are competitively superior; if this is true, it is true only in the long run; they harbor inherent short-term vulnerabilities, which, however, may be partially mitigated by conservatism.

Paul A. Rahe, classist and historian, writes, "nowhere in the world can a polity as economically stagnant as despotism manage to sustain itself against the forces brought to bear against it by technologically dynamic and commercially vibrant states of the sort that have appeared in Western Europe" (Rahe, 2009, p. 185). Rahe cites Montesquieu's faith in the superior competitiveness of modern, liberal republics over and above absolute monarchies and ancient despotisms, applying this logic to contemporary liberal republics facing populist and totalitarian states. "Over the last three hundred years," Rahe states, "though they [liberal republics] may at times have bled, they did endure" (Rahe, 2009, p. 239).

Rahe is right, and for reasons outlined by Harrington.[17] For instance, financial innovations in Britain enabled increased war making capacity, just as American factories designed to make motor cars began making war planes. As illustrated by the Manhattan Project, liberal societies benefit, in the long run, from those minds, institutions, methods, and materials that individual freedoms allow to flourish, and which can be co-opted and adapted to the needs of state. The more *phenotypically plastic*[18] liberal state is able to adapt to innovations of enemies during wartime. After being severely threatened by Napoleonic France and Nazi Germany, Britain harnessed the power of its free, dynamic, and capitalistic economy to attain victory. The same might be said of South and North in the American Civil War. The more group selected South initially attained an aura of invincibility through its salient victories in the Eastern theatre, but the more individually selected North thereafter triumphed through a combination of enlisting immigrant and black regiments and capitalizing on its industrialized productivity and monetary system.

While liberal institutions are put into a war posture and while factories are retooled and the private sector leveraged, liberal societies remain especially vulnerable to initial onslaught from fascistic societies. This short-term vulnerability was recognized by John Fitzgerald Kennedy. In writing about the British reaction to totalitarian threat in his *Why England Slept*[19] Kennedy laid the blame for British paralysis on impersonal British institutions, thereafter generalizing about democracies as *peace-loving* institutions by nature ill prepared for the challenge of totalitarianism, which was geared for a "world at war." Kennedy recognized that "totalitarianism had significant short-run advantages, even though democracy was superior '*for the long-run*'" (Schlesinger, 1992, p. 85). Tocqueville was of the same mind: "An autocracy that does not defeat a democracy quickly always runs the risk of being defeated by it" (Runciman, 2018, p. 161).

The aforementioned broader degree of tolerance for a wider range of individual differences includes individual differences in social ideology. Although a liberal state has its share of individualistic social progressives, it also typically contains a subset of social conservatives within its "big tent." This ideological range affords the maintenance of conservatives, who can then be drawn upon in times of threat, instability, and war. When they restrict openness to demonstration, openness to free speech and openness to deliberation, when they suspend habeas corpus, when they impart dictatorial powers, and when, in other ways, they drift toward the more closed societies of their enemies, liberal republics often promote, elect, and take counsel with more conservative population members. This all happens in the throes of war. Prior to war, conservative minds, with their larger amygdalae, temperamental pessimism, and alertness to foreign threats, limit initial vulnerabilities to the aforementioned onslaught. Minds sensitive to themes of decline, original sin, and moral failings reflect a pessimism about human nature that will assert itself in policy, reducing somnambulant vulnerabilities.

From a biohistorical standpoint, liberal republics are still something of a novelty. As such, neither sanguine complacency nor foreboding pessimism should stand in for empirical data. Without the latter, it is ever harder to predict where liberal republics will end up on the international stage—as waning novelties, waxing behemoths, or stable competitors.

5 Is Liberal Republicanism Subject to Limits?

William Graham Sumner censured progressives, radicals, and reformers for their inability to distinguish between the *old* and the *bad* (Bannister, 1992). The reformer finds something that is old, benefits from its reformation because it is also bad; and then, condemning all that is old as if it were bad, expects the same benefit will derive from every subsequent change.[20] When mistaking the old for the bad, progressives, radicals, and reformers unwittingly tilt at social realities rooted in intransigent features of human nature that cannot be very easily eliminated or altered. Also, when seizing upon that which is both old and bad, progressives, radicals, and reformers sometimes pursue change at an unsustainable rate. As considered below, there are limits to the *scope* and *rate* of change.

With respect to the scope of change, we do well to recall Kant's quote: "Out of the crooked timber of humanity, no straight thing was ever made." This seizes on conservativism's temperamental fountainhead—an understanding of human nature rooted in original sin. Exemplifying the point, in his *A Defense of the Constitutions of Government of the United States of America,* John Adams wrote:

> In the institution of government, it must be remembered, that although reason ought always to govern individuals, it certainly never did since the Fall, and never will till the Millennium; and human nature must be taken as it is, as it has been, and will be. (Diggins, 1986, p. 94)

By contrast, those on the liberal end of the ideological spectrum not only repudiate negative views of human nature, but are also apt to deny its fixedness, and even its existence. For example, reformers like Colonel Edward M. House had "a vision of human nature as something infinitely malleable" (Lasch, 1965, p. 229). "If we turn now to the opinions of the Left on this matter," Bongie (2000, p. 93) states, "we will see that the revolutionary ideologues disagreed passionately with the basic assumption on which such historical conservatism rested: namely, the idea of a stable human nature, of an inflexible moral 'nature of things.'" Of course, it is in the nature of ideology to distort a more subtle reality where, in evolutionary parlance, there is a human nature, some aspects of which are *canalized* and other aspects of which are *phenotypically plastic.* Canalized[21] or biologically prepared features of human nature, when they become targets of reform, prove obstinately persistent, as when the French tried to dispense with religion, and the Russians tried to dispense with selfishness. The record of liberal republicanism's achievement, chronicled in prior chapters, provides grounds for pride and optimism, but also for melioristic overconfidence. Liberals might fail to see when they pass an inflection point, and begin to betray their own principles or otherwise undermine foundations of realizable progress. As Karl Popper (2011, p. 442) said, "…the attempt to make heaven on earth invariably produces hell."

Recall from previous chapters that, as ancient republics were transformed into liberal republics, citizens changed from being perceived as means to being perceived as ends. Recall also that these modern, liberal republicans were, in some ways, stronger and more stable than their ancient counterparts. Freedom wrought dynamism. Just as the shift from centralized planning to capitalist economies allowed markets to adjust, grow, and diversify; so the shift from ancient republics to liberal republics allowed individuals to adapt, innovate, and specialize. After their creation in the seventeenth century and proliferation in the eighteenth century, liberal republics continued down this Enlightenment course of development, as benchmarked, for example, by England's abolition of the slave trade, or by America's Civil Rights Movement. All of this change, we aver, conforms to classical liberal republicanism's record of reform, which is marked by many instances of power and stability felicitously following fairness and freedom. However, in the 1970s and increasingly thereafter, *individual freedom* and *social justice* became uncoupled, as documented in *Empire of Democracy: The Remaking of the West Since the Cold War*. The author, Simon Reid-Henry (2020, p. 128), speaks of a "loss of complementarity between the ideals of freedom and those of social justice." Some such policies are logical extremes of liberal republicanism, including the legalization of drugs and prostitution, articulated in modern libertarian tracts. These policies raise moral concerns, but are rationalized nonetheless as expressions of individual choice. But such choices can weaken a state, spreading homelessness, joblessness, and disease, the consequences of which are reliably externalized from the individual to the state. Other such policies, though originating from the political left, are antithetical to classical liberal republicanism. This *may* happen in the transition from negative to positive reform that reliably divides classical liberal republicanism from its modern heirs—in transitioning from removing obstacles to open fairness, to imposing barriers to natural inequalities; from abolishing nonrepresentative taxation to extremes of redistribution that drive wealth out of nations or suppress its creation; from outlawing cruel and unusual punishment to freeing offenders from consequences; or when proceeding from attacking unjust hierarchies based on caste, race, or religion, to attacking hierarchy itself even in its most basic, meritocratic form. Also, paternalist policies can be added to this list. These weaken the state in two ways. First, paternalism, though it preserves the individual as the end of government, changes the role of government *vis-à-vis* the individual, sometimes to the end of weakening the nation, just as a family would be weakened if an adult breadwinner were somehow changed to a dependent child. Furthermore, expanded bureaucracies, and the corruption that follows from them, are secondary consequences of paternalist policies that compound its detrimental effects.

Finally, some modern policies originating on the left undermine the Westphalian order of national sovereignty, thus altering relations between individuals and states that created such dynamism at the inception of liberal republics. For example, some rhetoric and laws advance lower-level groups within nations to the end, inadvertent though it might be, of eroding national identity. And then, globalist policies increasingly actualize the transnational cosmopolitanism evident on the political left since the Enlightenment. This has the effect of subtly shifting power and allegiance from the nation to supranational organizations. "Traditional concepts of citizenship and sovereignty," Linklater (1996) observed, "have come under pressure from the combined challenge of globalization and the subnational revolt." Whatever might be the ultimate end of social evolution, this is at least dangerous when pursued as a unilateral policy. It is akin to one-sided demilitarization. Reorientation away from a nation of individuals to globalism and subgroups arguably undermines external power and internal order—citizens are less categorically united together against rival nations, as they devolve into subgroups, pursuing their ends within the backdrop of transnationalism.

With respect to the rate of change, we invoke an analogy to *evolutionary mismatch*, the decline or extinction of a population or species unable to cope with rapid change. A loss of fitness follows *mismatch* between the present environment with which an organism must contend and the past environment to which it is adapted (Li et al., 2018). Humans, because of our numbers and newfound industrial technologies, are eminent drivers of evolutionary mismatch, as has been experienced by dodo birds, pepper moths, and sea turtles,[22] respectively experiencing hunting, air pollution, and light pollution. Often, humans simultaneously drive the process, and suffer its consequences. The rise in autoimmune disorders coinciding with hygienic environments created in the wake of industrialization illustrates the point. Likewise, from a combination of clothing (Buyukuslu et al., 2014), sunscreen (Norval & Wulf, 2009),[23] indoor living (Hyun Kim & Soo Jae Moon, 2000), and migration (Martin et al., 2016), natural Vitamin D synthesis through sunlight exposure seems suppressed (Mithal, et al., 2009), especially in the northern hemisphere (Kimlin, 2008), to the end of increasing cancer (Zmijewski, 2019), osteoporosis (Holick, 2007), cardiovascular disease (Judd & Tangpricha, 2009), and myriad autoimmune diseases (Szodoray et al., 2008). This is then a case of evolutionary mismatch: Reduced sun exposure of modern environments thwarts Vitamin D synthesis.

Unlike systems of governance slowly accreted from the bottom up, which are generally expressions of biocultural realities, revolutionary systems of governance rapidly imposed from the top down can create the conditions for mismatch. The most revolutionary of all revolutions, the French Revolution of 1789, illustrates the point in its abrupt creation of calendars and administrative districts, and more obviously in its abolition of religion and monarchy, both of which had to be reinstated after decades of sanguinary instability. Relatedly, liberal reformers might fail to appreciate that some cultural developments are made feasible by evolutionary trends; and are therefore more accurately outcomes of gene-culture coevolution. For example, reductions in global violence (Pinker, 2012) might be a biocultural outcome enabled, not only by laws and rules, but by millennia of self-domestication, accelerating through modernity, wherein a society's most aggressive individuals were systematically incarcerated or executed to the end of lowering their fecundity and survival.

6 Conclusions

Excessively group selected societies are hivelike, mounting vigorous external defenses, though not without the compensatory costs of austere and restrictive mores and policies, as seen in ancient Sparta. Lopsidedly group selected regimes may effect an efflorescence of power that is correspondingly evanescent. With an emphasis on hierarchy, displays of martial valor, and group loyalty, moderating elements are silenced, and such societies tend toward unsustainable militarism. There is insufficient balance, restraint, and moderation upon which to build an enduring polity with adaptable institutions and pragmatic policy. The restricted range of tolerance for internal diversity thus limits such a society's aggregate plasticity.

Liberal republicanism offered an alternative. Classical Liberal republics are something of an experiment of modernity, the success of which can be cautiously gauged by their persistence and proliferation. They seem to have successfully placed the individual at the forefront, guarding his rights, protecting his person and property, while at the same time yoking his interest to the fate of the state. Liberal freedoms developed in the context of the British Agricultural and Industrial Revolutions, which augmented taxable wealth, demographic increase, and sustained trade. Mandeville was vindicated; ultrasocial communities could be wrought from selfish individuals. Yet, in the wake of this success, some see liberal democracy as progressing to fixation, inevitably overtaking other forms of governance. This was true of Thomas Paine in the eighteenth century as it is true of Francis Fukuyama in the twenty first century.

Yet, classical Liberal republicanism is neither inevitable nor invulnerable. It is more like an orchid than an invasive weed. Classical Liberal republicanism correlates in time with demographic increase, higher agricultural yields, trade, printing, and mass literacy, and thus, may be reliant on these as preconditions. Also, as has been seen, liberal republicanism benefits from, and may be reliant on, the foundation of earlier group selected societies, and even the inertia of Puritanism and Calvinism.[24] Further still, the Channel and the Atlantic respectively buffered the United Kingdom and the United States from incursions that might otherwise have smothered liberal republicanism in the cradle. And then we must note the short-term vulnerability to totalitarian blitzkrieg, which is difficult to cope with given the checks and balances, constitutional restraints, and individual freedoms that can collectively, though temporarily, render the liberal state supine. From within also, liberal societies are increasingly challenged by industrial and technological combinations, electoral corruption, and lower-level agglomerations of self-interested behaviour that do not fortuitously or inadvertently strengthen the state.

Also of note, while liberal republicanism has succeeded, it has done so within particular cultures and ecologies, and it is not entirely clear if it can be transplanted to wholly different cultures and ecologies.[25] Further still, we should recognize the baroque complexity of liberal societies and recall that entropy is a formidable foe to all things complex—a lesson seen in the fossil record's preservation of exquisitely adapted organisms that are nonetheless extinct, as well as in the historical record's narration of enormously powerful empires that are nonetheless collapsed.[26] Modern liberal societies are prey to geopolitical perturbations, climatic change, and other forms of instability in systems that extend outside their territorial boundaries and possible control.

A historically informed biological perspective frames liberal societies as complex systems, viable within a bounded set of parameters, with some of these parameters relating to ecological conditions, economic supports, and historical trajectories. Still, there are other parameters that are partially kept within viable ranges only where liberals are balanced against conservatives. Liberal republicanism can amount to an evolutionarily stable strategy, we tentatively conclude, only with the check imposed by conservative minds. Conservativism is an important brake on liberal excess, keeping liberal societies orbiting in pragmatic balance: Change that might otherwise be extreme and rapid is rendered partial and slow enough to observe consequences and reverse course where necessary; institutions that might otherwise be discarded or replaced are retained long enough for their unacknowledged functions to be conserved or substituted; resources that might otherwise be consumed are saved; defenses that might otherwise be relaxed are upheld. There was a reason why, within the primate troop and hunter-gatherer band, under the demands of negative frequency-dependent balancing selection, a temperamental range evolved, not an optimum. That range, as has been reviewed, served a bet hedging function within families and groups by balancing opportunities for innovation, expansion, outbreeding, trade, and cultural innovation with risks of insubordination, extinction, dissolution, exploitation, and cultural degeneration. As these temperaments are mapped onto a modern political process, this bet hedging functioning plays out at the societal level.

Notes

1. Multilevel selection theory posits that there is always a balance between the frequently opposing forces of individual and group selection.
2. We also do not mean that group and individual selection somehow work in isolation. As we clarify elsewhere, to say someone is "more group-selected" or "more individually selected" merely characterizes the relative forces of these two selective pressures historically within each lineage.
3. This term, "'classical liberal republic' refers to Enlightenment Era republics which insinuated classical liberal values into ancient republican models. Thus, we use the term classical to differentiate classical liberalism from modern liberalism, and classical liberal republics to differentiate these from ancient republics.
4. Mandeville was previously mentioned; again this refers to Bernard Mandeville, a Dutch philosopher. In his fable of the bees, Mandeville discussed how private self-interest could be good for the group. This was

similar to Adam Smith's argument in the Wealth of Nations, where he described the invisible hand allowing individual economic interests aligning with national economic interests.

5. We find some support for this also in the literature on ultrasocial institutions, which are supra-ethnic aggregates of unrelated individuals stitched together into sprawling states or expansive empires.
6. However, one must also be cognizant of inertia. It is probable that the success of liberal societies owes much to preceding centuries of biocultural evolution under different sociopolitical regimes, such as feudalism and monarchy.
7. This is reminiscent of Thomas Paine, who saw, in the outbreak of the French Revolution, the spread of the American Revolution, which was then to topple every monarch in Europe.
8. Of note, flexibility and plasticity can evolve, which are adaptive responses to fluctuating environments.
9. As discussed in Chap. 4, personalities evolve under the pressures of negative frequency-dependent balancing selection wherein a trait becomes less adaptive as it becomes more common; by contrast, blood types are not precisely frequency dependent, but are balanced because diversity hedges against risk; in this case, biodiversity among immune systems hedges against extinction within lineages.
10. "Some male salmon develop large bodies and fight over territory, while others maintain small bodies and develop large testes allowing them to clandestinely wait for the female to lay a clutch of eggs, which they can then rapidly fertilize with a shower of sperm" (Hertler, 2015).
11. Recall, *ultrasocial*, a term discussed in Chap. 5, refers to the presence of social institutions fostering the cooperation and division of labor and aggregation of very large groups of unrelated individuals, as discussed by Turchin (2016) and Gowdy (2021).
12. Turner's thesis is related to the sociological concept of *caging* (Mann, 1997), which is discussed in multiple forms. For example, ecological caging refers to the role of the environment in keeping people aggregated, as the surrounding desert kept people near the Nile River Valley, possibly facilitating the rise of Egyptian civilizations. Caging more generally relates to binding individuals into societies so they no longer can leave and lead self-sufficient lives. For example, a maltreated or underpaid frontiersman in early America could readily find other employment because of high wages, or move West and establish a homestead because

of free lands. By contrast, generations later, his counterpart had fewer options, and so could be said to be subject to caging.

13. https://www.britannica.com/place/English-Channel
14. Malthusian refers to Thomas Robert Malthus, who famously wrote *An Essay on the Principle of Population* (1798). This work broadly discussed the implications of reproduction outstripping food production to the end that more organisms will be born than can be fed. So, Malthusian constraints are constraints on demographic growth; or limits on births because of limits on food.
15. The term *mineral economy* and its counterpart, *organic economy*, are terms used by David Landers (2003) in *The Field and the Forge*. Energy within organic economies comes from the photosynthetic power of plants and the muscles of men and animals. The Industrial Revolution brought about mineral economies, releasing the power of fossil fuels.
16. The great chain of being (*scala naturae*) was a Medieval Christian hierarchy, with God on top, kings and high church officials below, down to nobleman, tradesman, animals and plants, etc.
17. Harrington's work was discussed previously. It centered on the strength deriving from trade and freedom.
18. An organism is said to be phenotypically plastic when it can manifest a wide range of behaviors in response to changing environmental demands. Here we are using the term to describe liberal societies.
19. This is a play on Churchill's book entitled *While England Slept*.
20. "But radicalism misses the distinction," Sumner claims, "It behaves as if antiquity were the cause or the sure sign of evil and it rages wildly against all antiquity and tradition" (Bannister, 1992, p. 18).
21. Canalized traits unfold by genetic prescription effectively in any environment. For example, assuming normal genetics, a person will grow five fingers on each hand whether they are raised in one or another culture, abused or coddled, educated or neglected.
22. https://en.wikipedia.org/wiki/Evolutionary_mismatch
23. As these authors mention, normal use may not have an appreciable effect on vitamin *D* production, and of course vitamin *D* production must be weighed against skin cancer. This balance follows the evolution of skin color, which is typically balanced against skin cancer and vitamin *D* production.
24. John Adams to Massachusetts Militia, 11 October 1798: "Our constitution was made only for a moral and religious people. It is wholly inadequate to the government of any other." https://founders.archives.gov/documents/Adams/99-02-02-3102

25. This is an important question to be asked within a *social biogeography* framework, which looks for links between ecology, evolution, and civilization; however, we pass over this question to pursue the durability of societies as a function of the interaction between liberal and conservative minds.
26. This refers to the literature on civilizational cycling, and, more recently, *cliodynamics*. It also alludes to Tainter and his book, *The Collapse of Complex Societies,* which discusses the sclerosis and burdens that afflict complex societies. Nevertheless, we do not pursue this important theme because these entropic forces affect all societies, rather than being unique to liberal societies.

Integrative Metacommentary: Part VII. The End of Tolerated Opposition

Aurelio José Figueredo, Steven Charles Hertler, and Mateo Peñaherrera-Aguirre

The End of Tolerated Opposition

Part VII presented three chapters on the various destabilizing influences that may plague the precarious balance that had been established between opposing factions. After having described the evolution of the mechanisms within a constitutional republic that maintain social order in spite of tolerating peaceful dissent, this section examines various threats that may arise to the maintenance that order.

Chapter 18 reviewed the possibility that the relaxation of external threats to the existence of the state, presumably from other states, might permit a greater resurgence of dissension between rival political factions within the state. Long periods of relative peace, such as the *Pax Americana*, may serve to lull competing interests within a constitutional republic into a false sense of security regarding the seeming nonexistence of external

A. J. Figueredo • M. Peñaherrera-Aguirre
University of Arizona, Tucson, AZ, USA
e-mail: ajf@arizona.edu; mpeaher@email.arizona.edu

S. C. Hertler (✉)
Saint Elizabeth University, Morristown, NJ, USA

threats. Within a framework of perceived relative safety, those rival factions may believe that they can afford the luxury of heightened within-group competition without endangering the continued existence of the constitutional republic itself.

Chapter 19 analyzed the role of political polarization as one among many of these possible destabilizing influences. Four major facets of the intersection between polarization and the decline of constitutional republics were considered: (1) Inefficiency; (2) Demagogues and Dependents; (3) Disintegrating Bonds; and (4) Liberalism, interpreted by some as a form of Decadence at the societal level. Nevertheless, it was concluded that political polarization is likely to be more of a *symptom* than a *cause* of societal decline.

Chapter 20 brought us back from a narration of evolutionary history to consider the adaptive function of Classical Liberal republics in relation to their sociopolitical environment of origin and of survival. The Classical Liberal republic was thus considered as a sociopolitical adaptation from the standpoint of behavioral ecology, rather than the teleological end, purpose, or goal of human evolution analogous to Teilhard de Chardin's (1959) *Omega Point* in *Le Phénomène Humain*. As such, it is historically contingent and restricted to a circumscribed ecological niche within a certain bounded hypervolume of time and space. It is thus not the Aristotelian *Final Cause* of all previous history, as the Whigs no doubt would have had it. The Classical Liberal republic is neither inevitable nor necessarily invincible.

Epilogue

> How do we move from theories that identify evolutionary pressures as the origin of behavioral adaptation to theories that use evolutionary reasoning to explain variation across individuals, cultures, and historical context?
> —Settle and Detert (2023, p. 281)

At the end of this 20-chapter pilgrimage through prehistory and political history, allow us to narrate a final story, one too condensed and synthetic to be told before now:

Humans branched off from the other great apes about five million years ago; we are now hominids capable of coordinated violence and coalitional action accentuated by high intelligence, initially expressed within small, hunter-gatherer bands composed mostly of genetically related individuals. The ability of individuals to compete for resources and mates within groups affected individual survival and reproduction, just as did the ability to compete between such coalitions. Nevertheless, both ingroup and outgroup aggression and competition could imperil survival and reproduction. In such small groups, defeat could entail complete group annihilation, but victory could result in more prestige, status,

S. C. Hertler et al., *The Evolution of Political Ideology*,
https://doi.org/10.1007/978-3-031-83941-2

allies, and mates for individuals. Small size, genetic relatedness, internal social controls, and between-group competition all kept individual survival closely yoked to group survival, favoring group characteristics such as altruism and cooperation. Much of Pleistocene human evolution, starting about fifty thousand years ago, selected for tribalistic ingroup altruism and cooperation, only allowing for a stunted substrate of ideological variation.

As group sizes increased, evolution accelerated across many ecological, demographic, and sociocultural transitions. Population sizes increased, changing the selective pressures to which humans were subject. Although ingroup and outgroup aggression and competition could still imperil survival and reproduction, in warfare between larger groups, such as state societies, the defeated risked becoming conquered and enslaved instead of annihilated. In contrast, victorious states stood to gain wealth and labor extracted from conquered peoples. Thus, there arose both individual selective pressures and group selective pressures, and these forces were sometimes mutually supporting and at other times opposing each other; the latter happening more frequently in state societies than within ancestral hunter-gatherer bands. These ecological, demographic, and sociocultural transitions acting on larger populations over generations also multiplied the social niches within the adaptive landscape. Warrior classes arose, along with other differentiated social roles among noncombatants: shamans and healers, priests and seers, kings and queens, farmers, serfs, artisans, tradesman, and merchants, not to mention the high bourgeoisie who wrested power from the old feudal aristocracy during the Age of Revolutions. In more complex societies, often at ecological carrying capacity, individuals became more specialized and occupied ever more novel social niches that were generated by the increased sociopolitical complexity, and these larger groups aggregated into increasingly stable and secure territories.

This diversification of social niches generated a wider range of genetically encoded and neurocognitively represented temperaments both within and between populations, which imparted the rudiments of ideological differences. There was greater variation in risk aversion, orientation to novelty, reactions to outgroups, and ideologically relevant ideas of justice and fairness. Nevertheless, selective pressures did not foster the evolution, development, or expression of ideological diversity or complexity. Individually selected persons and ideologies especially, would not easily evolve under these selective regimes, because these small-scale societies could not generate the degree of social complexity that would later evoke ideological variation. So, it is not as if the Pleistocene served as the single *environment of evolutionary adaptedness*, and then the mind produced by it was dropped off in modernity unchanged. Group selective pressures were relaxed in burgeoning post-Neolithic societies, which were larger, less directly genetically related, and more complex and differentiated. This created a more variegated fitness environment, evoking the evolution of a broader array of temperamental and behavioral variation, including more fully individually selected types. This process hastened on as human societies passed through multiple stages of social evolution, not excepting eighteenth-and nineteenth-century agricultural, industrial, and political revolutions. This allowed the evolution and expression of more individually selected persons, and consequently culminated in creating a full complement of temperamental variation from which a full panoply of ideological variation derives.

Political ideology could then be more rightly understood as the ideational subset of this long-evolving temperamental variation that imparted heuristic worldviews, biasing predilections and perceptions on human nature, sexuality, fairness, hierarchy, conformity, and much else of relevance to relations between individuals and groups. Of course, these many milestones in social evolution did not only create the foregoing temperamental and ideological variation, but placed persons in some social role, class, town, nation, and occupation. Consequently, individuals in different social circumstances would tend to adopt self-serving stances on political issues. They would then act politically, not only on the basis of their inherited temperamentally informed ideological disposition, but also as a partisan of their group and position within that group. For example, during the English Civil War, it was probably no coincidence that the hereditary landed aristocracy tended to support the Royalist cause, whereas the emerging high bourgeoisie tended to support the Parliamentarian cause; as these causes promoted their respective political and economic interests. Of course, ideology and interest can coincide, as in the person of a conservative defense contractor who lines his pockets while assuaging his fears of foreign conquest. On the other hand, ideology and interest can conflict, as when a liberal finds himself in that same position. Thus, there are, as discussed, two sides of the political, of which ideology is not uncommonly the weaker.

Only with such distinctions made, and with some knowledge of its evolutionary origins and temperamental foundations, can one understand how political ideology interfaces with society; and not just some static view of society in the present, but society as it changed demographically and materially across the transitions from small-scale societies to chiefdoms, from ancient *poleis* to medieval fiefdoms, to early modern monarchies to late modern democracies; and as political ideology interacts with societies as they traveled through cyclical arcs of birth, decadence, senescence, and dissolution. All of this makes ideology an elusive concept, and that elusiveness hopefully justifies our long pilgrimage.

Reference

Settle, J., & Detert, L. (2023). Biology and politics. In L. Huddy, D. O. Sears, J. S. Levy, & J. Jerit (Eds.), *The Oxford handbook of political psychology* (pp. 280–309). Oxford University Press

References

Aarøe, L., Appadurai, V., Hansen, K. M., Schork, A. J., Werge, T., Mors, O., Børglum, A. D., Hougaard, D. M., Nordentoft, M., Mortensen, P. B., Thompson, W. K., Buil, A., Agerbo, E., & Petersen, M. B. (2021). Genetic predictors of educational attainment and intelligence test performance predict voter turnout. *Nature Human Behaviour, 5*(2), 281–291.

Aarøe, L., Petersen, M. B., & Arceneaux, K. (2017). The behavioral immune system shapes political intuitions: Why and how individual differences in disgust sensitivity underlie opposition to immigration. *American Political Science Review, 111*(2), 277–294.

Achenbach, A., & Foitzik, S. (2009). First evidence for slave rebellion: Enslaved ant workers systematically kill the brood of their social parasite *Protomognathus Americanus. Evolution, 63*(4), 1068–1075.

Ackerman, B. (2000). *We the people.* Harvard University Press.

Adams, J. Q. (1847). *Letters on the masonic Institution.* Press of TR Marvin.

Adcock, R., Bevir, M., & Stimson, S. C. (2007). *Modern political science: Anglo-American exchanges since 1880.* Princeton University Press.

Ágoston, G. (2014). Firearms and military adaptation: The Ottomans and the European military revolution, 1450–1800. *Journal of World History, 25*(1), 85–124.

Ågren, J. A. (2016). Selfish genetic elements and the gene's-eye view of evolution. *Current Zoology, 62*(6), 659–665.

S. C. Hertler et al., *The Evolution of Political Ideology*,
https://doi.org/10.1007/978-3-031-83941-2

Ågren, J. A., & Clark, A. G. (2018). Selfish genetic elements. *PLoS genetics, 14*(11), e1007700.

Alemu, A. M. (2016). The Impacts of ethnic, linguistic and religious fractionalization on state fragility in Africa. *Journal of Marketing & Management, 7*(2), 47.

Alesina, A. F., Giuliano, P., & Nunn, N. (2011). *On the Origins of Gender Roles: Women and the Plough.* National Bureau of Economic Research Working Paper 17098.

Alfani, G., & Gráda, C. Ó. (Eds.). (2017). *Famine in European history.* Cambridge University Press.

Alford, C. F. (2001). Group psychology is the state of nature. In *Political psychology* (pp. 201–214). Psychology Press.

Alford, J. R., Funk, C. L., & Hibbing, J. R. (2005). Are political orientations genetically transmitted? *American Political Science Review, 99*, 153–167.

Alford, J. R., Funk, C. L., & Hibbing, J. R. (2008). Beyond liberals and conservatives to political genotypes and phenotypes. *Perspectives on Politics, 6*(2), 321–328.

Allen, M. G. (2013). Melanesia's violent environments: Towards a political ecology of conflict in the western Pacific. *Geoforum, 44*, 152–161.

Allen, M. W., & Jones, T. L. (2016). *Violence and warfare among hunter-gatherers.* Left Coast Press.

Alloway, T. M. (1980). The origins of slavery in Leptothoracine ants (Hymenoptera: Formicidae). *The American Naturalist, 115*(2), 247–261.

Altemeyer, B. (2004). Highly dominating, highly authoritarian personalities. *The Journal of Social Psychology, 144*(4), 421–448.

Altemeyer, B., & Hunsberger, B. (2005). Fundamentalism and authoritarianism. In *Handbook of the psychology of religion and spirituality* (pp. 378–393). The Guilford Press.

Altemeyer, R. A., & Altemeyer, B. (1996). *The authoritarian specter.* Harvard University Press.

Amodio, D. M., Jost, J. T., Master, S. L., & Yee, C. M. (2007). Neurocognitive correlates of liberalism and conservatism. *Nature neuroscience, 10*(10), 1246–1247.

Anderson, J. A. M., Franklin, B., & Royster, P. (1734). *The constitutions of the Free-Masons.* An Online Electronic Edition. (1734). Lincoln, NE: Faculty Publications, UNL Libraries 25. https://digitalcommons.unl.edu/libraryscience/25

Anderson, P. N. (2011). John and Qumran: Discovery and Interpretation over Sixty Years. In M. L. Coloe & T. Thatcher (Eds.), *John, Qumran, and the Dead Sea Scrolls: Sixty years of discovery and debate* (Early Judaism and its Literature, Number 32) (pp. 15–50). SBL Press.

Andrewes, A. (2002). The government of classical Sparta. In M. Whitby (Ed.), *Sparta* (pp. 49–68). New York.

Ansell, B. W., & Samuels, D. J. (2014). *Inequality and democratization.* Cambridge University Press.

Anspach, N. M. (2023). Afraid of whom?: Threat sensitivity's influence changes with perceived source of threat. *Politics and the Life Sciences, 42*(1), 17–31.

Arbatlı, C. E., Ashraf, Q. H., Galor, O., & Klemp, M. (2020). Diversity and conflict. *Econometrica, 88*(2), 727–797.

Arden, R., & Adams, M. J. (2016). A general intelligence factor in dogs. *Intelligence, 55*, 79–85.

Arendzen, J. (1910). Manichæism. In *The Catholic Encyclopedia* (p. 9). The Encyclopedia Press.

Arie, S. (2004, June 15). Historians say inquisition wasn't that bad. *Guardian US.* Retrieved April 16, 2024, from https://www.theguardian.com/world/2004/jun/16/artsandhumanities.internationaleducationnews

Aristotle. (2002). *On the parts of animals* (J. G Lennox, Trans.). Clarendon Press.

Arnold, J. H. (2001). *Inquisition and power: Catharism and the confessing subject in medieval Langedoc.* University of Pennsylvania Press.

Arnold, J. H. (2016). *The* Cathar middle ages as a methodological and historiographical problem. In A. Sennis (Ed.), *Cathars in Question, Heresy and Inquisition in the Middle Ages* (pp. 53–78). York Medieval Press.

Arnold, J. H., & Biller, P. (2016). *Inquisition in France, 1200–1300.* Manchester University Press.

Arvan, M. (2013). A lot more bad news for conservatives, and a little bit of bad news for liberals? Moral judgments and the dark triad personality traits: A follow-up study. *Neuroethics, 6*, 51–64.

Asbridge, T. (2010). *The Crusades: The authoritative history of the war for the Holy Land.* Simon & Schuster.

Ashby, R. W., Aubrey, N., Benbow, T., Catton, C., & Swanston, A. (2014). *Encyclopedia of warfare.* Amber Books.

Auslander, L. (2009). *Cultural revolutions: Everyday life and politics in Britain, North America, and France.* University of California Press.

Avital, E., & Jablonka, E. (2000). *Animal traditions: Behavioural inheritance in evolution*. Cambridge University Press.

Bachleda, S., Neuner, F. G., Soroka, S., Guggenheim, L., Fournier, P., & Naurin, E. (2020). Individual-level differences in negativity biases in news selection. *Personality and Individual Differences, 155*, 109675.

Bahar, S. (2017). *The essential tension: Competition, cooperation and multilevel selection in evolution*. Springer.

Bailey, M. (2014). *The decline of serfdom in late medieval England: From bondage to freedom*. Boydell & Brewer Ltd.

Bakker, B. N., Schumacher, G., Gothreau, C., & Arceneaux, K. (2020). Conservatives and liberals have similar physiological responses to threats. *Nature Human Behaviour, 4*(6), 613–621.

Baldwin, M. W. (1969). The decline and fall of Jerusalem, 1147–1189. In M. W. Baldwin (Ed.), *The Crusades. Volume I. The first hundred years* (pp. 590–621). The University of Wisconsin Press.

Ball, T. (2001). Discordant voices: American histories of political thought. In D. Castiglione & I. Hampsher-Monk (Eds.), *The history of political thought in national context* (pp. 107–133). Cambridge University Press.

Ballout, M. H., Briggs, A., Armstrong, J., & Clark, C. B. (2023). Assessing the relative contribution of Moral Foundation Theory, Right-Wing Authoritarianism, and Social Dominance Orientation in the prediction of political orientation. *Revista Interamericana de Psicología/Interamerican Journal of Psychology, 57*(3), e1756–e1756.

Bannister, R. C. (1992). *On liberty, society, and politics: The essential essays of William Graham Sumner*. Liberty Fund.

Barber, M. (2012). *The crusader states*. Yale University Press.

Barber, M., & Pope, J. C. (2024). The crucial role of race in twenty-first century US political realignment. *Public Opinion Quarterly, 88*, nfad063.

Barker, M. (1992). *The great angel: A study of Israel's second god*. Westminster/John Knox Press.

Barker, M. (1995). The secret tradition. *Journal of Higher Criticism, 2*(1), 31–67.

Bartelson, J. (2009). *Visions of world community*. Cambridge University Press.

Bartelson, J. (2023). *Becoming international*. Cambridge University Press.

Baum, D. (1978). Know-Nothingism and the Republican majority in Massachusetts: The political realignment of the 1850s. *The Journal of American History, 64*(4), 959–986.

Baumard, N. (2010). Has punishment played a role in the evolution of cooperation?A critical review. *Mind & Society, 9*(2), 171–192.

Baumgarten, J. M. (1986). The Book of Elkesai and Merkabah Mysticism. *Journal for the Study of Judaism in the Persian, Hellenistic, and Roman Period, 17*(2), 212–223.

Bayliss, A. J. (2022). *The spartans: A very short introduction.* Oxford University Press.

Bell, E., Kowalski, C. M., Vernon, P. A., & Schermer, J. A. (2021). Political Hearts of Darkness: The Dark Triad as predictors of political orientations and interest in politics. *Behavioral Sciences, 11*(12), 169.

Bell, E., Schermer, J. A., & Vernon, P. A. (2009). The origins of political attitudes and behaviours: An analysis using twins. *Canadian Journal of Political Science, 42*(04), 855–879.

Benjamin, D. J., Cesarini, D., Van Der Loos, M. J., Dawes, C. T., Koellinger, P. D., Magnusson, P. K., Chabris, C. F., Conley, D., Laibson, D., & Visscher, P. M. (2012). The genetic architecture of economic and political preferences. *Proceedings of the National Academy of Sciences, 109*(21), 8026–8031.

Bennett, G. (2004). *The battle of Trafalgar.* Pen and Sword.

Bennett, M. (2005). *The Hutchinson dictionary of ancient and medieval warfare.* Helicon Publishing.

Berggren, M., Akrami, N., Bergh, R., & Ekehammar, B. (2019). Motivated social cognition and authoritarianism: Is it all about closed-mindedness? *Journal of Individual Differences, 40*(4), 204.

Bergner, C. L. (2023). *The biological bases of political attachment: Neurobiological correlates of ideology and partisanship.* (Doctoral dissertation,. The University of Wisconsin-Milwaukee).

Bernard, G. (2006). *The inquisitor's guide: A medieval manual on heretics* (J. Shirley, Trans.). Welwyn Garden City.

Berry, J. M., & Sobieraj, S. (2011). Understanding the rise of talk radio. *Political Science & Politics, 44*(4), 762–767.

Betz, O. (1990). Was John the Baptist an Essene? *Bible Review, 6*, 20–25.

Bevir, M. (2006). Political studies as narrative and science, 1880–2000. *Political Studies, 54*(3), 583–606.

Bevir, M. (2015). Historicism and critique. *Philosophy of the Social Sciences, 45*(2), 227–245.

Bevir, M., & Rhodes, R. A. W. (2010). *The state as cultural practice.* Oxford University Press.

Bhushan, L. I. (1982). Validity of California F-scale: A review of studies. *Indian Psychological Review, 23*(Spec Issue No 1), 1–11.

Birdsong, D., Ramsey, N., & Gwon, M. (2014, August). Cable News: Audience Autonomy and Political Polarization. In *APSA 2014 Annual Meeting Paper*.

Black, J. (2006). *George III: America's last king*. Yale University Press.

Blanchard, A. E., Kibowski, F., & Dunn, T. J. (2020). Existential threats of immigration and terrorism predict voting for Brexit and Trump. *Evolutionary Psychological Science, 6*, 367–379.

Blanton, H., Strauts, E., & Perez, M. (2012). Partisan identification as a predictor of cortisol response to election news. *Political Communication, 29*(4), 447–460.

Blaydes, L., & Paik, C. (2015). The impact of Holy Land crusades on state formation: War mobilization, trade integration, and political development in medieval Europe. *International Organization, 70*(3), 551–586.

Boardman, J. D. (2011). Gene-environment interplay for the study of political behaviors. In P. K. Hatemi & R. McDermott (Eds.), *Man is by nature a political animal: Evolution, biology, and politics* (pp. 185–206). Chicago University Press.

Boas, F. (1896). The limitations of the comparative method of anthropology. *Science, 4*(103), 901–908.

Boehm, C. (1999). *Hierarchy in the forest: Egalitarianism and the evolution of human altruism*. Harvard University Press.

Boehm, C., Barclay, H. B., Dentan, R. K., Dupre, M. C., Hill, J. D., Kent, S., Knauft, B. M., Otterbein, K. F., & Rayner, S. (1993). Egalitarian behavior and reverse dominance hierarchy. *Current Anthropology, 34*(3), 227–254.

Bongie, L. L. (2000). *David Hume: Prophet of the counter-revolution* (2nd ed.). Liberty Fund.

Bonikowski, B., Feinstein, Y., & Bock, S. (2021). The partisan sorting of "America": How nationalist cleavages shaped the 2016 US Presidential election. *American Journal of Sociology, 127*(2), 492–561.

Bornet, V. D. (1983). *The Presidency of Lyndon B. Johnson* (Vol. 9). University Press of Kansas.

Borowiec, M. L., Cover, S. P., & Rabeling, C. (2021). The evolution of social parasitism in *Formica* ants revealed by a global phylogeny. *Proceedings of the National Academy of Sciences USA, 118*(38), e2026029118.

Borrello, M. E. (2010). *Evolutionary restraints: The contentious history of group selection*. Chicago University Press.

Borromeo, A. (2003). *L'Inquisizione: atti del Simposio internazionale: Città del Vaticano, 29–31 ottobre 1998*. Biblioteca Apostolica Vaticana.

Bouchard Jr, T. J., Segal, N. L., Tellegen, A., McGue, M., Keyes, M., & Krueger, R. (2003). Evidence for the construct validity and heritability of the Wilson–Patterson conservatism scale: A reared-apart twins study of social attitudes. *Personality and Individual Differences, 34*(6), 959–969.

Bowyer, B., & Kahne, J. (2017). Facing facts in an era of political polarization: Young people's learning and knowledge about economic inequality. *Political Science & Politics, 50*(4), 1056–1061.

Boyd, R., Gintis, H., Bowles, S., & Richerson, P. J. (2003). The evolution of altruistic punishment. *Proceedings of the National Academy of Sciences, 100*(6), 3531–3535.

Boyd, R., & Richerson, P. J. (1990). Group selection among alternative evolutionarily stable strategies. *Journal of Theoretical Biology, 145*(3), 331–342.

Boyd, R., & Richerson, P. J. (2005). *The origin and evolution of cultures.* Oxford University Press.

Braendle, C., Heyland, A., & Flatt, T. (2011). Integrating mechanistic and evolutionary analysis of life history variation. In T. Flatt & A. Heyland (Eds.), *Mechanisms of life history evolution: The genetics and physiology of life history traits and trade-offs* (pp. 3–10). Oxford University Press.

Brailsford, H. N. (1961). *The Levellers and the English revolution.* Stanford University Press.

Brandt, M. J., Wetherell, G., & Reyna, C. (2014). Liberals and conservatives can show similarities in negativity bias. *Behavioral & Brain Sciences, 37*(3), 307–308.

Brennan, J. (2017). *Against democracy.* Princeton University Press.

Brenner, R. (1996). Booms and declines of serfdom in Europe during the Middle Ages and the Modern Age. *Hispania, 56*(192), 173–201.

Brenner, R. (2014). The rises and declines of serfdom in medieval and early modern Europe. In M. L. Bush (Ed.), *Serfdom and slavery* (pp. 247–276). Routledge.

Brick, H. & Phelps, C. (2015). *Radicals in America: The U.S. left since the Second World War.* Cambridge University Press.

Brookmyre, C. (2011). *Not the end of the world.* Abacus.

Brooks, V. W. (1936). *The flowering of New England: 1815–1865.* E. P. Dutton & Company, Inc.

Brown, J. R., & Enos, R. D. (2021). The measurement of partisan sorting for 180 million voters. *Nature Human Behaviour, 5*(8), 998–1008.

Brown, R. (2022). *The conservative counter-revolution in Britain and America, 1980–2020.* Palgrave Macmillan.

Bruchmann, K., Koopmann-Holm, B., & Scherer, A. (2018). Seeing beyond political affiliations: The mediating role of perceived moral foundations on the partisan similarity-liking effect. *PloS One, 13*(8), e0202101.
Brumbach, B. H., Figueredo, A. J., & Ellis, B. J. (2009). Effects of harsh and unpredictable environments in adolescence on development of life history strategies. *Human Nature, 20*(1), 25–51.
Buechner, B. M., Clarkson, J. J., Otto, A. S., Hirt, E. R., & Ho, M. C. (2021). Political ideology and executive functioning: The effect of conservatism and liberalism on cognitive flexibility and working memory performance. *Social Psychological and Personality Science, 12*(2), 237–247.
Bugaric, B. (2019). The two faces of populism: Between authoritarian and democratic populism. *German Law Journal, 20*(3), 390–400.
Burgess, J. W. (1915). *The reconciliation of government with liberty*. Charles Scribner's Sons.
Burkart, J. M., Schubiger, M. N., & van Schaik, C. P. (2017). The evolution of general intelligence. *Behavioral and Brain Sciences, 40*, e195.
Burke, E. (1790/1992). *Reflections on the revolution in France and on the proceedings in certain societies in London relative to that event*. The Classics of Liberty Library.
Burner, D. (1979). *Herbert Hoover, a public life*. Alfred a Knopf Incorporated.
Burt, A., & Trivers, R. (2006). *Genes in conflict: The biology of selfish genetic elements*. The Belknap Press of Harvard University Press.
Buschinger, A. (2009). Social parasitism among ants: A review (Hymenoptera Formicidae). *Myrmecological News, 12*, 219–235.
Buss, D. M. (1987). Selection, evocation, and manipulation. *Journal of Personality and Social Psychology, 53*, 1214–1221.
Butler, M. (Ed.). (1984). *Burke, Paine, Godwin, and the revolution controversy*. Cambridge University Press.
Butterfield, H. (1931). *The Whig Interpretation of History*. WW Norton and Company.
Buyukuslu, N., Esin, K., Hizli, H., Sunal, N., Yigit, P., & Garipagaoglu, M. (2014). Clothing preference affects vitamin D status of young women. *Nutrition Research, 34*(8), 688–693.
Byrne, R. W. (2018). Machiavellian intelligence retrospective. *Journal of Comparative Psychology, 132*(4), 432–436.
Calder, W. A. (1984). *Size, function and life history*. Harvard University Press.
Calhoun, J. C. (1851/1995). *A disquisition on government and a discourse on the constitution and government of the United States*. The Classics of Liberty Library.

Campbell, D. B. (2012). *Spartan warrior 735–331 BC.* Bloomsbury Publishing.

Campbell, D. T., & Fiske, D. W. (1959). Convergent and discriminant validation by the multitrait-multimethod matrix. *Psychological Bulletin, 56*(2), 81–105.

Caplan, B. (2011). *The myth of the rational voter. In The Myth of the Rational Voter*. Princeton University Press.

Carr, J. (2012). *Sparta's Kings*. Pen and Sword.

Carraro, L., Castelli, L., & Negri, P. (2016). The hand in motion of liberals and conservatives reveals the differential processing of positive and negative information. *Acta Psychologica, 168*, 78–84.

Casey, J. (1989). *The history of the family*. Basil Blackwell.

Casey, M. (2010). *Jesus of Nazareth: An Independent Historian's Account of His Life and Teaching*. Clark.

Castiglione, D., & Hampsher-Monk, I. (2001). *The history of political thought in national context*. Cambridge University Press.

Cavanagh, W. (2012). Central and Southern Peloponnese. In E. H. Cline (Ed.), *The Oxford Handbook of the Bronze Age Aegean* (pp. 631–642). New York. Oxford University Press.

Cavanagh, W. (2018). An archaeology of ancient Sparta with reference to Laconia and Messenia. In A. Powel (Ed.), *A companion to Sparta* (pp. 59–92). John Wiley & Sons.

Cawkwell, G. L. (2002). The decline of Sparta. In M. Whitby (Ed.), *Sparta* (pp. 236–257). Routledge.

Cervantes, F. (1994). *The Devil in the New World: The impact of diabolism in New Spain*. Yale University.

Chaker, A. (2017). Rising income inequality increasing political polarization? A state-level analysis over two decades. *Journal of Applied Business and Economics, 19*(2), 20–34.

Chapais, B. (1995). Alliances as a means of competition in primates: Evolutionary, developmental, and cognitive aspects. *American Journal of Physical Anthropology, 38*(S21), 115–136.

Charney, E. (2008). Genes and ideologies. *Perspectives on. Politics, 6*(2), 299–319.

Chen, H., Tao, X., Li, N., & Han, Z. (2018, May). Political polarization analysis using random matrix theory: Case study for USA biparty public view. In *2018 IEEE International Conference on Communications (ICC)* (pp. 1–6). IEEE.

Chisholm, J. S. (1999). Attachment and time preference: Relations between early stress and sexual behavior in a sample of American University Women. *Human Nature, 10*(*1*), 51–83.

Christ, O., Schmid, K., & Green, E. G. T. (2022). Macro-diversity and intergroup attitudes. In D. Osborne & C. G. Sibley (Eds.), *The Cambridge handbook of political psychology* (pp. 243–257). Cambridge University Press.

Cikara, M., Van Bavel, J. J., Ingbretsen, Z. A., & Lau, T. (2017). Decoding "us" and "them": Neural representations of generalized group concepts. *Journal of Experimental Psychology: General, 146*(5), 621.

Ciuk, D. J. (2023). Value disagreement and partisan sorting in the American mass public. *Political Research Quarterly, 76*(1), 60–74.

Claassen, R. L., Djupe, P. A., Lewis, A. R., & Neiheisel, J. R. (2021). Which party represents my group? The group foundations of partisan choice and polarization. *Political Behavior, 43*, 615–636.

Claessens, S., Chaudhuri, A., Sibley, C., & Atkinson, Q. (2022). The evolutionary basis of political ideology. In D. Osborne & C. G. Sibley (Eds.), *The Cambridge handbook of political psychology* (pp. 22–36). Cambridge University Press.

Clark, G. (2007). *A farewell to alms: A brief economic history of the world.* Princeton University Press.

Clark, G., & Jacks, D. (2007). Coal and the industrial revolution, 1700–1869. *European Review of Economic History, 11*(1), 39–72.

Clemit, P. (2011). *The Cambridge companion to British literature of the French Revolution in the 1790s.* Cambridge University Press.

Cline, E. H. (2024). *After 1177 B.C.: The Survival of Civilizations* (Turning Points in Ancient History) (Vol. 12). Princeton University Press.

Cohen, F., Solomon, S., Maxfield, M., Pyszczynski, T., & Greenberg, J. (2004). Fatal attraction: The effects of mortality salience on evaluations of charismatic, task-oriented, and relationship-oriented leaders. *Psychological Science, 15*(12), 846–851.

Cohen, R. (1984). Warfare and state formation: Wars make states and states make wars. In *Warfare, culture, and environment* (pp. 329–358). Academic Press.

Conder, C. R. (2013). *Latin kingdom of Jerusalem.* Routledge.

Connolly, W. E. (2002). *Neuropolitics: Thinking, culture, speed* (Vol. 23). University of Minnesota Press.

Contreras, J., & Henningsen, G. (1986). Forty-four thousand cases of the Spanish Inquisition (1540–1700): Analysis of a historical data bank. In *The inquisition in early Modern Europe: Studies on sources and methods* (pp. 100–129). Northern Illinois University Press.

Converse, P. E. (1963). A major political realignment in the south? In A. P. Sindler (Ed.), *Change in the contemporary south* (pp. 195–222). Duke University Press.

Conway, L. G., III, Houck, S. C., Gornick, L. J., & Repke, M. A. (2018). Finding the Loch Ness monster: Left-wing authoritarianism in the United States. *Political Psychology, 39*(5), 1049–1067.

Conway, L. G., III, & McFarland, J. D. (2019). Do right-wing and left-wing authoritarianism predict election outcomes?: Support for Obama and Trump across two United States presidential elections. *Personality and Individual Differences, 138*, 84–87.

Corr, P. J., & Matthews, G. (2009). *The Cambridge handbook of personality psychology*. Cambridge University Press.

Costello, T. H., Bowes, S. M., Stevens, S. T., Waldman, I. D., Tasimi, A., & Lilienfeld, S. O. (2022). Clarifying the structure and nature of left-wing authoritarianism. *Journal of personality and social psychology, 122*(1), 135.

Cotlar, S. (2011). *Tom Paine's America: The rise and fall of transatlantic radicalism in the early republic.* University of Virginia Press.

Cotterill, B. F. (2023). *Personality psychology, ideology, and voting behavior: Beyond the ballot.* Springer Nature.

Craig, J. L., & Douglas, M. E. (1986). Resource distribution, aggressive asymmetries and variable access to resources in the nectar feeding bellbird. *Behavioral Ecology and Sociobiology, 18*, 231–240.

Craig, M. A., & Richeson, J. A. (2014). On the precipice of a "majority-minority" America: Perceived status threat from the racial demographic shift affects White Americans' political ideology. *Psychological Science, 25*(6), 1189–1197.

Creasy, E. (2013). *The fifteen decisive battles of the world.* Classics of Liberty Library.

Critchlow, D. T. (2015). *American political history: A very short introduction* (Vol. 420). Oxford University Press.

Cromartie, A. (2006). *The constitutionalist revolution: an essay on the history of England*, 1450–1642 (Vol. 75). Cambridge University Press.

Cronbach, L. J., & Meehl, P. E. (1955). Construct validity in psychological tests. *Psychological Bulletin, 52*(4), 281–302.

Cullen, P. K., Gilman, T. L., Winiecki, P., Riccio, D. C., & Jasnow, A. M. (2015). Activity of the anterior cingulate cortex and ventral hippocampus underlie increases in contextual fear generalization. *Neurobiology of Learning and Memory, 124*, 19–27.

Curta, F. (2019). *Eastern Europe in the Middle Ages (500–1300)* (Brill's Companions to European History 19) (Vol. 2). Brill.
Curtis, V. (2011). Why disgust matters. *Philosophical Transactions of the Royal Society B: Biological Sciences, 366*(1583), 3478–3490.
Czechowski, W., & Godzinska, E. J. (2015). Enslaved ants: Not as helpless as they were thought to be. *Insectes Sociaux, 62*, 9–22.
D'Avray, D. (2016). The Cathars from non-Catholic sources. In A. Sennis (Ed.), *Cathars in Question* (pp. 177–184). York Medieval Press.
D'Ettore, P., & Heinze, J. (2001). Sociobiology of slave-making ants. *Acta Ethologica, 3*, 67–82.
Dagger, R. (1997). *Civic virtues: Rights, citizenship, and Republican Liberalism.* Oxford University Press.
Dalberg-Acton, J. E. E. (1907/1993). *The history of freedom and other essays.* The Classics of Liberty Library.
Daly, M., & Wilson, M. (1988). *Homicide*. Aldine de Gruyter.
Danford, J. W. (2006). Riches valuable to all times to all men: Hume and the eighteenth century debate on commerce and liberalism. In D. Womersley (Ed.). *Liberty and American experience in the eighteenth century* (pp. 319–347). Liberty Fund.
Dangerfield, G. (1986). *The Era of Good Feelings*. Easton Press.
Darwin, C. (1859). *On the origin of species by means of natural selection, or the preservation of favoured races in the struggle for life*. John Murray.
Darwin, C. (1871). *The descent of man and selection in relation to sex*. John Murray.
Davydov, D. A. (2023). The left-wing ideology of the new elites: From luxury beliefs to universal humanism. *Obŝestvennye nauki i sovremennost', 4*, 7–21.
Dawidowicz, L. S. (1986). *The war against the Jews: 1933–1945*. Bantam.
Dawkins, R. (1976). *The selfish gene*. Oxford University Press.
Dawkins, R. (1981). *The extended phenotype: The long reach of the gene*. Oxford University Press.
Day, D. (2008). *Conquest: How societies overwhelm others*. Oxford University Press.
de Jouvenel, B. (1993). *On power: The natural history of its growth*. Liberty Fund.
de Lima, M. A. X., Baldo, M. V. C., Oliveira, F. A., & Canteras, N. S. (2022). The anterior cingulate cortex and its role in controlling contextual fear memory to predatory threats. *ELife, 11*, e67007.
de Waal, F. B. (2000). Primates: A natural heritage of conflict resolution. *Science, 289*(5479), 586–590.
de Waal, F. B. (2007). *Chimpanzee politics: Power and sex among apes*. JHU Press.
Deaner, R. O., Van Schaik, C. P., & Johnson, V. (2006). Do some taxa have better domain-general cognition than others? A meta-analysis of nonhuman

primate studies. *Evolutionary Psychology, 4*(1). https://doi.org/10.1177/147470490600400114

Degner, J., & Dalege, J. (2013). The apple does not fall far from the tree, or does it? A meta-analysis of parent–child similarity in intergroup attitudes. *Psychological Bulletin, 139*(6), 1270.

Den Otter, S. M. (2007). The origins of a historical political science in late Victorian and Edwardian Britain. In R. Adcock, M. Bevir, & S. C. Stimson (Eds.), *Modern political science: Anglo-American exchanges since 1880* (pp. 37–65). Princeton University Press.

Desan, S. (2006). *The family on trial in revolutionary France.* University of California Press.

DeVries, K. (2017). Gunpowder weaponry and the rise of the early modern state. In K. DeVries (Ed.), *Medieval Warfare 1300–1450* (pp. 457–475). Routledge.

DeWitte, S. N. (2015). Setting the stage for medieval plague: Pre-black death trends in survival and mortality. *American Journal of Physical Anthropology, 158*(3), 441–451.

Dickie, J. (2020). *The craft: How the Freemasons made the Modern World.* Public Affairs.

Dickinson, H. T. (2011). The political context. In P. Celmit (Ed.), *The Cambridge companion to British literature of the French Revolution in the 1790s* (pp. 1–15). Cambridge University Press.

Diggins, J. P. (1986). *The lost soul of American politics: Virtue, self-interest, and the foundations of liberalism.* Chicago University Press.

Dimont, M. (1994). *Jews, God, and history.* Penguin.

Dingemanse, N. J., Kazem, A. J., Réale, D., & Wright, J. (2010). Behavioural reaction norms: Animal personality meets individual plasticity. *Trends in Ecology & Evolution, 25*(2), 81–89.

Dingemanse, N. J., & Réale, D. (2013). What is the evidence for natural selection maintaining animal personality variation?. In *Animal personalities: Behavior, physiology, and evolution* (pp. 201–220). University of Chicago Press.

Djeriouat, H., & Trémolière, B. (2014). The Dark Triad of personality and utilitarian moral judgment: The mediating role of Honesty/Humility and Harm/Care. *Personality and Individual Differences, 67*, 11–16.

do Rosario Petrucci, B., Januario, M., & Quental, T. (2022). Paleobuddy: An R package for flexible simulations of diversification and fossil sampling. *Methods in Ecology and Evolution, 13*(12), 2692–2698.

Dobrzanska, J., & Dobrzanski, J. (1989). Controversies on the subject of slave-raids in Amazon ants (Genus Polyergus). *Acta Neurobiologiae Experimentalis, 49*, 367–379.

Dodd, M. D., Balzer, A., Jacobs, C. M., Gruszczynski, M. W., Smith, K. B., & Hibbing, J. R. (2012). The political left rolls with the good and the political right confronts the bad: Connecting physiology and cognition to preferences. *Philosophical Transactions of the Royal Society B: Biological Sciences, 367*(1589), 640–649.

Donohue, R. N., Hewitt, D. G., Fulbright, T. E., Deyoung, C. A., Litt, A. R., & Draeger, D. A. (2013). Aggressive behavior of white-tailed deer at concentrated food sites as affected by population density. *The Journal of Wildlife Management, 77*(7), 1401–1408.

Drower, E. S. (1960). *The secret Adam, a study of Nasoraean gnosis*. Clarendon Press.

Duca, J. V., & Saving, J. L. (2017). Income inequality, media fragmentation, and increased political polarization. *Contemporary Economic Policy, 35*(2), 392–413.

Ducat, J. (2018). The Perioikoi. In A. Powel (Ed.), *A companion to Sparta* (pp. 596–614). John Wiley & Sons.

Duckitt, J. (2022). Authoritarianism: Conceptualization, research, and new developments. In D. Osborne & C. G. Sibley (Eds.), *The Cambridge handbook of political psychology* (pp. 177–197). Cambridge University Press.

Duff, D. (2011). Burke and Paine: Contrasts. In P. Celmit (Ed.), *The Cambridge companion to British literature of the French Revolution in the 1790s* (pp. 47–70). Cambridge University Press.

Dunbar, R. I. (2013). The origin of religion as a small-scale phenomenon. In S. Clarke, R. Powell, & J. Savulescu (Eds.), *Religion, intolerance, and conflict: A scientific and conceptual investigation* (pp. 48–66). Oxford University Press.

Dunbar, R. I., Clark, A., & Hurst, N. L. (1995). Conflict and cooperation among the Vikings: Contingent behavioral decisions. *Ethology and Sociobiology, 16*(3), 233–246.

Dunn, S. (1999). *Sister Revolutions: French lightning, American light*. Faber & Faber, Incorporated.

Durant, W. (1942). *The story of civilization: Vol. 1: Our oriental heritage* (Vol. 1). Simon & Schuster.

Durkheim, E. (1964/1893). *The Division of labor in society*. Collier Macmillan.

Easton, M. J., & Holbein, J. B. (2021). The democracy of dating: How political affiliations shape relationship formation. *Journal of Experimental Political Science, 8*(3), 260–272.

Eaves, L., Heath, A., Martin, N., Maes, H., Neale, M., Kendler, K., Kirk, K., & Corey, L. (1999). Comparing the biological and cultural inheritance of personality and social attitudes in the Virginia 30 000 study of twins and their relatives. *Twin Research and Human Genetics, 2*(2), 62–80.

Eaves, L. J., & Eysenck, H. J. (1974). Genetics and the development of social attitudes. *Nature, 249*, 288–289.

Eaves, L. J., & Hatemi, P. K. (2008). Transmission of attitudes toward abortion and gay rights: Effects of genes, social learning and mate selection. *Behavior Genetics, 38*, 247–256.

Eaves, L. J., & Hatemi, P. K. (2011). Do we choose our spouse based on our in-laws? Resolving the effects of family background and spousal choice for educational attainment, religious practice, and political preference. *Social Science Quarterly, 92*(5), 1253–1278.

Eaves, L. J., Martin, N., Heath, A., Schieken, R., Meyer, J., Silberg, J., Neale, M., & Corey, L. (1997). Age changes in the causes of individual differences in conservatism. *Behavioral Genetics, 27(2)*, 121–124.

Ebstein, R. P., Monakhov, M. V., Lu, Y., Jiang, Y., Lai, P. S., & Chew, S. H. (2015). Association between the dopamine D4 receptor gene exon III variable number of tandem repeats and political attitudes in female Han Chinese. *Proceedings of the Royal Society B: Biological Sciences, 282*(1813), 20151360.

Edbury, P. (1991). *The kingdom of Cyprus and the Crusades, 1191–1374*. Cambridge University Press.

Edgerton, R. B. (1971). *The individual in cultural adaptation: A study of four East African peoples*. University of California Press.

Edgington, S. (2019). *Baldwin I of Jerusalem, 1100–1118*. Routledge.

Edwards, J. S. S., & Ogilvie, S. C. (2018). *Did the Black Death Cause Economic Development by "Inventing" Fertility Restriction?* CESifo Working Paper no. 7016.

Eggenberger, D. (2012). *An encyclopedia of battles: Accounts of over 1,560 battles from 1479 BC to the present*. Courier Corporation.

Ehrman, B. D. (2003). *Lost christianities: The battles for scripture and the faiths we never knew*. Oxford University Press.

Ehrman, B. D. (2005). *Lost Christianities: The Battles for scripture and the faiths we never knew*. Oxford University Press.

Elkins, S., & McKitrick, E. (1993). *The age of federalism*. Oxford University Press.

Ellerbe, H. (1995). *The dark side of Christian history*. Morningstar Books.

Elms, A. C., & Milgram, S. (1966). Personality characteristics associated with obedience and defiance toward authoritative command. *Journal of Experimental Research in Personality, 1*(4), 282–289.

Emery, C. (1909). Über den Ursprung der dulotischen, parasitischen und myrmekophilen Ameisen. *Biologisches Centralblatt, 29*, 352–362.

Engels, W. (1990). *Social insects an evolutionary approach to castes and reproduction*. Springer-Verlag.

Enroth, H. (2022). *Political science and the problem of social order*. Cambridge University Press.

Estevez, I., Newberry, R. C., & Keeling, L. J. (2002). Dynamics of aggression in the domestic fowl. *Applied Animal Behaviour Science, 76*(4), 307–325.

Evans, C. (2006). *Debating the Revolution: Britain in the 1790s*. I.B. Taurus.

Evans, G., & Opacic, A. (2022). How social class influences political choices. In D. Osborne & C. G. Sibley (Eds.), *The Cambridge handbook of political psychology* (pp. 382–398). Cambridge University Press.

Evrigenis, I. D. (2007). *Fear of enemies and collective action*. Cambridge University Press.

Eysenck, H. J. (1981). Left-wing authoritarianism: Myth or reality? *Political Psychology, 3*(1/2), 234–238.

Eysenck, H. J., Gudjonsson, G. H., Eysenck, H. J., & Gudjonsson, G. H. (1989). Crime and personality. In *The causes and cures of criminality* (pp. 43–89). Springer.

Fagan, B. (2000). *The little ice age: How climate made history, 1300–1850*. Perseus Books Group.

Faria, F. N. (2019). *The evolutionary limits of liberalism: Democratic problems, market solutions and the ethics of preference satisfaction*. Palgrave Macmillan.

Farr, J., & Williams, D. L. (Eds.). (2015). *The General Will*. Cambridge University Press.

Farrenkopf, J. (2001). *Prophet of decline: Spengler on world history and politics*. LSU Press.

Fatke, M. (2017). Personality traits and political ideology: A first global assessment. *Political Psychology, 38*(5), 881–899.

Fawcett, E. (2022). *Conservatism: The fight for a tradition*. Princeton University Press.

Featherston, J., & Durand, P. M. (2012). Cooperation and conflict in cancer: An evolutionary perspective. *South African Journal of Science, 108*(9), 1–7.

Federico, C. M. (2022). The personality basis of political preferences. In D. Osborne & C. G. Sibley (Eds.), *The Cambridge handbook of political psychology* (pp. 68–88). Cambridge University Press.

Feezell, J. T., Wagner, J. K., & Conroy, M. (2021). Exploring the effects of algorithm-driven news sources on political behavior and polarization. *Computers in human behavior, 116*, 106626.

Fehr, E., Fischbacher, U., & Gächter, S. (2002). Strong reciprocity, human cooperation, and the enforcement of social norms. *Human Nature, 13*, 1–25.

Feist, G. J., & Brady, T. R. (2004). Openness to experience, non-conformity, and the preference for abstract art. *Empirical Studies of the Arts, 22*(1), 77–89.

Fenton, J. H. (1982). 'Cycle Theory' and political realignment in the United States. *Political Studies, 30*(3), 426–431.
Ferling, J. (1992/2001). *John Adams: A life*. Easton Press.
Ferling, J. (2003). *A leap in the dark: The struggle to create the American Republic*. Oxford University Press.
Ferling, J. (2004). *Adams vs. Jefferson: The tumultuous election of 1800*. Oxford University Press.
Fernandes, H. B., Peñaherrera-Aguirre, M., of Menie, M. A. W., & Figueredo, A. J. (2020). Macroevolutionary patterns and selection modes for general intelligence (G) and for commonly used neuroanatomical volume measures in primates. *Intelligence, 80*, 101456.
Fernandes, H. B., Woodley, M. A., & te Nijenhuis, J. (2014). Differences in cognitive abilities among primates are concentrated on G: Phenotypic and phylogenetic comparisons with two meta-analytical databases. *Intelligence, 46*, 311–322.
Figueira, T. J. (1986). Population patterns in late archaic and classical Sparta. *Transactions of the American Philological Association (1974–), 116*, 165–213.
Figueira, T. J. (2018). Helotage and the Spartan economy. In A. Powel (Ed.), *A companion to Sparta* (pp. 565–595). John Wiley & Sons.
Figueredo, A. J. (1999). Separation and its discontents: Toward an evolutionary theory of anti-Semitism (Kevin B. MacDonald). Book review. *Politics and the Life Sciences, 18*(1), 136–137.
Figueredo, A. J., & Wolf, P. S. (2009). Assortative pairing and life history strategy. *Human Nature, 20*(3), 317–330.
Figueredo, A. J., Cabeza de Baca, T., Fernandes, H. B. F., Black, C. J., Peñaherrera, M., Hertler, S., García, R. A., Meisenberg, G., & Woodley of Menie, M. A. (2017). A sequential canonical cascade model of social biogeography: Plants, parasites, and people. *Evolutionary Psychological Science, 3*, 40–61.
Figueredo, A. J., Corral-Verdugo, V., Frıas-Armenta, M., Bachar, K. J., White, J., McNeill, P. L., Kirsner, B. R., & del PilarCastell-Ruiz, I. (2001). Blood, solidarity, status, and honor: The sexual balance of power and spousal abuse in Sonora. *Mexico. Evolution and Human Behavior, 22*(5), 295–328.
Figueredo, A. J., Fernandes, H. B. F., & Woodley, M. A. (2017). The quantitative theoretical ecology of life history strategies. *Mankind Quarterly, 57*(3), 305.
Figueredo, A. J., Fernandes, H. B. F., Peñaherrera-Aguirre, M., & Hertler, S. C. (2018). The evolution of personality: Building on Buss. In P. Corr (Ed.), *Personality and individual differences: Revisiting the classic studies*. Sage.

Figueredo, A. J., & Jacobs, W. J. (2010). Aggression, risk-taking, and alternative life history strategies: The behavioral ecology of social deviance. In *Bio-psychosocial perspectives on interpersonal violence* (pp. 3–28). Nova Science Publishers, Inc.

Figueredo, A. J., Jacobs, W. J., Burger, S. B., Gladden, P. R., & Olderbak, S. G. (2011). The biology of personality. In *Information and living systems: Essays in philosophy of biology* (pp. 371–406). MIT Press.

Figueredo, A. J., Jacobs, W. J., Wolf, P. S. A., Hohman, Z. J., & Fernandes, H. B. F. (2020). Preface: A pan-selectionist model of evolution and development. In S. Hertler, A. J. Figueredo, & M. Peñaherrera Aguirre (Eds.), *Multilevel selection: Theoretical foundations, historical examples, and empirical evidence*. Palgrave Macmillan.

Figueredo, A. J., Peñaherrera-Aguirre, M., Fernandes, H. B. F., Lomayesva, S. L., Woodley, M. A., Hertler, S. C., & Sarraf, M. A. (2019). The ecology of empire: The dynamics of strategic differentiation-integration in two competing Western European biocultural groups. *Politics and the Life Sciences, 38*, 210–225.

Figueredo, A. J., Tal, I. R., McNeil, P., & Guillén, A. (2004). Farmers, herders, and fishers: The ecology of revenge. *Evolution and Human Behavior, 25*(5), 336–353.

Figueredo, A. J., Vasquez, G., Brumbach, B. H., & Schneider, S. M. (2004). The heritability of life history strategy: The k-factor, covitality, and personality. *Social Biology, 51*(3–4), 121–143.

Figueredo, A. J., Vásquez, G., Brumbach, B. H., Schneider, S. M., Sefcek, J. A., Tal, I. R., Hill, D., Wenner, C. J., & Jacobs, W. J. (2006). Consilience and life history theory: From genes to brain to reproductive strategy. *Developmental Review, 26*(2), 243–275.

Figueredo, A. J., Vásquez, G., Brumbach, B. H., Sefcek, J. A., Kirsner, B. R., & Jacobs, W. J. (2005). The K-factor: Individual differences in life history strategy. *Personality and Individual Differences, 39*(8), 1349–1360.

Figueredo, A. J., Wolf, P. S. A., Gladden, P. R., Olderbak, S., Andrzejczak, D. J., & Jacobs, W. J. (2011). Ecological approaches to personality. In D. Buss & P. Hawley (Eds.), *The evolution of personality and individual differences* (pp. 210–238). Oxford University Press.

Figueredo, A. J., Wolf, P. S. A., Olderbak, S. G., Gladden, P. R., Fernandes, H. B. F., Wenner, C., Hill, D., Andrzejczak, D. J., Sisco, M. M., Jacobs, W. J., Hohman, Z. J., Sefcek, J. A., Kruger, D., Howrigan, D. P., MacDonald, K., & Rushton, J. P. (2014). The psychometric assessment of human life history strategy: A meta-analytic construct validation. *Evolutionary Behavioral Sciences, 8*(3), 148–185.

Figueredo, A. J., Woodley, M. A., Brown, S. D., & Ross, K. C. (2013). Multiple successful tests of the Strategic Differentiation-Integration Effort (SD-IE) hypothesis. *Journal of Social, Evolutionary, and Cultural Psychology, 7*(4), 361–383.

Fioroni, S. B., Lotz, A. D., Soroka, S., & Hiaeshutter-Rice, D. (2022). Political sorting in US entertainment media. *Popular Communication, 20*(2), 117–132.

Fischer, D. H. (1996). *The great wave: Price revolutions and the rhythms of history*. Oxford University Press.

Fischer, D. H. (2005). *Liberty and freedom*. Oxford University Press.

Fischer, R., Hanke, K., & Sibley, C. G. (2012). Cultural and institutional determinants of social dominance orientation: A cross-cultural meta-analysis of 27 societies. *Political Psychology, 33*(4), 437–467.

Fishman, L., & Jainike, J. (2014). Selfish genetic elements and genetic conflict. In J. B. Losos (Ed.), *the Princeton guide to evolution* (pp. 347–355). Princeton University Press.

Flannery, K., & Marcus, J. (2012). *The creation of inequality: How our prehistoric ancestors set the stage for monarchy, slavery, and empire*. Harvard University Press.

Fogel, R. W. (1992). Problems in modeling complex dynamic interactions: The political realignment of the 1850s. *Economics & Politics, 4*(3), 215–254.

Fogel, R. W. (1995). The fourth great awakening and the political realignment of the 1990s. *Brigham Young University Studies, 35*(3), 31–43.

Fournier, P., Petersen, M. B., & Soroka, S. (2021). The political phenotype of the disgust sensitive: Correlates of a new abbreviated measure of disgust sensitivity. *Electoral Studies, 72*, 102347.

Fournier, P., Soroka, S., & Nir, L. (2020). Negativity biases and political ideology: A comparative test across 17 countries. *American Political Science Review, 114*(3), 775–791.

Fowler, J. H., Baker, L. A., & Dawes, C. T. (2008). Genetic variation in political participation. *American Political Science Review, 102*(2), 233–248.

Fowler, J. H., & Kam, C. D. (2006). Patience as a political virtue: Delayed gratification and turnout. *Political Behavior, 28*, 113–128.

Fowler, J. H., Loewen, P. J., Settle, J., & Dawes, C. T. (2011). Genes, games, and political participation. In P. K. Hatemi & R. McDermott (Eds.), *man is by nature a political animal: Evolution, biology, and politics* (pp. 207–223). Chicago University Press.

Fowler, J. H., & Smirnov, O. (2007). *Mandates, parties, and voters: How elections shape the future*. Temple University Press.

Fox, H. S. A. (1995). Servants, cottagers and tied cottages during the later middle ages: Towards a regional dimension. *Rural History, 6*(2), 125–154.

Fox, J., Friendly, G. G., Graves, S., Heiberger, R., Monette, G., Nilsson, H., Ripley, B., Weisberg, S., Fox, M. J., & Suggests, M. A. S. S. (2007). The car package. *R Foundation for Statistical Computing, 1109*, 1431.

Fozdar, V. (2011). That grand primeval and fundamental religion: The transformation of Freemasonry into a British imperial cult. *Journal of World History, 22*(3), 493–525.

Frankland, P. W., Bontempi, B., Talton, L. E., Kaczmarek, L., & Silva, A. J. (2004). The involvement of the anterior cingulate cortex in remote contextual fear memory. *Science, 304*(5672), 881–883.

Fried, M. H. (1967). *The evolution of political society, an essay in political anthropology*. Random House.

Friesen, A., Gruszczynski, M., Smith, K. B., & Alford, J. R. (2020). Political attitudes vary with detection of androstenone. *Politics and the Life Sciences, 39*(1), 26–37.

Friesen, A., & Ksiazkiewicz, A. (2015). Do political attitudes and religiosity share a genetic path? *Political Behavior, 37*, 791–818.

Friesen, J. P., Kay, A. C., Eibach, R. P., & Galinsky, A. D. (2014). Seeking structure in social organization: Compensatory control and the psychological advantages of hierarchy. *Journal of Personality and Social Psychology, 106*(4), 590.

Frye, T., Reuter, O. J., & Szakonyi, D. (2014). Political machines at work voter mobilization and electoral subversion in the workplace. *World Politics, 66*(2), 195–228.

Fukuyama, F. (2011). *The origins of political order: From prehuman times to the French Revolution*. Farrar, Straus and Giroux.

Furnham, A., & Fenton-O'Creevy, M. (2018). Personality and political orientation. *Personality and Individual Differences, 129*, 88–91.

Gat, A. (2009). *War in human civilization*. Oxford University Press.

Gat, A. (2017). *The causes of war and the spread of peace*. Oxford University Press.

Gat, A. (2022). *Ideological fixation: From the stone age to today's culture wars*. Oxford University Press.

Gellner, E. (1987). *Culture, identity, and politics*. Cambridge University Press.

Gerber, J. S. (1992). *The Jews of Spain: A history of the Sephardic experience*. Free Press.

Gerring, J. (1997). Ideology: A definitional analysis. *Political Research Quarterly, 50*(4), 957–994.

Gerson, G. (2022). Liberalism, pluralism and the sphere division in Harold Laski. *Theoria, 69*(170), 35–60.

Gidengil, E., & Stolle, D. (2022). Populism. In D. Osborne & C. G. Sibley (Eds.), *The Cambridge handbook of political psychology* (pp. 442–457). Cambridge University Press.

Gidengil, E., Wass, H., & Valaste, M. (2016). Political socialization and voting: The parent–child link in turnout. *Political Research Quarterly, 69*(2), 373–383.

Gilbert, R. E. (2024). Civil rights and political realignment: The presidency of John F. Kennedy. *PCS–Politics, Culture and Socialization, 10.*

Gilboy, R. (1992). Crisis mortality in Civil War Oxford, 1642–1646. *Revolution, 132*, 21.

Gilmartin K. (2011). Counter-revolutionary culture. In P. Celmit (Ed.), *The Cambridge companion to British literature of the French Revolution in the 1790s* (pp. 129–144). Cambridge University Press.

Gilpin, M. E. (1975). *Group selection in predator-prey communities* (Monographs in Population Biology 9). Princeton University Press.

Gintis, H. (2000). Strong reciprocity and human sociality. *Journal of Theoretical Biology, 206*(2), 169–179.

Gintis, H. (2017). *Individuality and entanglement: The moral and material bases of social life*. Princeton University Press.

Gitlitz, D. M. (1996). *Secrecy and deceit: The religion of the Crypto-Jews*. Jewish Publication Society.

Giuseffi, K. E. (2012). *Processing facial emotions: An eeg study of the differences between conservatives and liberals and across political participation.*

Given, J. B. (2018). *Inquisition and medieval society: Power, discipline, and resistance in Languedoc.* Cornell University Press.

Gladden, P. R., & Cleator, A. M. (2018). Slow life history strategy predicts six moral foundations. *EvoS Journal: The Journal of the Evolutionary Studies Consortium, 9*, 43–63.

Godefroidt, A. (2023). How terrorism does (and does not) affect citizens' political attitudes: A meta-analysis. *American Journal of Political Science, 67*(1), 22–38.

Goldman, L. (2003). Civil society in nineteenth-century Britain and Germany: J. M. Ludlow, Lujo Brentano, and the labour question. In J. Harris (Ed.), *Civil society in British history: Ideas, identities, institutions* (pp. 97–114). Oxford University Press.

Goldschmidt, W. (1965). Theory and strategy in the study of cultural adaptability. *American Anthropologist, 67*, 402–407.

Goldstone, J. A. (1991). *Revolution and rebellion in the early modern world: Population change and state breakdown in England, France, Turkey, and China, 1600–1850*. California University Press.

Goldstone, J. A. (2001). Toward a fourth generation of revolutionary theory. *Annual Review of Political Science, 4*(1), 139–187.

Gómez, J. M., Verdú, M., & González-Megías, A. (2021). Killing conspecific adults in mammals. *Proceedings of the Royal Society B, 288*(1955), 20211080.

Gómez, J. M., Verdú, M., González-Megías, A., & Méndez, M. (2016). The phylogenetic roots of human lethal violence. *Nature, 538*(7624), 233–237.

González, F. J. N., Vidal, J. J., Jurado, J. M. L., McLean, A. K., & Bermejo, J. V. D. (2019). Dumb or smart asses? Donkey's (Equus asinus) cognitive capabilities share the heritability and variation patterns of human's (Homo sapiens) cognitive capabilities. *Journal of Veterinary Behavior, 33*, 63–74.

Gonzalez-Voyer, A., & von Hardenberg, A. (2014). An introduction to phylogenetic path analysis. In *Modern phylogenetic comparative methods and their application in evolutionary biology: Concepts and practice* (pp. 201–229). Springer.

Gooch, D. M. (2018). From Consensus to Conflict: Political Polarization, the Culture War, and Gay Rights. In C. Rackaway & L. L. Rice (Eds.), *American Political Parties under Pressure* (pp. 77–120). Palgrave Macmillan.

Gorsuch, R. L. (1983). *Factor analysis*. Lawrence Erlbaum.

Gorsuch, R. L. (1991). *UniMult for univariate and multivariate data analysis*. Unimult.

Gøtzsche-Astrup, O. (2021). Dark triad, partisanship and violent intentions in the United States. *Personality and Individual Differences, 173*, 110633.

Gowdy, J., & Krall, L. (2013). The ultrasocial origin of the Anthropocene. *Ecological Economics, 95*, 137–147.

Gowdy, J., Mazzucato, M., van den Bergh, J. C. J. M., van der Leeuw, S. E., & Wilson, D. S. (2016). Shaping the Evolution of Complex Societies. In D. S. Wilson & A. Kirman (Eds.), *complexity and evolution: Toward a new synthesis for economics* (pp. 185–199). The MIT Press.

Gowdy, J. M. (2021). *Ultrasocial: The evolution of human nature and the quest for a sustainable future*. Cambridge University Press.

Grafen, A. (1984). Natural selection, kin selection and group selection. *Behavioural ecology: An evolutionary approach, 2*, 62–84.

Graham, J., Englander, Z., Morris, J., Hawkins, C. B., Haidt, J., & Nosek, B. A. (2012). Warning bell: Liberals implicitly respond to group morality before rejecting it explicitly. *Available at SSRN 2071499*.

Grande, E. (2008). Comparative perspective. In H. Kriesi, E. Grande, R. Lachat, M. Dolezal, S. Bornschier, & T. Frey (Eds.), *West European politics in the age of globalization* (pp. 320–344). Cambridge University Press.

Greene, J., & Haidt, J. (2002). How (and where) does moral judgment work? *Trends in Cognitive Sciences, 6*(12), 517–523.

Griskevicius, V., Tybur, J. M., Delton, A. W., & Robertson, T. E. (2011). The influence of mortality and socioeconomic status on risk and delayed rewards: A life history theory approach. *Journal of Personality and Social Psychology, 100*(6), 1015.

Gros-Louis, J., Perry, S., & Manson, J. H. (2003). Violent coalitionary attacks and intraspecific killing in wild white-faced capuchin monkeys (Cebus capucinus). *Primates, 44*, 341–346.

Guibernau, M. (1999). *Nations without states: Political communities in a global age*. Polity.

Guillem, M., Drijfhout, F., & Martin, S. J. (2014). Chemical deception among ant social parasites. *Current Zoology, 60*(1), 62–75.

Gus, D. (2001). *Pagans & Christians: The personal spiritual experience*. Llewellyn.

Haas, I. J. (2016). Political neuroscience. In J. R. Absher & J. Cloutier (Eds.), *Neuroimaging personality, social cognition, and character* (pp. 355–370). Academic Press.

Haas, I. J., Baker, M. N., & Gonzalez, F. J. (2017). Who can deviate from the party line? Political ideology moderates evaluation of incongruent policy positions in insula and anterior cingulate cortex. *Social Justice Research, 30*(4), 355–380.

Haas, I. J., Warren, C., & Lauf, S. J. (2020). Political neuroscience: Understanding how the brain makes political decisions. In *Oxford research encyclopedia of politics*. Oxford University Press.

Hague, W. (2005). *William Pitt the Younger*. Alfred A. Knoff.

Haidt, J. (2001). The emotional dog and its rational tail: A social intuitionist approach to moral judgment. *Psychological Review, 108*(4), 814.

Haidt, J. (2003). The moral emotions. *Handbook of affective sciences, 11*, 852–870.

Haidt, J. (2007). The new synthesis in moral psychology. *Science, 316*(5827), 998–1002.

Haidt, J. (2013). *The Righteous mind: Why good people are divided by politics and religion*. Vintage.

Haidt, J., & Joseph, C. (2007). The moral mind: How five sets of innate intuitions guide the development of many culture-specific virtues, and perhaps even modules. In P. Carruthers, S. Laurence, & S. Stich (Eds.), *The Innate Mind* (pp. 367–391). Oxford University Press.

Halperin, T. D. (2016). *The Alien and Sedition Acts of 1798*. Johns Hopkins University Press.

Hamilton, C. D. (1970). Spartan politics and policy, 405–401 BC. *The American Journal of Philology, 91*(3), 294–314.

Hamilton, W. D. (1964). The genetical evolution of social behaviour. *II. Journal of theoretical biology, 7*(1), 17–52.

Hamilton, W. D. (1975). Innate social aptitudes of man: An approach from evolutionary genetics. *Biosocialanthropology, 133*, 115–132.

Hamilton, B. (2000). *The leper king and his heirs: Baldwin IV and the Crusader Kingdom of Jerusalem*. Cambridge University Press.

Harbottle, T. B. (1981). *Harbottle's Dictionary of Battles*. Van Nostrand Reinhold.

Harcourt, A. H., & Stewart, K. J. (2007). *Gorilla society: Conflict, compromise, and cooperation between the sexes*. Chicago University Press.

Harden, K. P. (2021). *The genetic lottery: Why DNA matters for social equality*. Princeton University Press.

Hare, C., & Poole, K. T. (2014). The polarization of contemporary American politics. *Polity, 46*(3), 411–429.

Harland-Jacobs, J. (1999). "Hands across the sea": The Masonic network, British Imperialism, and the North Atlantic World. *The Geographical Review, 89*(2), 237–253.

Harland-Jacobs, J. (2003). All in the Family: Freemasonry and the British Empire in the Mid-Nineteenth Century. *Journal of British Studies, 42*(4), 448–482.

Harris, E., Pärnamets, P., Sternisko, A., Robertson, C., & Van Bavel, J. J. (2022). The psychology and neuroscience of partisanship. In D. Osborne & C. G. Sibley (Eds.), *The Cambridge Handbook of Political Psychology* (pp. 50–67). Cambridge University Press.

Harris, J. (2003). *Civil society in British history: Ideas, identities, institutions*. Oxford University Press.

Harrison, G. W., Lau, M. I., & Williams, M. B. (2002). Estimating individual discount rates in Denmark: A field experiment. *American Economic Review, 92*(5), 1606–1617.

Harvey, P. (2012). *Freedom's coming: Religious culture and the shaping of the South from the Civil War through the civil rights era*. University of North Carolina.

Hastings, A. (1997). *The construction of nationhood: Ethnicity, religion, and nationalism*. Cambridge University Press.

Hatemi, P. K., Crabtree, C., & McDermott, R. (2017). The relationship between sexual preferences and political orientations: Do positions in the bedroom affect positions in the ballot box? *Personality and Individual Differences, 105*, 318–325.

Hatemi, P. K., Dawes, C. T., Frost-Keller, A., Settle, J. E., & Verhulst, B. (2011). Integrating social science and genetics: News from the political front. *Biodemography and Social Biology, 57*(1), 67–87.

Hatemi, P. K., & Fazekas, Z. (2018). Narcissism and political orientations. *American Journal of Political Science, 62*(4), 873–888.

Hatemi, P. K., Funk, C. L., Medland, S. E., Maes, H. M., Silberg, J. L., Martin, N. G., & Eaves, L. J. (2009). Genetic and environmental transmission of political attitudes over a life time. *The Journal of Politics, 71*(3), 1141–1156.

Hatemi, P. K., Hibbing, J. R., Medland, S. E., Keller, M. C., Alford, J. R., Smith, K. B., Martin, N. G., & Eaves, L. J. (2010). Not by twins alone: Using the extended family design to investigate genetic influence on political beliefs. *American Journal of Political Science, 54*(3), 798–814.

Hatemi, P. K., & McDermott, R. (2011). Evolution as a theory for political behavior. In P. K. Hatemi & R. McDermott (Eds.), *Man is by nature a political animal: Evolution, biology, and politics* (pp. 13–46). Chicago University Press.

Hatemi, P. K., & McDermott, R. (2012a). Broadening political psychology. *Political Psychology, 33*(1), 11–25.

Hatemi, P. K., & McDermott, R. (2012b). The genetics of politics: Discovery, challenges, and progress. *Trends in Genetics, 28*(10), 525–533.

Hatemi, P. K., McDermott, R., Eaves, L. J., Kendler, K. S., & Neale, M. C. (2013). Fear as a disposition and an emotional state: A genetic and environmental approach to out-group political preferences. *American Journal of Political Science, 57*(2), 279–293.

Hatemi, P. K., Medland, S. E., Klemmensen, R., Oskarsson, S., Littvay, L., Dawes, C. T., Verhulst, B., McDermott, R., Nørgaard, A. S., Klofstad, C., Christensen, K., Johannesson, M., Magnusson, P. K. E., Eaves, L. J., & Martin, N. G. (2014). Genetic influences on political ideologies: Twin analyses of 19 measures of political ideologies from five democracies and genome-wide findings from three populations. *Behavior genetics, 44*, 282–294.

Hawkins, J. W. (2023). The Royal Family and Freemasonry. *Ars Quatuor Coronatorum, 136*, 1–34.

Heaven, P. C. (1985). Construction and validation of a measure of authoritarian personality. *Journal of Personality Assessment, 49*(5), 545–551.

Hechter, M., & Horne, C. (2009). *Theories of social order, a reader*. Stanford University Press.

Helanterä, H. (2021). Pathways to parasitic strategies in ants. *Proceedings of the National Academy of Sciences, 118*(41), e2115607118.

Henrich, J. (2004). Cultural group selection, coevolutionary processes and large-scale cooperation. *Journal of Economic Behavior & Organization, 53*(1), 3–35.

Henrich, J. (2006). Cooperation, punishment, and the evolution of human institutions. *Science, 311*(5769), 60–61.

Hermann, M. G. (1986). *Political psychology: Contemporary problems and issues.* Jossey-Bass.

Herrera, J. P. (2017). Primate diversification inferred from phylogenies and fossils. *Evolution, 71*(12), 2845–2857.

Hertler, S., Figueredo, A. J., & Peñaherrera-Aguirre, M. (2020). *Multilevel selection: Theoretical foundations, historical examples, and empirical evidence.* Palgrave Macmillan.

Hertler, S. C. (2015). Crossing the paradigmatic divide in personality: Coming to an evolutionary understanding of personality variation. *Journal for Philosophy and Psychiatry.*

Hertler, S. C. (2017). Closing on a note of conciliation: On the attempt to reconcile science and religion at the American Museum of Natural History's Hall of Human Origins. *Museums & Social Issues, 12*(2), 140–149.

Hertler, S. C. (2019). A second Bonaparte: Searching for the character of Alexander Hamilton. In D. N. Hagist (Ed.), *Journal of the American Revolution: Annual volume 2019.* Westholme Publishing.

Hess, R. S. (2007). *Israelite religions: An archaeological and biblical survey.* Baker Academic.

Hesseltine, W. B. (1957). *Rise and fall of third parties: From Anti-Masonry to Wallace.* Peter Smith.

Heywood, A. (2021). *Political ideologies: An introduction.* Bloomsbury Academic.

Hibbing, J. R., Smith, K. B., & Alford, J. R. (2014). Differences in negativity bias underlie variations in political ideology. *Behavioral and Brain Sciences, 37*(3), 297–307.

Hirschi, C. (2011). *The origins of nationalism: An alternative history from ancient Rome to early modern Germany.* Cambridge University Press.

Hirst, F. W. (1935). *Liberty and tyranny.* Duckworth.

Hiruta, K. (2021). *Hannah Arendt and Isaiah Berlin: Freedom, politics and humanity.* Princeton University Press.

Ho, L. S. T., Ane, C., Lachlan, R., Tarpinian, K., Feldman, R., Yu, Q., van der Bijl, W., Maspons, J., Vos, R., Bastide, P., & Ho, M. L. S. T. (2016). Package

'phylolm'. Retrieved February, 2018, from http://cran.r-project.org/web/packages/phylolm/index.html
Hobbes, T. (1998). *Leviathan*. Oxford University Press.
Hodkinson, S. (2002). Social order and the conflict of values in classical Sparta. In M. Whitby (Ed.), *Sparta* (pp. 104–130). Routledge.
Hodkinson, S. (2006). Was classical Sparta a military society. In S. Hodkinson & A. Powel (Eds.), *Sparta and war* (pp. 111–162). The Classical Press of Wales.
Hodson, G. (2014). Is it impolite to discuss cognitive differences between liberals and conservatives? *Behavioral & Brain Sciences, 37*(3), 313–314.
Hoeller, S. A. (2002). *Gnosticism: New light on the ancient tradition of inner knowing*. Quest Books.
Hofstadter, R. (1969). *The idea of a party system: The rise of legitimate opposition in the United States, 1780–1840*. University of California Press.
Holick, M. F. (2007). Optimal vitamin D status for the prevention and treatment of osteoporosis. *Drugs & Aging, 24*(12), 1017–1029.
Hölldobler, B. (1976). Tournaments and slavery in a desert ant. *Science, 192*(4242), 912–914.
Holldobler, B., & Wilson, E. O. (1990). *The ants*. Springer-Verlag.
Holt, M. F. (1992). *Political parties and American political development from the age of Jackson to the Age of Lincoln*. Louisiana State University Press.
Holt, P. M. (2004). *The Crusader States and their Neighbours: 1098–1291*. Routledge.
Hopkins, K. (2001). *A world full of gods: The strange triumph of Christianity*. Plume.
Hoppitt, W., & Laland, K. N. (2013). *Social learning: An introduction to mechanisms, methods, and models*. Princeton University Press.
Hout, M., Perrett, S., & Cowan, S. K. (2022). Stasis and sorting of Americans' abortion opinions: Political polarization added to religious and other differences. *Socius, 8*. https://doi.org/10.1177/23780231221117648
Howe, D. W. (2009). *Making the American self: Jonathan Edwards to Abraham Lincoln*. Oxford University Press.
Hsü, K. J. (1989). Evolution, ideology, Darwinism and science. *Klinische Wochenschrift, 67*(17), 923–928.
Hudson, J. (2012). *The Oxford history of the laws of England Volume II: 871–1216*. Oxford University Press.
Hume, M. (1907). *The court of Philip IV: Spain in decadence*. Eveleigh Nash.
Humphrey, N. K. (1976). The social function of intellect. In P. P. G. Bateson & R. A. Hinde (Eds.), *Growing points in ethology* (pp. 303–317). Cambridge University Press.

Humphreys, M. (2003). *Economics and violent conflict* (p. 31). Harvard Humanitarian Initiative.

Hundert, E. J. (1994). *The Enlightenment's fable: Bernard Mandeville and the discovery of society*. Cambridge University Press.

Hurst, A. L. (2010). Languages of class in US party platforms, 1880–1936. *Journal of Historical Sociology, 23*(4), 542–569.

Hurst, L. (1995). Selfish genetic elements and their role in evolution: The evolution of sex and some of what that entails. *Philosophical Transactions of the Royal Society of London. Series B: Biological Sciences, 349*(1329), 321–332.

Huston, J. L. (2011). The Illinois political realignment of 1844–1860: Revisiting the analysis. *Journal of the Civil War Era, 1*(4), 506–535.

Hutchinson, R. (2014). *The Spanish Armada*. Macmillan.

Hyun Kim, J., & Moon, S. J. (2000). Time spent outdoors and seasonal variation in serum concentrations of 25–hydroxyvitamin D in Korean women. *International Journal of Food Sciences and Nutrition, 51*(6), 439–451.

Ikenberry, G. J. (2011). *Liberal Leviathan*. Princeton University Press.

Inglehart, R. F. (2018). *Cultural evolution: People's motivations are changing, and reshaping the world*. Cambridge University Press.

Ingvarsson, P. K., & Taylor, D. R. (2002). Genealogical evidence for epidemics of selfish genes. *Proceedings of the National Academy of Sciences, 99*(17), 11265–11269.

Innes, J. (2003). Central government 'interference': Changing concepts, practices, and concerns, *c.*1700–1850. In J. Harris (Ed.), *Civil society in British history: Ideas, identities, institutions* (pp. 39–60). Oxford University Press.

Irwin, R. (2018). *Ibn Khaldun: An intellectual biography*. Princeton University Press.

Iyengar, S., & McGuire, W. J. (1993). *Explorations in political psychology*. Duke University Press.

Jackson, C. (2005). The Rage of Parliaments: The House of Commons, 1690–1715. *The Historical Journal, 48*(2), 567–587.

James, S. C. (2000). *Presidents, parties, and the state: A party system perspective on democratic regulatory choice, 1884–1936*. Cambridge University Press.

Jaques, T. (2007). *Dictionary of battles and sieges: A guide to 8,500 battles from antiquity through the twenty-first century* (3 Vols.). Bloomsbury Publishing.

Jefferson, T. (1984). *Writings: Autobiography, notes on the State of Virginia, public and private papers, addresses, letters*. Library of America.

Jenner, R. E. (2010). *FDR's republicans: Domestic political realignment and American Foreign Policy*. Rowman & Littlefield.

Johnson, W. (2007). Genetic and environmental influences on behavior: Capturing all the interplay. *Psychological Review, 114*, 423–440.

Johnston, C. D., & Madson, G. J. (2022). Negativity bias, personality and political ideology. *Nature Human Behaviour, 6*(5), 666–676.

Jolly, A. (1966). Lemur social behavior and primate intelligence: The step from prosimian to monkey intelligence probably took place in a social context. *Science, 153*(3735), 501–506.

Jones, D. S. (2014). *Masters of the universe. Hayek, Friedman, and the birth of neoliberal politics*. Princeton University Press.

Jones, G. (2015). *Hive mind: How your nation's IQ matters so much more than your own*. Stanford University Press.

Jordan, D. (1993). *The king's trial: The French Revolution vs. Louis XVI*. Gryphon Press.

Jordan, W. C. (1997). *The great famine: Northern Europe in the early fourteenth century*. Princeton University Press.

Jost, J. T. (2006). The end of the end of ideology. *American Psychologist, 61*(7), 651.

Jost, J. T. (2013). *Left & right: The psychological significance of a political distinction*. Oxford University Press.

Jost, J. T., & Amodio, D. M. (2012). Political ideology as motivated social cognition: Behavioral and neuroscientific evidence. *Motivation and Emotion, 36*, 55–64.

Jost, J. T., & Sidanius, J. (2004). *Political psychology*. Psychology Press.

Jost, J. T., Sterling, J., & Stern, C. (2017). Getting closure on conservatism, or the politics of epistemic and existential motivation. In C. E. Kopetz & A. Fishbach (Eds.), *The motivation-cognition interface* (pp. 56–87). Routledge.

Jost, J. T., Stern, C., Rule, N. O., & Sterling, J. (2017). The politics of fear: Is there an ideological asymmetry in existential motivation? *Social Cognition, 35*(4), 324–353.

Jotischky, A. (2013). *Crusading and the Crusader states*. Routledge.

Judd, S. E., & Tangpricha, V. (2009). Vitamin D deficiency and risk for cardiovascular disease. *The American Journal of the Medical Sciences, 338*(1), 40–44.

Kalmoe, N. P., & Johnson, M. (2022). Genes, ideology, and sophistication. *Journal of Experimental Political Science, 9*(2), 255–266.

Kalyvas, A. (2008). *Democracy and the politics of the extraordinary: Max Weber, Carl Schmitt, and Hannah Arendt*. Cambridge University Press.

Kalyvas, A., & Katznelson, I. (2008). *Liberal beginnings: Making a republic for the moderns*. Cambridge University Press.

Kamen, H. (1998). *The Spanish Inquisition: A historical revision.* Yale University Press.

Kanai, R., Feilden T., Firth C., & Rees G. (2011). Political orientations are correlated with brain structure in young adults. *Current Biology, 21*(8), 677–80.

Kandler, C., Bleidorn, W., & Riemann, R. (2012). Left or right? Sources of political orientation: The roles of genetic factors, cultural transmission, assortative mating, and personality. *Journal of Personality and Social Psychology, 102*(3), 633.

Kandola, S. S., & Egan, V. (2014). Individual differences underlying attitudes to the death penalty. *Personality and Individual Differences, 66*, 48–53.

Kaplan, J. M. (2023). The biology of politics: Some limitations and reasons for caution. In H. Kincaid & J. Van Bouwel (Eds.), *The Oxford Handbook of Philosophy of Political Science* (pp. 17–35). Oxford University Press.

Kaplan, J. T., Freedman, J., & Iacoboni, M. (2007). Us versus them: Political attitudes and party affiliation influence neural response to faces of presidential candidates. *Neuropsychologia, 45*(1), 55–64.

Karnane, P., & Quinn, M. A. (2019). Political instability, ethnic fractionalization and economic growth. *International Economics and Economic Policy, 16*, 435–461.

Karpenko, O. V. (2019). The "Overton Window" as manipulative mechanism of public values transformation. *Вісник Національної академії керівних кадрів культури і мистецтв, 1*, 51–54.

Kavanagh, D. (1994). Harold Laski: A political biography. *Parliamentary Affairs, 47*(1), 146–154.

Kavanagh, P. S., & Kahl, B. L. (2016). Life history theory. In *Encyclopedia of evolutionary psychological science* (pp. 1–12). Springer.

Kawamoto, T., Van der Linden, D., & Dunkel, C. S. (2017). The General Factor of Personality (GFP) and moral foundations. *Personality and Individual Differences, 119*, 78–82.

Kebbell, P. (2009). *The changing face of English Freemasonry, 1640–1740* (Doctoral Dissertation, University of Bristol).

Keeley, L. H. (1997). *War before civilization.* Oxford University Press.

Kelemen, L., Szabó, Z. P., Mészáros, N. Z., László, J., & Forgas, J. P. (2014). Social cognition and democracy: The relationship between system justification, just world beliefs, authoritarianism, need for closure, and need for cognition in Hungary. *Journal of Social and Political Psychology, 2*(1), 197–219.

Kelly, R. C. (2005). *Warless societies and the origin of war.* University of Michigan Press.

Kennell, N. M. (2010). *Spartans: A new history* (Vol. 8). John Wiley & Sons.

Kenrick, D. T., Griskevicius, V., Neuberg, S. L., & Schaller, M. (2010). Renovating the pyramid of needs: Contemporary extensions built upon ancient foundations. *Perspectives on Psychological Science: A Journal of the Association for Psychological Science, 5*(3), 292–314.

Kern, F. D. (1915). The genetic relationship of parasites. *American Journal of Botany, 2*(3), 116–131.

Kerr, B. (2009). Theoretical and experimental approaches to the evolution of altruism and the levels of selection. In T. Garland & M. R. Rose (Eds.), *Experimental evolution: Concepts, methods, and applications of selection experiments* (pp. 585–630). California University Press.

Kevins, A., & Soroka, S. N. (2018). Growing apart? Partisan sorting in Canada, 1992–2015. *Canadian Journal of Political Science/Revue canadienne de science politique, 51*(1), 103–133.

Kimlin, M. G. (2008). Geographic location and vitamin D synthesis. *Molecular Aspects of Medicine, 29*(6), 453–461.

Kirby, K. R., Gray, R. D., Greenhill, S. J., Jordan, F. M., Gomes-Ng, S., Bibiko, H. J., Blasi, D. E., Botero, C. A., Bowern, C., Ember, C. R., Leehr, D., Low, B. S., McCarter, J., Divale, W., & Gavin, M. C. (2016). D-PLACE: A global database of cultural, linguistic and environmental diversity. *PloS one, 11*(7), e0158391.

Kirkwood, T. B. (1977). Evolution of ageing. *Nature, 270*, 301–304.

Kirkwood, T. B. L. (1987). Immortality of the germ-line versus disposability of the soma. In A. D. Woodhead & K. H. Thompson (Eds.), *Evolution of longevity in animals: A comparative approach* (pp. 209–218). Plenum Press.

Klebl, C., Donner, M., & Bishnoi, I. (2022). Disliking: From adaptive disgust to ugliness. In M. Skov & M. Nadal (Eds.), *The Routledge International Handbook of Neuroaesthetics* (pp. 71–88). Routledge.

Klein, E. (2020). *Why we're polarized*. Simon and Schuster.

Klemmensen, R., Hatemi, P. K., Hobolt, S. B., Petersen, I., Skytthe, A., & Nørgaard, A. S. (2012). The genetics of political participation, civic duty, and political efficacy across cultures: Denmark and the United States. *Journal of Theoretical Politics, 24*(3), 409–427.

Klemmensen, R., Hatemi, P. K., Hobolt, S. B., Skytthe, A., & Nørgaard, A. S. (2012). Heritability in political interest and efficacy across cultures: Denmark and the United States. *Twin Research and Human Genetics, 15*(1), 15–20.

Kleppestø, T. H. (2021). The political animal: The roles of genetics, environments, evolution, attachment, and personality in the formation of political attitudes (Doctoral dissertation, University of Oslo).

Kleppestø, T. H., Czajkowski, N. O., Vassend, O., Røysamb, E., Eftedal, N. H., Sheehy-Skeffington, J., Kunst, J. R., & Thomsen, L. (2019). Correlations between social dominance orientation and political attitudes reflect common genetic underpinnings. *Proceedings of the National Academy of Sciences, 116*(36), 17741–17746.

Klofstad, C. A., McDermott, R., & Hatemi, P. K. (2012). Do bedroom eyes wear political glasses? The role of politics in human mate attraction. *Evolution and Human Behavior, 33*(2), 100–108.

Klofstad, C. A., McDermott, R., & Hatemi, P. K. (2013). The dating preferences of liberals and conservatives. *Political Behavior, 35*(3), 519–538.

Klosko, G. (2012). *History of political theory: An introduction, volume I: Ancient and Medieval* (2nd ed.). Oxford University Press.

Knight, C. (2005). Darwinism and collective representations. In *The archaeology of human ancestry* (pp. 312–325). Routledge.

Koenig, A. (2002). Competition for resources and its behavioral consequences among female primates. *International Journal of Primatology, 23*, 759–783.

Kofoed, L., & MacMillan, J. (1986). Alcoholism and Antisocial Personality: The Sociobiology of an Addiction. *The Journal of nervous and mental disease, 174*(6), 332–335.

Kowaleski, M. (2014). Medieval people in town and country: New perspectives from demography and bioarchaeology. *Speculum, 89*(3), 573–600.

Krasnow, M. M., Delton, A. W., Cosmides, L., & Tooby, J. (2015). Group cooperation without group selection: Modest punishment can recruit much cooperation. *PloS one, 10*(4), e0124561.

Krispenz, A., & Bertrams, A. (2023). Understanding left-wing authoritarianism: Relations to the dark personality traits, altruism, and social justice commitment. *Current Psychology, 43*, 1–17.

Krosch, A. R., Jost, J. T., & Van Bavel, J. J. (2021). The neural basis of ideological differences in race categorization. *Philosophical Transactions of the Royal Society B, 376*(1822), 20200139.

Kruse, K. M., & Zelizer, J. E. (2019). *Fault lines: A history of the United States since 1974*. WW Norton & Company.

Ksiazkiewicz, A. (2020). Conservative larks, liberal owls: The relationship between chronotype and political ideology. *The Journal of Politics, 82*(1), 367–371.

Ksiazkiewicz, A., & Friesen, A. (2017). Genes and politics. In S. A. Peterson & A. Somit (Eds.), *The handbook of biology and politics* (pp. 85–105). Edward Elgar Publishing.

Ladd, E. C., Hadley, C., & King, L. (1971). A new political realignment?. *The Public Interest, 23*, 46.

Lamb, P. (2004). *Harold Laski*. Palgrave Macmillan.

Lammers, J., & Baldwin, M. (2022). Two sides of the same coin: A new look at differences and similarities across political ideology. In D. Osborne & C. G. Sibley (Eds.), *The Cambridge Handbook of political psychology* (pp. 674–686). Cambridge University Press.

Landau, M. J., Kay, A. C., & Whitson, J. A. (2015). Compensatory control and the appeal of a structured world. *Psychological Bulletin, 141*(3), 694.

Landers, J. (2003). *The field and the forge: Population, production, and power in the pre-industrial West*. Oxford University Press.

Lang, C., & Pearson-Merkowitz, S. (2015). Partisan sorting in the United States, 1972–2012: New evidence from a dynamic analysis. *Political Geography, 48*, 119–129.

Larson, E. J. (2007). *A magnificent catastrophe: The tumultuous election of 1800, America's first presidential campaign*. Simon and Schuster.

Lasch, C. (1965). *The new radicalism in America*. Knopf.

Laski, H. J. (1917). *The problems of sovereignty*. Yale University Press.

Laski, H. J. (1936). *The Rise of European Liberalism*. George Allen & Unwin, Ltd.

Lau, R. R., & Sears, D. O. (1986). *Political cognition*. Erlbaum.

Lazenby, J. F. (1985). *The Spartan Army*. Aris and Phillips.

Le Billon, P. (2004). *The geopolitics of resource wars*. Routledge.

Le Moli, F. (1980). On the origin of slaves in dulotic ant societies. *Italian Journal of Zoology, 47*(1–2), 207–212.

LeBlanc, S. A., & Register, K. E. (2004). *Constant battles: The myth of the peaceful, noble savage*. Macmillan.

Lee, S. W., & Ma, C. (2023). Pain sensitivity predicts support for moral and political views across the aisle. *Journal of Personality and Social Psychology, 125*(6), 1239–1264.

Leeds, E. J. (1979). *No Man's Land: Combat and identity in World War I*. Cambridge University Press.

Leffler, M. P. (2010). *The emergence of an American grand strategy, 1945–1952*. In M. Leffler & A. Westad (Eds.), *Cambridge history of the cold war* (pp. 67–89). Cambridge University Press.

Legendre, P., & Oksanen, M. J. (2018). Package 'lmodel2'. https://CRAN.R-project.org/package=lmodel2

Lehmann, T. C., & Zhukov, Y. M. (2019). Until the Bitter End? The Diffusion of Surrender Across Battles. *International Organization, 73*(1), 133–169.

Lemkin, R. (2012). *Lemkin on genocide*. Rowman & Littlefield.

Lepage, J., Bègue, L., Zerhouni, O., Dambrun, M., Vezirian, K., Besson, T., Bonneterre, S., & Mermillod, M. (2022). Authoritarian attitudes are associated with higher autonomic reactivity to stress and lower recovery. *Emotion, 22*(3), 526.

Levack, B., & P. (1987). *The witch-hunt in Early Modern Europe* (2nd ed.). Longman.

Levendusky, M. (2009). *The partisan sort: How liberals became Democrats and conservatives became Republicans*. University of Chicago Press.

Levendusky, M. (2023). *Our common bonds: Using what Americans share to help bridge the partisan divide*. Chicago University Press.

Levinson, E. (1937). The right to break strikes. *Current History (1916–1940), 45*(5), 77–82.

Lewis, E. E., Campbell, J. F., & Sukhedeo, M. V. K. (2002). *The behavioural ecology of parasites*. CABI Publishing.

Lewis, G. J. (2016). Hans Eysenck and the first wave of socio-political genetics. *Personality and Individual Differences, 103*, 135–139.

Lewis, G. J., & Bates, T. C. (2010). Genetic evidence for multiple biological mechanisms underlying in-group favoritism. *Psychological Science, 21*(11), 1623–1628.

Lewis, G. J., & Bates, T. C. (2014). Common heritable effects underpin concerns over norm maintenance and in-group favoritism: Evidence from genetic analyses of right-wing authoritarianism and traditionalism. *Journal of Personality, 82*(4), 297–309.

Lewis, K. J. (2017). *The Counts of Tripoli and Lebanon in the twelfth century: Sons of Saint-Gilles*. Routledge.

Lewis, V. (2019). *Ideas of power: The politics of American party ideology development*. Cambridge University Press.

Li, N. P., van Vugt, M., & Colarelli, S. M. (2018). The evolutionary mismatch hypothesis: Implications for psychological science. *Current Directions in Psychological Science, 27*(1), 38–44.

Lightfoot, J. B. (1875). *On Some Points Connected with the Essenes. St. Paul's epistles to the Colossians and to Philemon: A revised text with introductions, notes, and dissertations*. Macmillan Publishers.

Lilienfeld, S. O., & Latzman, R. D. (2014). Threat bias, not negativity bias, underpins differences in political ideology. *Behavioral & Brain Sciences, 37*(3), 318–319.

Lindholm, A. K., & Price, T. A. (2016). The evolutionary consequences of selfish genetic elements. *Current Zoology, 62*(6), 655–658.

Line, N. D., & Hanks, L. (2016). The effects of environmental and luxury beliefs on intention to patronize green hotels: The moderating effect of destination image. *Journal of Sustainable Tourism, 24*(6), 904–925.

Linklater, A. (1996). Citizenship and sovereignty in the post-Westphalian state. *European Journal of International Relations, 2*(1), 77–103.

Linquist, S. (2015). Which evolutionary model best explains the culture of honour? *Biology & Philosophy, 31*, 213–235.

Little, A. C. (2017). Evolutionary explanations for financial and prosocial biases: Beyond mating motivation. *Behavioral and Brain Sciences, 40*, e34.

Littvay, L. (2012). Do heritability estimates of political phenotypes suffer from an equal environment assumption violation? Evidence from an empirical study. *Twin Research and Human Genetics, 15*(1), 6–14.

Liuzza, M. T., Lindholm, T., Hawley, C. B., Gustafsson Sendén, M., Ekström, I., Olsson, M. J., & Olofsson, J. K. (2018). Body odour disgust sensitivity predicts authoritarian attitudes. *Royal Society Open Science, 5*(2), 171091.

Llewellyn, A. (1994). *Witchcraze: A new history of the European witch hunts.* Pandora Press.

Lock, P. (2009). *The Routledge companion to the Crusades.* Routledge.

Locke, J. (1962). *Second treatise on civil government in social contract.* Oxford University Press.

Lockyer, A., Hatemi, P. K., & Hopcroft, R. (2018). Genetics and politics: A review for the social scientist. In R. L. Hopcroft (Ed.), *Oxford handbook of evolution, biology, and society* (pp. 281–304). Oxford University Press.

Lovestone, J. (1923). *The Government--strikebreaker: A Study of the Role of the Government in the Recent Industrial Crisis* (Vol. 2). Workers party of America.

Lovett, F. (2015). The republican critique of liberalism. In S. Wall (Ed.) *The Cambridge companion to liberalism* (pp. 381–400). Cambridge University Press.

Lubell, S. (1970). *The hidden crisis.* Norton.

Ludeke, S., Johnson, W., & Bouchard, T. J., Jr. (2013). "Obedience to traditional authority:" A heritable factor underlying authoritarianism, conservatism and religiousness. *Personality and Individual Differences, 55*(4), 375–380.

Lumsden, C. J., & Wilson, E. O. (1981). *Genes, mind, and culture: The coevolutionary process.* Harvard University Press.

Luomanen, P. (2011). *Recovering Jewish-Christian sects and gospels.* Brill.

Lupi, M. (2006). Amompharetos, the lochos of Pitane and the Spartan system of villages. In S. Hodkinson & A. Powel (Eds.), *Sparta and war* (pp. 185–218). The Classical Press of Wales.

MacDonald, K. B. (1994). *A people that shall dwell alone: Judaism as a group evolutionary strategy*. Praeger.

MacDonald, K. B. (1998). *Separation and its discontents: Toward an evolutionary theory of antisemitism*. Praeger.

Macfarlan, S. J., Walker, R. S., Flinn, M. V., & Chagnon, N. A. (2014). Lethal coalitionary aggression and long-term alliance formation among Yanomamö men. *Proceedings of the National Academy of Sciences, 111*(47), 16662–16669.

Macfarlane, A. (2003). *The savage wars of peace: England, Japan and the Malthusian trap*. Palgrave Macmillan.

Machiavelli, N. (1993). *The prince (1513)*. Wordsworth Editions.

MacKay, K. C. (1972). *The Progressive movement of 1924*. Octagon Books.

Mackey, A. G. (1946). *Encyclopedia of freemasonry* (Vol. 1). The Masonic History Company.

Malka, A. (2022). Religiosity and openness to authoritarian governance. In D. Osborne & C. G. Sibley (Eds.), *The Cambridge handbook of political psychology* (pp. 285–297). Cambridge University Press.

Malka, A., & Soto, C. J. (2014). How encompassing is the effect of negativity bias on political conservatism. *Behavioral and Brain Sciences, 37*(3), 320–321.

Malthus, T. R. (1798/1992). *An essay on the principle of population*. Easton Press.

Mann, M. (1997). Has globalization ended the rise and rise of the nation-state? *Review of International Political Economy, 4*(3), 472–496.

Mansell, J., Harell, A., Gidengil, E., & Stewart, P. A. (2021). Psychophysiology, cognition, and political differences: Guest editors' introduction to the special issue. *Politics and the Life Sciences, 40*(2), 137–141.

Marcion of Sinope. (1891). *The Gospel of the Lord* (J. H. Hill, Trans.). Whitehead.

Martin, C. A., Gowda, U., & Renzaho, A. M. (2016). The prevalence of vitamin D deficiency among dark-skinned populations according to their stage of migration and region of birth: A meta-analysis. *Nutrition, 32*(1), 21–32.

Martin, N. G., Eaves, L. J., Heath, A. C., Jardine, R., Feingold, L. M., & Eysenck, H. J. (1986). Transmission of social attitudes. *Proceedings of the National Academy of Sciences, 83*(12), 4364–4368.

Martin, S. (2006). *The Gnostics: The first Christian heretics*. Pocket Essentials.

Mason, R. (2011). Political realignment. In M. Small (Ed.), *A Companion to Richard M. Nixon* (pp. 252–269). Blackwell Publishing.

Mason, S. N., & Honora, C. (2008). *Flavius Josephus: Translation and commentary: Volume 1B: Judean War 2*. Brill.

Maxwell, A. (2020). Primordialism for scholars who ought to know better: Anthony D. Smith's critique of modernization theory. *Nationalities Papers, 48*(5), 826–842.

Mayhew, D. (2000). Electoral realignments. *Annual Review of Political Science, 3*, 449–474.

McAdams, D. P., & Mayukha, A. (2023). Hiding in plain view: An historical perspective on the study of morality in personality psychology. *Journal of Personality, 92(3)*, 666–682.

McCann, S. J. (2014). Neuroticism and state differences in partisanship in the USA: Emotional stability, ideological orientation, and Republican preference. *Journal of Social and Political Psychology, 2*(1), 242–267.

McCarthy, C. (1901). *The Antimasonic Party: A Study of Political Antimasonry in the United States, 1827–1840*. US Government Printing Office.

McCarty, N., Poole, K. T., & Rosenthal, H. (2013). *Political bubbles: Financial crises and the failure of American democracy*. Princeton University Press.

McCrae, R. R. (1996). Social consequences of experiential openness. *Psychological Bulletin, 120*(3), 323.

McDermott, R., & Hatemi, P. K. (2017). The relationship between physical aggression, foreign policy and moral choices: Phenotypic and genetic findings. *Aggressive Behavior, 43*(1), 37–46.

McDermott, R., Tingley, D., & Hatemi, P. K. (2014). Assortative mating on ideology could operate through olfactory cues. *American Journal of Political Science, 58*(4), 997–1005.

McGlynn, S. (2015). *Kill Them All: Cathars and Carnage in the Albigensian Crusade*. The History Press.

McGuire, W. J. (2004). The poly-psy relationship: Three phases of a long affair. In J. T. Jost & J. Sidanius (Eds.), *Political psychology* (pp. 22–32). Psychology Press.

McKinstry, L. (2014). *Operation Sea Lion*. Abrams.

McLachlan, S. (2010). *Medieval Handgonnes*. Osprey Publishing.

McNeil, W. H. (1963). *The rise of the West: A history of the human community*. Chicago University Press.

McNeill, W. H. (1979). *A world history* (3rd ed.). Oxford University Press.

McRae, T. (2022). Neuropolitics: The state of the literature & its implications for political science. (Doctoral dissertation, University of Ottawa).

Melchionna, M., Mondanaro, A., Serio, C., Castiglione, S., Di Febbraro, M., Rook, L., Diniz-Filho, J. A. F., Manzi, G., Profico, A., Sansalone, G., & Raia, P. (2020). Macroevolutionary trends of brain mass in Primates. *Biological Journal of the Linnean Society, 129*(1), 14–25.

Mendel, J. G. (1866). "Versuche über Pflanzenhybriden", *Verhandlungen des naturforschenden Vereines in Brünn*, Bd. IV für das Jahr, 1865, *Abhandlungen*, 3–47.

Mendez, M. F. (2017). A neurology of the conservative-liberal dimension of political ideology. *The Journal of neuropsychiatry and clinical neurosciences, 29*(2), 86–94.

Meyer, D., Aguiar, V. R. C., Bitarello, B. D., Brandt, D. Y. C., & Nunes, K. (2018). A genomic perspective on HLA evolution. *Immunogenetics, 70*(1), 5–27.

Milad, M. R., Quirk, G. J., Pitman, R. K., Orr, S. P., Fischl, B., & Rauch, S. L. (2007). A role for the human dorsal anterior cingulate cortex in fear expression. *Biological psychiatry, 62*(10), 1191–1194.

Miles, M. R., & Haider-Markel, D. P. (2019). Personality and genetic associations with military service. *Armed Forces & Society, 45*(4), 637–658.

Millan, M., & Saluppo, A. (2021). *Corporate policing, yellow unionism, and strikebreaking, 1890–1930: In defence of freedom* (p. 298). Routledge.

Miller, & Bakar, K. S. (2023). Conflict events worldwide since 1468BC: Introducing the historical conflict event dataset. *The Journal of Conflict Resolution, 67*(2–3), 522–554.

Milojev, P., Osborne, D., Greaves, L. M., Bulbulia, J., Wilson, M. S., Davies, C. L., Liu, J. H., & Sibley, C. G. (2014). Right-wing authoritarianism and social dominance orientation predict different moral signatures. *Social Justice Research, 27*, 149–174.

Mitchell, B. P. (2007). *Eight ways to run the country: A new and revealing look at left and right.* Greenwood Publishing Group.

Mithal, A., Wahl, D. A., Bonjour, J. P., Burckhardt, P., Dawson-Hughes, B., Eisman, J. A., El-Hajj Fuleihan, G., Josse, R. G., Lips, P., & Morales-Torres, J. (2009). Global vitamin D status and determinants of hypovitaminosis D. *Osteoporosis International, 20*(11), 1807–1820.

Mondak, J. J. (2010). *Personality and the foundations of political behavior.* Cambridge University Press.

Moore, R. I. (2012). *The War on Heresy*. Belknap Press of Harvard University Press.

Morison, S. E. (1969). *Harrison Gray Otis 1765–1848, the urbane Federalist.* Houghton Mifflin Company.

Morris, M. (1998). *The British Monarchy and the French Revolution.* Yale University Press.

Morris, I. (2009). The greater Athenian state. In I. Morris & W. Scheidel (Eds.), *The dynamics of ancient empires: State power from Assyria to Byzantium* (pp. 99–177). Oxford University Press.

Morrissey, M. B., & Liefting, M. (2016). Variation in reaction norms: Statistical considerations and biological interpretation. *Evolution, 70*(9), 1944–1959.

Morton, N. (2018). *The field of blood: The battle for Aleppo and the remaking of the medieval Middle East.* Basic Books.

Morton, N. (2020). *The crusader states and their neighbours: A military history, 1099–1187.* Oxford University Press.

Morton Smith, J. (1966). *Freedom's fetters: The Alien and Sedition Laws and American civil liberties.* Cornell University Press.

Moss, J., & O'Connor, P. J. (2020). The Dark Triad traits predict authoritarian political correctness and alt-right attitudes. *Heliyon, 6*(7), e04453.

Motyl, M. (2012). Party evolutions in moral intuitions: A text-analysis of US political party platforms from 1856–2008. SSRN Electronic Journal. https://doi.org/10.2139/ssrn.2158893

Mudde, C., & Kaltwasser, C. R. (2017). *Populism: A very short introduction.* Oxford University Press.

Muir, W. M. (1996). Group selection for adaptation to multiple-hen cages: Selection program and direct responses. *Poultry Science, 75*(4), 447–458.

Murphey, D. D. (2015). A gentle look at race, evolution and genetics: Nicholas wade's elevation of science over ideology. *The Journal of Social, Political, and Economic Studies, 40*(1/2), 117.

Murray, A. V., & Nicholson, H. (2006). Latin kingdom of Jerusalem. In A. V. Murray (Ed.), *The Crusades: An encyclopedia* (pp. 662–672). ABC-CLIO.

Murray, D. R., Schaller, M., & Suedfeld, P. (2013). Pathogens and politics: Further evidence that parasite prevalence predicts authoritarianism. *PloS One, 8*(5), e62275.

Mustafaj, M., Madrigal, G., Roden, J., & Ploger, G. W. (2022). Physiological threat sensitivity predicts anti-immigrant attitudes. *Politics and the Life Sciences, 41*(1), 15–27.

Nagayoshi, T., Ishikawa, R., & Kida, S. (2022). Anterior cingulate cortex projections to the dorsal hippocampus positively control the expression of contextual fear generalization. *Learning & Memory, 29*(3), 77–82.

Nail, P. R., McGregor, I., Drinkwater, A. E., Steele, G. M., & Thompson, A. W. (2009). Threat causes liberals to think like conservatives. *Journal of Experimental Social Psychology, 45*(4), 901–907.

Nam, H. H., Jost, J. T., Kaggen, L., Campbell-Meiklejohn, D., & Van Bavel, J. J. (2018). Amygdala structure and the tendency to regard the social system as legitimate and desirable. *Nature Human Behaviour, 2*(2), 133–138.

Navarrete, C. D., & Fessler, D. M. (2006). Disease avoidance and ethnocentrism: The effects of disease vulnerability and disgust sensitivity on intergroup attitudes. *Evolution and Human Behavior, 27*(4), 270–282.

Nawata, K. (2020). A glorious warrior in war: Cross-cultural evidence of honor culture, social rewards for warriors, and intergroup conflict. *Group Processes & Intergroup Relations, 23*(4), 598–611.

Needham, J. (1986). *Science and civilisation in China, Volume 5: Chemistry and chemical technology, Part 7, military technology: The gunpowder epic*. Cambridge University Press.

Nelkin, D. (2001). A brief history of the political work of genetics. *Jurimetrics, 42*, 121.

Netanyahu, B. (1995). *The origins of the Inquisition in 15th Century Spain*. Random House.

Newman, M. (1993). *Harold Laski: A political biography*. Springer.

Newton, M. (1998). *Holy homicide: An encyclopedia of those who go with their god and kill*. Loompanics.

Nisbett, R. E., & Cohen, D. (1996). *Culture of honor: The psychology of violence in the South*. Westview Press.

Niven, J. (1983). *Martin Van Buren: The Romantic age of American politics*. Oxford University Press.

Noel, H. (2013). *Political ideologies and political parties in America*. Cambridge University Press.

Nolan, D. (1971). Classifying and analyzing politico-economic systems. *The Individualist, 1*, 5–11.

Norenzayan, A. (2013). *Big gods: How religion transformed cooperation and conflict*. Princeton University Press.

Norenzayan, A., & Shariff, A. F. (2008). The origin and evolution of religious prosociality. *Science, 322*(5898), 58–62.

Norval, M., & Wulf, H. C. (2009). Does chronic sunscreen use reduce vitamin D production to insufficient levels? *British Journal of Dermatology, 161*(4), 732–736.

O'Grady, J. (1994). *Early Christian heresies*. Barnes & Noble.

Oakeshott, M. (2001). *The voice of liberal learning*. Liberty Fund.

Obolensky, D. (1972). *The Bogomils: A study in Balkan Neo-Manichaeism*. Hall.

Odling-Smee, F. J., Laland, K. N., & Feldman, M. W. (2003). *Niche construction: The neglected process in evolution*. Princeton University Press.

Okasha, S. (2006). *Evolution and levels of selection*. Oxford University Press.

Okasha, S. (2009). Individuals, groups, fitness and utility: Multi-level selection meets social choice theory. *Biology & Philosophy, 24*(5), 561–584.

Olson, L. C. (2004). *Benjamin Franklin's vision of American community: A study in rhetorical iconography*. University of South Carolina Press.

Ordabayeva, N., & Fernandes, D. (2018). Better or different? How political ideology shapes preferences for differentiation in the social hierarchy. *Journal of Consumer Research, 45*(2), 227–250.

Orey, B. D. A., & Park, H. (2012). Nature, nurture, and ethnocentrism in the Minnesota Twin Study. *Twin Research and Human Genetics, 15*(1), 71–73.

Orme, D., Freckleton, R., Thomas, G., Petzoldt, T., Fritz, S., Isaac, N., & Pearse, W. (2013). The caper package: Comparative analysis of phylogenetics and evolution in R. *R Package Version, 5*(2), 1–36.

Ormrod, W. M. (1990). The Peasants' Revolt and the Government of England. *Journal of British Studies, 29*(1), 1–30.

Osborne, D., & Sibley, C. G. (2015). Within the limits of civic training: Education moderates the relationship between openness and political attitudes. *Political Psychology, 36*(3), 295–313.

Osborne, D., & Sibley, C. G. (2022). *The Cambridge handbook of political psychology*. Cambridge University Press.

Osmundsen, M., Hendry, D. J., Laustsen, L., Smith, K. B., & Petersen, M. B. (2022). The psychophysiology of political ideology: Replications, reanalyses, and recommendations. *The Journal of Politics, 84*(1), 50–66.

Ostiguy, P. (2017). Populism: A sociocultural approach. In C. R. Kaltwasser, P. A. Taggart, P. O. Espejo, & P. Ostiguy (Eds.), *The Oxford handbook of populism* (pp. 73–97). Oxford University Press.

Oswald, M. (Ed.). (2022). *The Palgrave handbook of populism* (p. 3). Palgrave Macmillan.

Otterbein, K. F. (1989). *The evolution of war: A cross-cultural study* (3rd ed.). HRAF.

Otterbein, K. F. (2004). *How war began* (Vol. 10). Texas A&M University Press.

Otterbein, K. F., & Otterbein, C. S. (1965). An eye for an eye, a tooth for a tooth: A cross-cultural study of feuding 1. *American Anthropologist, 67*(6), 1470–1482.

Overton, M. (1996). *Agricultural revolution in England: The transformation of the agrarian economy 1500–1850* (Vol. 23). Cambridge University Press.

Oxley, D. R., Smith, K. B., Alford, J. R., Hibbing, M. V., Miller, J. L., Scalora, M., Hatemi, P. K., & Hibbing, J. R. (2008). Political attitudes vary with physiological traits. *Science, 321*(5896), 1667–1670.

Pagels, E. (1979). *The Gnostic Gospels.* Vintage Books.

Paine, T. (1879). *The political works of Thomas Paine.* Belfords, Clarke & Company.

Pamminger, T., Leingärtner, A., Achenbach, A., Kleeberg, I., Pennings, P. S., & Foitzik, S. (2012). Geographic distribution of the anti-parasite trait 'slave rebellion'. *Evolutionary Ecology, 27*, 39–49.

Panish, A., & Nam, H. H. (2024). The neurobiology of political ideology: Theories, findings, and future directions. *Social and Personality Psychology Compass, 18*(1), e12916.

Paradis, E., Blomberg, S., Bolker, B., Brown, J., Claude, J., Cuong, H. S., et al. (2019). Package 'ape'. *Analyses of Phylogenetics and Evolution, Version, 2*(4), 47.

Park, J. H. (2007). Persistent misunderstandings of inclusive fitness and kin selection: Their ubiquitous appearance in social psychology textbooks. *Evolutionary Psychology, 5*(4), https://doi.org/10.1177/147470490700500414.

Parker, G. (1988). *The military revolution: Military innovation and the rise of the West, 1500–1800.* Cambridge University Press.

Parker, G. (2008). Crisis and Catastrophe: The Global Crisis of the Seventeenth Century. *The American Historical Review, 113*(4), 1053–1079.

Parkinson, P. (2022). Marriage and luxury beliefs at the United Nations. *Quadrant, 66*(1–2), 34–39.

Parmentier, T. (2020). Guests of social insects. In C. Starr (Ed.), *Encyclopedia of Social Insects* (pp. 1–15). Springer.

Passmore, J. (2000). *The perfectibility of man.* Liberty Fund.

Pedersen, W. S., Muftuler, L. T., & Larson, C. L. (2017). Conservatism and the neural circuitry of threat: Economic conservatism predicts greater amygdala-BNST connectivity during periods of threat vs. safety. *Social Cognitive and Affective Neuroscience, 13*(1), 43–51.

Pegg, M. G. (2001). *The corruption of Angels: The great inquisition of 1245–124.* Princeton University Press.

Pegg, M. G. (2008). *A most holy war: The Albigensian Crusade and the Battle for Christendom* (Pivotal Moments in World History). Oxford University Press.

Peñaherrera Aguirre, M., & Fernandes, H. B. (2021). *Phylogenetic analysis within comparative psychology* (*Encyclopedia of Evolutionary Psychological Science*, 5967–5975). Springer.

Peñaherrera-Aguirre, M., Figueredo, A. J., & Hertler, S. C. (2020). Lethal intergroup competition in non-state societies: From small-scale raids to large-scale battles. In *Multilevel selection*. Palgrave Macmillan.

Peñaherrera-Aguirre, M., Figueredo, A. J., Jurgensen, J., Woodley of Menie, M. A., & Hertler, S. C. (2022). Macrohistorical and evolutionary dynamics of between-group competition, sociopolitical complexity, and differentiation-integration effects. *Evolutionary Psychological Science, 9*(1), 90–118.

Peñaherrera-Aguirre, M., Sarraf, M. A., of Menie, M. A. W., & Miller, G. F. (2023). The ten-million-year explosion: Paleocognitive reconstructions of domain-general cognitive ability (G) in extinct primates. *Intelligence, 101*, 101795.

Penke, L., & Jokela, M. (2016). The evolutionary genetics of personality revisited. *Current Opinion in Psychology, 7*, 104–109.

Perry, G. (2018). *The lost boys: Inside Muzafer Sherif's Robbers Cave experiment*. Scribe Publications.

Perry, S. L. (2022). American religion in the era of increasing polarization. *Annual Review of Sociology, 48*, 87–107.

Pestritto, R. J., & Atto, W. J. (2008). *American progressivism*. Lexington Books.

Peter of Vaux de Cernay. (1998). *The History of the Albigensian Crusade* (W. A. Sibly, & M. D. Sibly, Trans.). Woodbridge, SFK: Boydell Press.

Peterson, N. (1993). Demand sharing: Reciprocity and the pressure for generosity among foragers. *American Anthropologist, 95*(4), 860–874.

Philip, M. (2014). *Reforming ideas in Britain: Politics and language in the shadow of the French Revolution, 1789–1815*. Cambridge University Press.

Phillips, K. P. (2014). *The emerging republican majority: Updated edition*. Princeton University Press.

Pijper, F. (1909). The Christian church and slavery in the middle ages. *The American Historical Review, 14*(4), 675–695.

Pinker, S. (2011). *The better angels of our nature*. Viking.

Plavcan, J. M. (2001). Mating system, intrasexual competition and sexual dimorphism in primates. In P. C. Lee (Ed.), *Comparative primate socioecology* (pp. 241–270). Cambridge University Press.

Plavcan, J. M. (2012). Sexual size dimorphism, canine dimorphism, and male-male competition in primates: Where do humans fit in? *Human Nature, 23*, 45–67.

Plavcan, J. M., & van Schaik, C. P. (1992). Intrasexual competition and canine dimorphism in anthropoid primates. *American Journal of Physical Anthropology, 87*(4), 461–477.

Plavcan, J. M., & Van Schaik, C. P. (1997a). Interpreting hominid behavior on the basis of sexual dimorphism. *Journal of Human Evolution, 32*(4), 345–374.

Plavcan, J. M., & Van Schaik, C. P. (1997b). Intrasexual competition and body weight dimorphism in anthropoid primates. *American Journal of Physical Anthropology: The Official Publication of the American Association of Physical Anthropologists, 103*(1), 37–68.

Plumb, J. H. (1967). *The growth of political stability in England*, 1675-1725. Houghton Mifflin.

Pocock, J. G. A. (1975). *The Machiavellian moment: Florentine political thought and the Atlantic republican tradition*. Princeton University Press.

Pollock, G., & Rissing, S. W. (1989). Intraspecific brood raiding, territoriality, and slavery in ants. *The American Naturalist, 133*(1), 61–70.

Popper, K. (2011). *The open society and its enemies*. Routledge.

Postan, M. M. (1975). *The medieval economy and society: An economic history of Britain in the Middle Ages*. Penguin Books.

Pournelle, J. E. (1964). *The American political continuum: An examination of the validity of the left-right model as an instrument for studying contemporary American Political 'isms'*. University of Washington.

Prentice, P. M., Thorton, A., & Wilson, A. J. (2021). A multivariate view of cognitive differences reveals domain-general correlation structure in the Trinidadian Guppy (*Poecilia reticulata*). *bioRxiv*, 2021-11.

Price, G. R. (1970). Selection and covariance. *Nature, 227*(5257), 520–521.

Price, G. R. (1972). Fisher's 'fundamental theorem' made clear. *Annals of Human Genetics, 36*(2), 129–140.

Primus, R. A. (1999). *The American language of rights* (Vol. 54). Cambridge University Press.

Pruitt, J. N., Goodnight, C. J., & Riechert, S. E. (2017). Intense group selection selects for ideal group compositions, but selection within groups maintains them. *Animal Behaviour, 124*, 15–24.

Purdy, J. (2009). *A tolerable anarchy: Rebels, reactionaries, and the making of American freedom*. Vintage.

Putnam, R. D. (2007). E pluribus unum: Diversity and community in the twenty-first century the 2006 Johan Skytte Prize Lecture. *Scandinavian Political Studies, 30*(2), 137–174.

Rabeling, C. (2020). Social Parasitism. In C. Starr (Ed.), *Encyclopedia of social insects* (pp. 1–23). Springer.
Rabin, J. S. (2021). Blue Brain-Red Brain: The Biopsychology of Political Beliefs and Behavior. *The Psychology of Political Behavior in a Time of Change*, 15–53.
Rahe, P. A. (2009). *Montesquieu and the logic of liberty: War, religion, commerce, climate, terrain, technology, uneasiness of mind, the spirit of political vigilance, and the foundations of the modern republic*. Yale University Press.
Rawlings, H. (2008). *The Spanish Inquisition*. John Wiley & Sons.
Ray, J. J. (1976). Do authoritarians hold authoritarian attitudes? *Human Relations, 29*, 307–325.
Ray, J. J. (1979). Does authoritarianism of personality go with conservatism? *Australian Journal of Psychology, 31*, 9–14.
Ray, J. J. (1984). Alternatives to the F scale in the measurement of authoritarianism: A catalog. *The Journal of Social Psychology, 122*(1), 105–119.
Reader, S. M., Hager, Y., & Laland, K. N. (2011). The evolution of primate general and cultural intelligence. *Philosophical Transactions of the Royal Society B: Biological Sciences, 366*(1567), 1017–1027.
Redlawsk, D. P., & Mattes, K. (2022). Emotions and politics. In D. Osborne & C. G. Sibley (Eds.), *The Cambridge handbook of political psychology* (pp. 139–158). Cambridge University Press.
Reid-Henry, S. (2020). *Empire of democracy: The remaking of the west since the Cold War*. Simon & Schuster.
Renshon, J., Lee, J. J., & Tingley, D. (2015). Physiological arousal and political beliefs. *Political Psychology, 36*(5), 569–585.
Revell, L. J. (2012). Phytools: An R package for phylogenetic comparative biology (and other things). *Methods in Ecology and Evolution, 2*, 217–223.
Revelle, M. W. (2015). Package 'psych'. *The Comprehensive R Archive Network, 337*(338).
Reznick, D. (1982). The impact of predation on life history evolution in Trinidadian guppies: Genetic basis of observed life history patterns. *Evolution, 36*(6), 1236–1250.
Reznick, D. (2014). Evolution of life histories. In J. B. Losos (Ed.), *The Princeton guide to evolution* (pp. 268–275). Princeton University Press.
Ribeiro, M. H., Calais, P. H., Almeida, V. A., & Meira Jr, W. (2017). "Everything I disagree with is# FakeNews": Correlating political polarization and spread of misinformation. *arXiv preprint arXiv:1706.05924*.
Richardson, G. B., Chen, C. C., Dai, C. L., Hardesty, P. H., & Swoboda, C. M. (2014). Life history strategy and young adult substance use.

Evolutionary Psychology, 12(5). https://doi.org/10.1177/147470491401200506

Richerson, P., Baldini, R., Bell, A. V., Demps, K., Frost, K., Hillis, V., Mathew, S., Newton, E. K., Naar, N., Newson, L., Ross, C., Smaldino, P. E., Waring, T. M., Zefferman, M., & Ross, C. (2016). Cultural group selection plays an essential role in explaining human cooperation: A sketch of the evidence. *Behavioral and Brain Sciences, 39*, e30.

Richerson, P., & Henrich, J. (2012). Tribal social instincts and the cultural evolution of institutions to solve collective action problems. *Cliodynamics, 3*(1), 38–80.

Riley-Smith, J. (1973). *The feudal nobility and the Kingdom of Jerusalem, 1174–1277*. Macmillan.

Riley-Smith, J. (Ed.). (1999). *The Oxford illustrated history of the Crusades*. Oxford University Press.

Rindermann, H. (2018). *Cognitive capitalism: Human capital and the wellbeing of nations*. Cambridge University Press.

Rindermann, H., Flores-Mendoza, C., & Woodley, M. A. (2012). Political orientations, intelligence and education. *Intelligence, 40*(2), 217–225.

Rio, A. (2017). *Slavery after Rome, 500–1100*. Oxford University Press.

Robbers, M. (2006). Tough-mindedness and fair play: Personality traits as predictors of attitudes toward the death penalty–an exploratory gendered study. *Punishment & Society, 8*(2), 203–222.

Robinson, B., Wiseman, N., Zhang, M., Staber, C. J., & Woodruff, R. C. (2008). Identification of selfish genetic elements in natural populations of Drosophila melanogaster. II. Segregation Distorter (SD). *Drosophila Information Service, 91*, 183–185.

Robinson, J. H. (1905). *Readings in European history*. Ginn.

Robinson, J. J. (2009). *Born in blood: The lost secrets of Freemasonry*. Rowman and Littlefield.

Robisheaux, T. (2013). The German witch trials. In B. P. Levak (Ed.), *The Oxford handbook of witchcraft in early modern Europe and Colonial America* (pp. 179–198). Oxford University Press.

Rodríguez-Sala, M. L. (2009). Cárcel del tribunal del Santo Oficio de la Inquisición. In M. L. Rodríguez-Sala, V. Ramírez, P. O. Alfonzo, M. d. J. Ángel, & C. Rivera (Eds.), *Cinco cárceles de la Ciudad de México, sus cirujanos y otros personajes: 1574–1820* (pp. 149–253). Universidad Nacional Autónoma de México.

Rogers, C. J., Caferro, W., & Reid, S. (2010). *The Oxford encyclopedia of Medieval warfare and military technology, Vol. I.* Oxford University Press.

Rokeach, M. (1973). *The nature of human values.* Free press.

Rose, M. A. (2021). *A world after liberalism: Philosophers of the radical right.* Yale University Press.

Rosenfeld, R. N. (1997). *American Aurora: A democratic-republican returns. The suppressed history of our nation's beginning and the heroic newspaper that tried to report it.* St. Martin's Press.

Rudolph, K. (1987). *Gnosis: The nature and history of Gnosticism.* HarperOne.

Ruisch, B. C., Anderson, R. A., Inbar, Y., & Pizarro, D. A. (2021). A matter of taste: Gustatory sensitivity predicts political ideology. *Journal of Personality and Social Psychology, 121*(2), 394.

Rule, N. O., Freeman, J. B., Moran, J. M., Gabrieli, J. D., Adams, R. B., Jr., & Ambady, N. (2010). Voting behavior is reflected in amygdala response across cultures. *Social Cognitive and Affective Neuroscience, 5*(2–3), 349–355.

Rummel, R. J. (1994). *Death by Government: Genocide and Mass Murder Since 1900* (1st ed.). Transaction Publishers.

Runciman, D. (1997). *Pluralism and the personality of the state.* Cambridge University Press.

Runciman, D. (2018). *The confidence trap: A history of democracy in crisis from World War I to the present.* Princeton University Press.

Rushton, J. P. (1986). Gene-culture, co-evolution, and Genetic Similarity Theory: Implications for ideology, ethnic nepotism, and geopolitics. *Politics and the Life Sciences, 4*(2), 144–148.

Rushton, J. P. (1989). Genetic similarity, human altruism, and group selection. *Behavioral and Brain Sciences, 12*, 503–559.

Rushton, P. (1998). Genetic similarity theory, ethnocentrism, and group selection. In *In ethnic conflict and indoctrination: Altruism and identity in evolutionary perspective.* Berghahn Books.

Ryan, M. (2020). *The genetics of political behavior: How evolutionary psychology explains ideology.* Routledge.

Sachs, A. (2011). Silencing the union movement. In C. R. Smith (Ed.), *Silencing the opposition: How the US government suppressed freedom of expression during major crises* (pp. 125–150). SUNY Press.

Sack, J. J. (1993). *From Jacobite to conservative.* Cambridge University Press.

Saler, M. (2019). Review of *Historicism and the Human Sciences in Victorian Britain; The Myth of Disenchantment: Magic, Modernity, and the Birth of the Human Sciences*, by M. Bevir & J. Ā. Josephson-Storm. *Victorian Studies, 61*(3), 531–534.

Salm, R. (2008). *The myth of Nazareth: The invented town of Jesus*. American Atheist Press.

Salter, A. W. (2018). Constitutional drift and political dysfunction: underappreciated maladies of the political commons. In *Economic freedom and prosperity* (pp. 32–47). Routledge.

Salter, M. E., Duymaç, F. Y., Yilmaz, O., Bahçekapili, H. G., & Harma, M. (2022). Is negativity bias intuitive for liberals and conservatives? *Current Psychology*, 1–13. https://doi.org/10.1007/s12144-021-02557-y

Sarraf, M. A., Pestow, R. N., & Fernandes, H. B. (2017). Social epistasis amplifies the fitness costs of deleterious mutations, engendering rapid fitness decline among modernized populations. *Evolutionary Psychological Science, 3*(2), 181–191.

Schaff, P., & Wace, H. (1890). *A select library of the Nicene and Post-Nicene fathers of the Christian Church, second series, volume I*. Eerdmans.

Schattschneider, E. E. (1960). *The semi-sovereign people*. Holt, Rinehart, and Winston.

Schechter, D. E., & Francis, C. M. (2010). A life history approach to understanding youth time preference: Mechanisms of environmental risk and uncertainty and attitudes toward risk behavior and education. *Human Nature, 21*(2), 140–164.

Schlenker, B. R., Chambers, J. R., & Le, B. M. (2012). Conservatives are happier than liberals, but why? Political ideology, personality, and life satisfaction. *Journal of Research in Personality, 46*(2), 127–146.

Schlesinger, A. M. (1992). *A thousand days: John F. Kennedy in the White House*. Easton Press.

Schnall, S., Haidt, J., Clore, G. L., & Jordan, A. H. (2008). Disgust as embodied moral judgment. *Personality and Social Psychology Bulletin, 34*(8), 1096–1109.

Schofield, N. (2006). Equilibria in the spatial stochastic model of voting with party activists. *Review of Economic Design, 10*, 183–203.

Schofield, N. (2007). Political equilibria with electoral uncertainty. *Social Choice and Welfare, 28*(3), 461–490.

Schofield, N. (2009). Switching equilibria. In W. B. Heller & C. Mershon (Eds.), *Political parties and legislative party switching* (pp. 55–79). Palgrave Macmillan.

Schofield, N., Miller, G., & Martin, A. (2003). Critical elections and political realignments in the USA: 1860–2000. *Political Studies, 51*(2), 217–240.

Schreiber, D. (2017). Neuropolitics: Twenty years later. *Politics and the Life Sciences, 36*(2), 114–131.

Schreiber, D., Fonzo, G., Simmons, A. N., Dawes, C. T., Flagan, T., Fowler, J. H., & Paulus, M. P. (2013). Red brain, blue brain: Evaluative processes differ in Democrats and Republicans. *PLoS One, 8*(2), e52970.

Schwabe, I., Jonker, W., & Van Den Berg, S. M. (2016). Genes, culture and conservatism-a psychometric-genetic approach. *Behavior Genetics, 46*, 516–528.

Sennis, A. (2016). Questions about the Cathars. In A. Sennis (Ed.), *Cathars in question, heresy and inquisition in the Middle Ages* (pp. 1–20). York Medieval Press.

Settle, J., & Detert, L. (2023). Biology and politics. In L. Huddy, D. O. Sears, J. S. Levy, & J. Jerit (Eds.), *The Oxford handbook of political psychology* (pp. 280–309). Oxford University Press.

Shanahan, T. (2004). *The evolution of Darwinism: Selection, adaptation and progress in evolutionary biology*. Cambridge University Press.

Sharman, J. C. (2018). Myths of military revolution: European expansion and Eurocentrism. *European Journal of International Relations, 23*(3), 491–513.

Shaw, C., & Mallett, M. (2018). *The Italian wars 1494–1559: War, state and society in Early Modern Europe*. Routledge.

Sherif, M. & Sherif, C. (1964). *Reference Groups: Exploration into conformity and deviation of adolescents*. Harper and Row.

Sherlock, J. M., Zietsch, B. P., Tybur, J. M., & Jern, P. (2016). The quantitative genetics of disgust sensitivity. *Emotion, 16*(1), 43.

Sherman, R. A., Figueredo, A. J., & Funder, D. C. (2013). The behavioral correlates of overall and distinctive life history strategy. *Journal of Personality and Social Psychology, 105*(5), 873.

Shi, Y., Mast, K., Weber, I., Kellum, A., & Macy, M. (2017). Cultural fault lines and political polarization. In *Proceedings of the 2017 ACM on Web Science Conference* (pp. 213–217). ACM.

Shipley, G. (2002). Perioecic society. In M. Whitby (Ed.), *Sparta* (pp. 182–189). Routledge.

Shirley, J. (2006). *The inquisitor's guide: A medieval manual on heretics*. Ravenhall Books.

Showalter, D. E. (2004). The Prussian military state. In G. Mortimer (Ed.), *Early Modern military history, 1450–1815* (pp. 118–134). Palgrave Macmillan.

Shultz, S., & Dunbar, R. I. (2022). Socioecological complexity in primate groups and its cognitive correlates. *Philosophical Transactions of the Royal Society B, 377*(1860), 20210296.

Sibly, W. A., & Sibly, M. D. (1998). *The history of the Albigensian crusade: Peter of les Vaux-de-Cernay's Historia Albigensis*. Boydell Press.

Skinner, B. F. (1953). *Science and human behavior*. Macmillan.

Skinner, Q. (1998). *Liberty before liberalism*. Cambridge University Press.

Skjönsberg, M. (2021). *The persistence of party: Ideas of harmonious discord in eighteenth-century Britain*. Cambridge University Press.

Skocpol, T. (1979). *States and social revolutions: A comparative analysis of France, Russia, & China*. Cambridge University Press.

Skorupski, J. (2015). The conservative critique of liberalism. In S. Wall (Ed.) *The Cambridge companion to liberalism* (pp. 401–422). Cambridge University Press.

Smail, R. C. (1995). *Crusading warfare, 1097–1193* (2nd ed.). Cambridge University Press.

Smith, J. M. (1964). Group selection and kin selection. *Nature, 201*(4924), 1145.

Smith, J. M. (1966). *Freedom's fetters: The Alien and Sedition Laws and American civil liberties*. Cornell University Press.

Smith, K. B., & Hibbing, J. R. (2011). The mind-body connection: Psychophysiology as an approach to studying political attitudes and behaviors. In P. K. Hatemi & R. McDermott (Eds.), *Man is by nature a political animal: Evolution, biology, and politics* (pp. 224–246). Chicago University Press.

Smith, K. B., Oxley, D. R., Hibbing, M. V., Alford, J. R., & Hibbing, J. R. (2011). Linking genetics and political attitudes: Reconceptualizing political ideology. *Political Psychology, 32*(3), 369–397.

Smith, M. S. (2003). *The origins of biblical monotheism: Israel's polytheistic background and the ugaritic texts*. Oxford University Press.

Smith, T. (2017). *Why Wilson matters: The origin of American liberal internationalism and its crisis today*. Princeton University Press.

Sober, E., & Wilson, D. S. (1998). *Unto others: The evolution and psychology of unselfish behavior*. Harvard University Press.

Soltis, J., Boyd, R., & Richerson, P. J. (1995). Can group-functional behaviors evolve by cultural group selection?: An empirical test. *Current Anthropology, 36*(3), 473–494.

Somit, A., & Peterson, S. (2005). *The failure of democratic nation building: Ideology meets evolution*. Springer.

Southern, L. M., Deschner, T., & Pika, S. (2021). Lethal coalitionary attacks of chimpanzees (Pan troglodytes troglodytes) on gorillas (Gorilla gorilla gorilla) in the wild. *Scientific Reports, 11*(1), 14673.

Sowell, T. (2002). *A conflict of visions: Ideological origins of political struggles*. Books.

Spear, B. (2014). Coal–Parent of the industrial revolution in Great Britain: The early patent history. *World Patent Information, 39*, 85–88.

Spengler, O. (1991). *The Decline of the West*. Oxford University Press.

Stam, A. C., Von Hagen-Jamar, A., & Worthington, A. B. (2012). Fear and attitudes towards torture and preventive war. *Twin Research and Human Genetics, 15*(1), 60–70.

Stankov, L. (2009). Conservatism and cognitive ability. *Intelligence, 37*(3), 294–304.

Stears, M. (2006). *Progressives, pluralists, and the problems of the state*. Oxford University Press.

Steklis, H. D., Steklis, N. G., Peñaherrera-Aguirre, M., & Figueredo, A. J. (2023). *Human-animal interactions from an evolutionary perspective: Symbioses as extended ultrasociality* (Human-Animal Interactions). CABI.

Stenner, K. (2005). *The authoritarian dynamic*. Cambridge University Press.

Stevens, J. R. (2017). Intertemporal choice and delayed gratification. In J. Call, G. M. Burghardt, I. M. Pepperberg, C. T. Snowdon, & T. Zentall (Eds.), *APA handbook of comparative psychology: Perception, learning, and cognition* (pp. 35–552). American Psychological Association.

Stevenson, D. (1988). *The origins of Freemasonry: Scotland's century, 1590–1710*. Cambridge University Press.

Stewart, A. J., McCarty, N., & Bryson, J. J. (2018). Explaining parochialism: A causal account for political polarization in changing economic environments. *arXiv preprint arXiv:1807.11477*.

Stewart, A. J., Plotkin, J. B., & McCarty, N. (2021). Inequality, identity, and partisanship: How redistribution can stem the tide of mass polarization. *Proceedings of the National Academy of Sciences, 118*(50), e2102140118.

Stone, W. F. (1980). The myth of left-wing authoritarianism. *Political Psychology, 2*(3/4), 3–19.

Strickler, R. (2016). A "sorted" America? Geographic polarization and value overlap in the American electorate. *Social Science Quarterly, 97*(2), 439–457.

Stroumsa, G. G. (2010). Gnosis. In A. Cohen & P. Mendes-Flohr (Eds.), *20th Century Jewish religious thought: Original essays on critical concepts, movements, and beliefs*. The Jewish Publication Society.

Sumner, C., Scofield, J. E., Buchanan, E. M., Evans, M. R., & Shearing, M. (2023). The role of personality, authoritarianism and cognition in the

United Kingdom's 2016 referendum on European Union membership. *Frontiers in Psychology, 14*, 1077354.
Sumner Maine, H. (1976). *Popular government.* Liberty Classics.
Sumpter, D. J. (2010). *Collective animal behavior*. Princeton University Press.
Sun, J., & Smillie, L. (2023). Why moral psychology needs personality psychology. *Journal of Personality, 92*(3), 653–665.
Sweetser, K. D. (2014). Partisan personality: The psychological differences between Democrats and Republicans, and independents somewhere in between. *American Behavioral Scientist, 58*(9), 1183–1194.
Szodoray, P., Nakken, B., Gaal, J., Jonsson, R., Szegedi, A., Zold, E., Brun, J. G., Gesztelyi, R., Zeher, M., & Bodolay, E. (2008). The complex role of vitamin D in autoimmune diseases. *Scandinavian Journal of Immunology, 68*(3), 261–269.
Tainter, J. A. (1988). *The collapse of complex societies*. Cambridge University Press.
Talebi, M. G., Beltrão-Mendes, R., & Lee, P. C. (2009). Intra-community coalitionary lethal attack of an adult male southern muriqui (Brachyteles arachnoides). *American Journal of Primatology: Official Journal of the American Society of Primatologists, 71*(10), 860–867.
Tang, J., Ko, S., Ding, H. K., Qiu, C. S., Calejesan, A. A., & Zhuo, M. (2005). Pavlovian fear memory induced by activation in the anterior cingulate cortex. *Molecular Pain, 1*, 1744–8069.
Tatz, C., & Higgins, W. (2016). *The magnitude of genocide*. Bloomsbury Publishing.
Teilhard de Chardin, P. (1959). *The phenomenon of man [Phénomène humain (English)]*. Harper.
Terrizzi, J. A., Jr., Shook, N. J., & Ventis, W. L. (2010). Disgust: A predictor of social conservatism and prejudicial attitudes toward homosexuals. *Personality and Individual Differences, 49*(6), 587–592.
Thayer, B. A. (2010). The failure of democratic nation building: Ideology meets evolution. *Politics and the Life Sciences, 29*(2), 102–105.
Thomas, R. (2007). The Israelite origins of the Mandaean people. *Studia Antiqua, 5*(2), 3–27.
Thornhill, R., & Fincher, C. L. (2007). What is the relevance of attachment and life history to political values? *Evolution and Human Behavior, 28*(4), 215–222.
Tibble, S. (2018). *The Crusader Armies: 1099–1187*. Yale University Press.
Tifferet, S., Agrest, S., & Benisti Shlomo, A. (2011). Problem gambling: An outcome of a life history strategy. *International Gambling Studies, 11*(2), 253–262.

Titus, H. E. (1968). F Scale Validity Considered Against Peer Nomination Criteria. *The Psychological Record, 18*, 395–403.

Toynbee, A. (2011). *Lectures on the Industrial Revolution in England: Popular addresses, notes and other fragments*. Cambridge University Press.

Toynbee, A. J. (1951). *A study of history* (Vol. I). Oxford University Press.

Trevor-Roper, H. R. (1959). The general crisis of the 17th century. *Past & Present, 16*, 31–64.

Trivers, R. L. (1971). The evolution of reciprocal altruism. *Quarterly Review of Biology, 46*, 35–57.

Tuchman, B. W. (1987). *A distant mirror: The calamitous 14th century*. Random House.

Tucker, S. (2010). *A global chronology of conflict*. ABC-CLIO.

Turchin, P. (2003a). *Complex population dynamics: A theoretical/empirical synthesis*. Princeton University Press.

Turchin, P. (2003b). *Historical dynamics: Why states rise and fall*. Princeton University Press.

Turchin, P. (2006a). Population dynamics and internal warfare: A reconsideration. *Social Evolution & History, 5*(2), 112–147.

Turchin, P. (2006b). *War and peace and war: The life cycles of imperial nations*. Pi Press.

Turchin, P. (2006c). *War and peace and war: The rise and fall of empires*. Plume.

Turchin, P. (2010). Warfare and the evolution of social complexity: A multilevel-selection approach. *Structure and Dynamics, 4*(3), 1–37.

Turchin, P. (2013a). Complex population dynamics: A theoretical/empirical synthesis (MPB-35). In *Complex population dynamics*. Princeton University Press.

Turchin, P. (2013b). How did large-scale complex societies evolve. In P. J. Richerson & M. H. Christiansen (Eds.), *Cultural evolution: Society, technology, language, and religion* (pp. 61–74). MIT Press.

Turchin, P. (2013c). Modeling social pressures toward political instability. *Cliodynamics, 4*(2), 281.

Turchin, P. (2015). *Ultrasociety: How 10,000 years of war made humans the greatest cooperators on earth*. Beresta Books.

Turchin, P. (2023). *End times: elites, counter-elites, and the path of political disintegration*. Penguin.

Turchin, P., Hoyer, D., Korotayev, A., Kradin, N., Nefedov, S., Feinman, G., Levine, J., Reddish, J., Cioni, E., Thorpe, C., Bennett, J. S., Francois, P., & Whitehouse, H. (2021). Rise of the war machines: Charting the evolution of military technologies from the Neolithic to the Industrial Revolution. *PLoS One, 16*(10), e0258161.

Turchin, P., & Nefedov, S. A. (2009). *Secular cycles*. Princeton University Press.

Turchin, P., Whitehouse, H., Gavrilets, S., Hoyer, D., François, P., Bennett, J. S., Feeney, K. C., Peregrine, P., Feinman, G., Korotayev, A., Kradin, N., Levine, J., Reddish, J., Cioni, E., Wacziarg, R., Mendel-Gleason, G., & Benam, M. (2022). Disentangling the evolutionary drivers of social complexity: A comprehensive test of hypotheses. *Science Advances, 8*(25), eabn3517.

Turkheimer, E. (2000). Three laws of behavior genetics and what they mean. *Current Directions in Psychological Science, 9*(5), 160–164.

Turner, F. J. (1921). *The frontier in American history*. Henry Holt & Company.

Tybur, J. M., Çınar, Ç., Karinen, A. K., & Perone, P. (2018). Why do people vary in disgust? *Philosophical Transactions of the Royal Society B: Biological Sciences, 373*(1751), 20170204.

Tyerman, C. (2006). *God's war: A new history of the Crusades*. Penguin.

Tyler, M. C. (1941/1897). *The literary history of the American Revolution*. Barnes & Noble.

Uskul, A. K., & Cross, S. E. (2020). Socio-ecological roots of cultures of honor. *Current Opinion in Psychology, 32*, 177–180.

Vallance, T. (2015). Political thought. In M. J. Braddick (ed.), *The Oxford handbook of the English Revolution* (pp. 430–446). Oxford University Press.

van der Bijl, W. (2018). Phylopath: Easy phylogenetic path analysis in R. *PeerJ, 6*, e4718.

Van Es, S., Nieuwenhuijsen, B. W., Lenouvel, F., van Deursen, E. M., & Schaap, P. (1994). Universal signals control slime mold stalk formation. *Proceedings of the National Academy of Sciences, 91*(17), 8219–8223.

Van Hiel, A., Onraet, E., & De Pauw, S. (2010). The relationship between social-cultural attitudes and behavioral measures of cognitive style: A meta-analytic integration of studies. *Journal of Personality, 78*(6), 1765–1800.

van Schaik, C. P., Pandit, S. A., & Vogel, E. R. (2006). Toward a general model for male-male coalitions in primate groups. In P. M. Kappeler & C. P. van Schaik (Eds.), *Cooperation in primates and humans* (pp. 151–171). Springer.

Van Vugt, M., & Schaller, M. (2008). Evolutionary approaches to group dynamics: An introduction. *Group Dynamics: Theory, Research, and Practice, 12, 1–6.*

Venning, T., & Frankopan, P. (2015). *A chronology of the crusades*. Routledge.

Verhoeven, W. (2013). *Americomania and the French Revolution debate in Britain, 1789–1802.* Cambridge University Press.

Verhulst, B., Hatemi, P. K., & Martin, N. G. (2010). The nature of the relationship between personality traits and political attitudes. *Personality and Individual Differences, 49*(4), 306–316.

Verweij, K. J., Shekar, S. N., Zietsch, B. P., Eaves, L. J., Bailey, J. M., Boomsma, D. I., & Martin, N. G. (2008). Genetic and environmental influences on individual differences in attitudes toward homosexuality: An Australian twin study. *Behavior Genetics, 38*, 257–265.

Verweij, K. J. H., Zietsch, B. P., Bailey, J. M., & Martin, N. G. (2009). Shared aetiology of risky sexual behaviour and adolescent misconduct: Genetic and environmental influences. *Genes, Brain and Behavior, 8*(1), 107–113.

Vigil, J. M. (2010). Political leanings vary with facial expression processing and psychosocial functioning. *Groups Processes & Intergroup Relations, 13*(5), 547–558.

Voigtländer, N., & Voth, H. J. (2013). How the West 'Invented' Fertility Restriction. *American Economic, Review, 103*(6), 2227–2264.

von Humboldt, W. (1993). *The limits of state action*. Liberty Fund.

Wade, M. J. (1977). An experimental study of group selection. *Evolution, 31*(1), 134–153.

Wade, M. J. (1978). A critical review of the models of group selection. *The Quarterly Review of Biology, 53*(2), 101–114.

Wajzer, M. (2020). Genopolitics: Introductory remarks. *Interdisciplinary Science Reviews, 45*(4), 508–524.

Wajzer, M. (2023). The reductionism of genopolitics in the context of the relationships between biology and political science. *Endeavour, 47*(3), 100874.

Wakefield, W. L., & Evans, A. P. (1991). *Heresies of the High Middle Ages*. Columbia University Press.

Walker, R. S., & Bailey, D. H. (2013). Body counts in lowland South American violence. *Evolution and Human Behavior, 34*(1), 29–34.

Wallace, G. (1908). *Human nature in politics*. A. Constable and co., limited.

Walsh, A. (2023). Eudaimonic happiness, temperament, and political ideology. *Journal of Ideology, 42*(1), 4.

Ward, L. (2004). *The politics of liberty in England and revolutionary America*. Cambridge University Press.

Ware, A. (2004). *Political parties and political systems*. Ediciones Akal.

Warne, R. T. (2020). *In the know: Debunking 35 myths about human intelligence*. Cambridge University Press.

Warren, C. (1942). *Odd byways in American history*. Harvard University Press.

Watts, D. P. (1997). Agonistic interventions in wild mountain gorilla groups. *Behaviour, 134*(1–2), 23–57.

Watts, E. J. (2021). *The eternal decline and fall of Rome: The history of a dangerous idea*. Oxford University Press.

Weber, M. (2019). *Economy and society*. Harvard University Press.

Weed, C. P. (1989). What Happened to the Republicans in the 1930s: Minority Party Dynamics during Political Realignment. *Polity, 22*(1), 5–23.

Wegenast, T. C., & Basedau, M. (2014). Ethnic fractionalization, natural resources and armed conflict. *Conflict Management and Peace Science, 31*(4), 432–457.

Weissflog, M., Choma, B. L., Dywan, J., van Noordt, S. J., & Segalowitz, S. J. (2013). The political (and physiological) divide: Political orientation, performance monitoring, and the anterior cingulate response. *Social Neuroscience, 8*(5), 434–447.

Wells, H. G. (1921). *The outline of history: Being a plain history of life and mankind*. Cassell.

Werren, J. H. (2011). Selfish genetic elements, genetic conflict, and evolutionary innovation. *Proceedings of the National Academy of Sciences, 108*, 10863–10870.

Wertheimer, J. (1993). *A people divided: Judaism in contemporary America*. Basic Books.

West-Eberhard, M. J. (1967). Dominance hierarchies and the evolution of social behaviour. *Science, 157*, 1585–1595.

West-Eberhard, M. J. (1970). *The wasps*. University of Michigan Press.

West-Eberhard, M. J. (1975). The evolution of social behavior by kin selection. *The Quarterly Review of Biology, 50*(1), 1–33.

West-Eberhard, M. J. (2003). *Developmental plasticity and evolution*. Oxford University Press.

Wezel, C. (2022). A cultural theory of autocracy vs. democracy: On the psychological foundations of political regimes. In D. Osborne & C. G. Sibley (Eds.), *The Cambridge handbook of political psychology* (pp. 458–475). Cambridge University Press.

Wheeler, J. (2017). Analyzing the role of income inequality in American political polarization. (Master's Thesis: Bemidji State University).

Wheeler, W. M. (1907). The origin of slavery among ants. *The Popular Science Monthly, 71*, 550–559.

Wheeler, W. M. (1910). *Ants*. Columbia University.

Wheeler, W. M. (1923). Social life among the insects. *The Scientific Monthly, 16*(1), 5–33.

Whitby, M. (2002). *Sparta*. Routledge.

White, M. (2012). Selected death tolls for wars, massacres and atrocities before the 20th Century. *Necrometrics*. Retrieved June 25, 2024, from https://necrometrics.com/pre1700a.htm#30YrW

Widengren, G. (1946). *Mesopotamian elements in Manichaeism (King and Saviour II): Studies in Manichaean, Mandaean, and Syrian-gnostic religion*. Lundequistska bokhandeln.

Williams, G. C. (1971/2008). *Group selection*. Aldine Transaction.

Willis, J. L. (2017). *Culture wars? Applying categorical variation measures to the study of sociocultural and political polarization* (Doctoral dissertation, Colorado State University. Libraries).

Wilson, D. (2002). *Darwin's cathedral: Evolution, religion, and the nature of society*. University of Chicago Press.

Wilson, D. S. (1983). The group selection controversy: History and current status. *Annual Review of Ecology and Systematics, 14*(1), 159–187.

Wilson, D. S. (2015). *Does altruism exist? Culture, genes and the welfare of others*. Yale University Press.

Wilson, D. S. (2019). *This view of life: Completing the Darwinian revolution*. Pantheon Books.

Wilson, D. S., Madhavan, G., Gelfand, M. J., Hayes, S. C., Atkins, P. W., & Colwell, R. R. (2023). Multilevel cultural evolution: From new theory to practical applications. *Proceedings of the National Academy of Sciences, 120*(16), e2218222120.

Wilson, E. O. (1971). *The insect societies*. Belknap Press of Harvard University Press.

Wilson, E. O. (1975a). Leptothorax duloticus and the beginnings of slavery in ants. *Evolution, 29*, 108–119.

Wilson, E. O. (1975b). Slavery in ants. *Scientific American, 232*(6), 32–36.

Wilson, E. O. (1975c). *Sociobiology: The new synthesis*. Cambridge, Massachusetts: Belknap Press of Harvard University.

Wilson, E. O. (2012). *The social conquest of earth*. WW Norton & Company.

Wilson, G. (2013). *The psychology of conservatism (Routledge revivals)*. Routledge.

Wilson, G. D., Ausman, J., & Mathews, T. R. (1973). Conservatism and art preferences. *Journal of Personality and Social Psychology, 25*(2), 286–288.

Wilson, M. L., Boesch, C., Fruth, B., Furuichi, T., Gilby, I. C., Hashimoto, C., Hobaiter, C. L., Hohmann, G., Itoh, N., Koops, K., Lloyd, J. N., Matsuzawa, T., Mitani, J. C., Mjungu, D. C., Morgan, D., Muller, M. N., Mundry, R., Nakamura, M., Pruetz, J., et al. (2014). Lethal aggression in Pan is better explained by adaptive strategies than human impacts. *Nature, 513*(7518), 414–417.

Wimmer, A. (2013). *Waves of war: Nationalism, state formation, and ethnic exclusion in the modern world.* Cambridge University Press.

Winter, U., & Buschinger, A. (1983). The reproductive biology of a slavemaker ant, Epimyrma ravouxi, and a degenerate slavemaker, E. kraussei (Hymenoptera: Formicidae). *Entomologia Generalis, 9*, 1–15.

Wolf, P. S., & Figueredo, A. J. (2011). Fecundity, offspring longevity, and assortative mating: Parametric tradeoffs in sexual and life history strategy. *Biodemography and Social Biology, 57*(2), 171–183.

Wolfinger, R. E., & Rosenstone, S. J. (1980). *Who votes?* Yale University Press.

Woodard, C. (2016). *American character: A history of the epic struggle between individual liberty and the common good.* Viking.

Woodley, M. A. (2011). The cognitive differentiation-integration effort hypothesis: A synthesis between the fitness indicator and life history models of human intelligence. *Review of General Psychology, 15*(3), 228–245.

Woodley of Menie, M. A., Figueredo, A. J., Peñaherrera-Aguirre, M., Jurgensen, J., & Sarraf, M. A. (2022). Haidt's moral foundations tracked over 200 years of lexicographic data, and their predictors. *Anthropological Review, 85*(2), 79–102.

Woodley of Menie, M. A., Figueredo, A. J., Sarraf, M. A., Hertler, S., Fernandes, H. B. F., & Aguirre, M. P. (2017). *The rhythm of the West: A biohistory of the Modern era, AD 1600 to present* (Journal of Social, Political and Economic Studies Monograph Series, Volume 37). Scott Townsend Pub.

Woodley of Menie, M. A., & Peñaherrera-Aguirre, M. (2022). General intelligence as a major source of cognitive variation among individuals of three species of Lemur, uniting g with G. *Evolutionary Psychological Science, 8*, 241–253.

Woodley of Menie, M. A., & Peñaherrera-Aguirre, M. (2023). Convergence between G and g in three monkey species (*Sapajus spp, Ateles geoffroyi, and Macaca fascicularis*). *Journal of Comparative Psychology, 137*(1), 62.

Woodley of Menie, M. A., Peñaherrera-Aguirre, M., & Sarraf, M. A. (2022). Signs of a Flynn effect in rodents? Secular differentiation of the manifold of general cognitive ability in laboratory mice (Mus musculus) and Norwegian rats (Rattus norvegicus) over a century—Results from two cross-temporal meta-analyses. *Intelligence, 95*, 101700.

Woodley of Menie, M. A., Peñaherrera-Aguirre, M., & Sarraf, M. A. (2023). Do cleaner fish (*Labroides dimidiatus*) have general cognitive ability? A reanalysis of individual differences data and consideration of phylogenetic context. *Evolutionary Psychological Science, 9*, 309–316.

Woodward, C. V. (1981). *Origins of the new South, 1877–1913: A History of the South* (Vol. 9). LSU Press.

Woodward, C. V. (2008). *The burden of Southern history*. LSU Press.

Wrangham, R. W. (1999). Evolution of coalitionary killing. *American Journal of Physical Anthropology, 110*(S29), 1–30.

Wrangham, R. W., Wilson, M. L., & Muller, M. N. (2006). Comparative rates of violence in chimpanzees and humans. *Primates, 47*, 14–26.

Wright, J. P., Beaver, K. M., Morgan, M. A., & Connolly, E. J. (2017). Political ideology predicts involvement in crime. *Personality and Individual Differences, 106*, 236–241.

Wrong, D. H. (1994). *The problem of social order: What unites and divides society*. Free Press.

Wyatt, D. (2021). Slavery in Northern Europe (Scandinavia and Iceland), and the British Isles, 500–1420. In C. Perry, D. Eltis, S. L. Engerman, & D. Richardson (Eds.), *The Cambridge world history of slavery* (AD 500–AD 1420) (Vol. 2, pp. 482–507). Cambridge University Press.

Wynne-Edwards, V. C. (1962). *Animal dispersal in relation to social behaviour*. Oliver & Boyd.

Xu, X., Burton, C. M., & Plaks, J. E. (2022). Three dimensions of American conservative political orientation differentially predict negativity bias and satisfaction with life. *Social Psychological and Personality Science, 13*(8), 1230–1245.

Xu, X., Mar, R. A., & Peterson, J. B. (2013). Does cultural exposure partially explain the association between personality and political orientation? *Personality and Social Psychology Bulletin, 39*(11), 1497–1517.

Xu, X., Soto, C. J., & Plaks, J. E. (2021). Beyond openness to experience and conscientiousness: Testing links between lower-level personality traits and American political orientation. *Journal of Personality, 89*(4), 754–773.

Yarchi, M., Baden, C., & Kligler-Vilenchik, N. (2021). Political polarization on the digital sphere: A cross-platform, over-time analysis of interactional, positional, and affective polarization on social media. *Political Communication, 38*(1–2), 98–139.

York, N. L. (1993). Freemasons and the American Revolution. *The Historian, 55*(2), 315–330.

Yutin, N., Raoult, D., & Koonin, E. V. (2013). Virophages, polintons, and transpovirons: A complex evolutionary network of diverse selfish genetic elements with different reproduction strategies. *Virology Journal, 10*(1), 1–15.

Závecz, G., & Littvay, L. (2022). Genopolitics. In T. J. Rudolph (Ed.), *Handbook on politics and public opinion* (pp. 71–82). Edward Elgar Publishing.

Zebarjadi, N., Adler, E., Kluge, A., Sams, M., & Levy, J. (2023). Ideological values are parametrically associated with empathy neural response to vicarious suffering. *Social Cognitive and Affective Neuroscience, 18*(1), nsad029.

Zeldis, L. (1998). Freemasonry's contribution to South American independence: A factual approach. *Ars Quatuor Coronatorum, 111*, 79–104.

Zhang, D. D., Lee, H. F., Wang, C., Li, B., Pei, Q., Zhang, J., & An, Y. (2011). The causality analysis of climate change and large-scale human crises. *Proceedings of the National Academy of Sciences, 108*, 17296–17301.

Zhang, J., & Zhong, Z. (2019). Life history and system justification: Higher individual fertility and lower provincial life expectancy correlate with stronger progovernment attitudes in China. *Political Psychology, 40*(2), 355–373.

Ziblatt, D. (2017). *Conservative parties and the birth of democracy*. Cambridge University Press.

Zietsch, B. P., Verweij, K. J. H., Bailey, J. M., Wright, M. J., & Martin, N. G. (2010). Genetic and environmental influences on risky sexual behaviour and its relationship with personality. *Behavior Genetics, 40*, 12–21.

Zmigrod, L. (2022). A psychology of ideology: Unpacking the psychological structure of ideological thinking. *Perspectives on Psychological Science, 17*(4), 1072–1092.

Zmijewski, M. A. (2019). Vitamin D and human health. *International Journal of Molecular Sciences, 20*(1), 145.

Zuckert (1999). *Natural rights and the new republicanism*. Princeton University Press.

Zwolinski, M., & Tomasi, J. (2023). *The Individualists: Radicals, reactionaries, and the struggle for the soul of libertarianism*. Princeton University Press.

Index[1]

[1] Note: Page numbers followed by 'n' refer to notes.

S. C. Hertler et al., *The Evolution of Political Ideology*,
https://doi.org/10.1007/978-3-031-83941-2

W

GPSR Compliance
The European Union's (EU) General Product Safety Regulation (GPSR) is a set of rules that requires consumer products to be safe and our obligations to ensure this.

If you have any concerns about our products, you can contact us on

ProductSafety@springernature.com

In case Publisher is established outside the EU, the EU authorized representative is:

Springer Nature Customer Service Center GmbH
Europaplatz 3
69115 Heidelberg, Germany

www.ingramcontent.com/pod-product-compliance
Lightning Source LLC
Chambersburg PA
CBHW051412031125
34911CB00002B/58
9783031839405